SURVEILLANCE AND GOVERNANCE: CRIME CONTROL AND BEYOND

SOCIOLOGY OF CRIME, LAW AND DEVIANCE

Series Editors: Mathieu Deflem (Volumes 6–10)
Jeffrey T. Ulmer (Volumes 1–5)

Recent Volumes:

SOCIOLOGY OF CRIME, LAW AND DEVIANCE VOLUME 10

SURVEILLANCE AND GOVERNANCE: CRIME CONTROL AND BEYOND

EDITED BY

MATHIEU DEFLEM

University of South Carolina, Columbia, USA

Emerald

JAI

United Kingdom – North America – Japan
India – Malaysia – China

JAI Press is an imprint of Emerald Group Publishing Limited
Howard House, Wagon Lane, Bingley BD16 1WA, UK

First edition 2008

British Library Cataloguing in Publication Data
A catalogue record for this book is available from the British Library

ISBN: 978-0-7623-1416-4
ISSN: 1521-6136 (Series)

To the memory of our colleague and friend Richard V. Ericson (1948–2007)

CONTENTS

PART IV: BEYOND CRIME CONTROL

LIST OF CONTRIBUTORS

Janet Chan	School of Social Sciences and International Studies, University of New South Wales, Sydney, Australia
Kirsten Christiansen	The Graduate Center of the City University of New York and John Jay College of Criminal Justice, New York, NY, USA
David Cunningham	Department of Sociology, Brandeis University, Waltham, MA, USA
Stephanie K. Decker	Department of Sociology, University of Kansas, Lawrence, KS, USA
Mathieu Deflem	Department of Sociology, University of South Carolina, Columbia, SC, USA
Benoît Dupont	Centre International de Criminologie Comparée, Université de Montréal, Montreal, Quebec, Canada
Richard V. Ericson†	Centre of Criminology, University of Toronto, Toronto, Canada
John Gilliom	Department of Political Science, Ohio University, Athens, OH, USA
Karen S. Glover	Department of Sociology, Criminology and Justice Studies Program, California State University, San Marcos, CA, USA
Kevin D. Haggerty	Department of Sociology, University of Alberta, Edmonton, Alberta, Canada
Nathan Harris	Regulatory Institutions Network, Australian National University, Canberra, Australia

Laura Huey	Department of Sociology, Concordia University, Montreal, Quebec, Canada
Fabien Jobard	Center of Sociological Research on Penal Institutions (CESDIP-CNRS), Guyancourt, France
Dominique Linhardt	Centre de sociologie de l'innovation (CSI-CNRS), École nationale supérieure des mines de Paris (ENSMP), Paris, France
Thomas Mathiesen	Department of Criminology and Sociology of Law, University of Oslo, Oslo, Norway
Michael McCahill	Department of Social Sciences, University of Hull, Hull, UK
John Noakes	Department of Sociology, Anthropology and Criminal Justice, Arcadia University, Glenside, PA, USA
Minas Samatas	Sociology Department, University of Crete, Rethymno, Crete, Greece
William G. Staples	Department of Sociology, University of Kansas, Lawrence, USA
Kevin Stenson	Crime and Conflict Research Centre, Middlesex University (London), Enfield, Middlesex, UK
James Walsh	Department of Sociology, University of California Santa Barbara, Santa Barbara, USA
Scott G. White	The Graduate Center, City University of New York and LaGuardia Community College, Long Island City, New York, USA
Jennifer Wood	Department of Criminal Justice, Temple University, Philadelphia, PA, USA

INTRODUCTION: AN EYE ON SURVEILLANCE AND GOVERNANCE

Mathieu Deflem

ABSTRACT

This volume presents recent insights in the sociological study of surveillance and governance in the context of criminal justice and other control strategies in contemporary societies. The collected chapters provide a varied set of theoretical perspectives and substantive research domains on the qualities and quantities of some of the most recent transformations of social control as well as their historical precursors in diverse social settings. Drawn from several quarters of the world, the contributors to this volume testify to the increasing relevance of surveillance and governance across the globe and, at the same time, demonstrate the cross-national spread of scholarly ideas on the study thereof.

Surveillance is not a conceptual invention of recent social science scholarship. In fact, the term has a very long history that appears more closely related to the multitude of policy functions that historically were increasingly monopolized by the state, including matters pertaining to health, deviance, poverty, geography, and economy (e.g., Forrest, 1896;

Surveillance and Governance: Crime Control and Beyond
Sociology of Crime Law and Deviance, Volume 10, 1–8

ISSN: 1521-6136/doi:10.1016/S1521-6136(07)00219-9

Langmuir, 1965). Surveillance is, in its origins, a concept of power. The term, however, has also acquired the status of a concept in social science scholarship, and it is here that the contributions in the present volume are situated to ponder on the patterns, dynamics, and implications of the social practices and institutions involved with the observing and monitoring of behavior and the collection of information thereon. Yet, even as a concept, surveillance has undergone changes in meaning and has, consequently, been applied in a variety of contexts. Surveillance can probably be less clearly defined and is better experienced when we see it, or when it is discovered to have violated our sense of trust and privacy when we did not. Further indicating conceptual and theoretical complexities there has been the more recent introduction in social science of the concept of governance, itself a term going back hundreds of years to its origins in the world of politics and policy. At a general level, governance can be defined as the administrative or application-oriented components of government, the latter broadly defined with respect not only to politics but power more generally, including other spheres of conduct, notably private and corporate action. Evidently, the study of social control in terms of both or either surveillance and governance introduces further complexities about their relations and meanings.

The authors contributing to this volume have many intelligent things to say, in theoretical and empirical respects, about surveillance and governance. In order to briefly situate these varied discussions, I wish to clarify a few broad strokes of the study of surveillance and governance on the basis of the work of the French philosopher Michel Foucault. Of course, modern surveillance scholars have made much effort in recent years to move beyond Foucault, and in many instances they surely have done so successfully (e.g., see the contributions in Haggerty & Ericson, 2006; Lyon, 2006; Zurawski, 2007). However, it remains instructive, also, just as much as we sociologists habitually remind ourselves that we are anchored in the 19th century, to briefly situate recent developments in the study of surveillance and governance with reference to Foucault's work. This review can be useful, minimally, for strategic reasons aimed at uncovering the historical centrality of Foucault's terms of discipline and governmentality in the development of what is now sometimes called the sub-disciplinary specialty of 'surveillance studies' (see Contemporary Sociology, 2007). Conceptually, furthermore, a case can be made that Foucault's twin notions correspond to surveillance and governance, respectively, and have informed central aspects of their study in the social and human sciences.

OF DISCIPLINE AND GOVERNMENTALITY

Foucault discussed the theme of surveillance, in his now famous work on the birth of the prison (Foucault, 1977), in a manner that ignited the sociological imagination to devote increasing attention to a variety of mechanisms and technologies of control. A study on the transformation of punishment in the modern era, Foucault's investigation is centrally involved in analyzing the disappearance of punishment as a public and violent spectacle centered on the infliction of pain (public torture) and the emergence of a meticulous surveillance of the soul. Since the second half of the 18th century, Foucault suggests, reform proposals were introduced in matters of punishment that proposed leniency only to enhance intervention and efficiency. Although the prison system originally did not fit this model, detention would become the most typical form of punishment. This peculiar development makes sense, according to Foucault, in terms of the spread of a new form of punishment called discipline. Oriented at the production of docile bodies, discipline involves a series of techniques of surveillance which emphasize a continuous supervision, examination, and normalization of behavior. Like other theaters of disciplinary power (the school, the clinic, the factory), the modern prison has the Panopticon as its most prototypical expression to economically keep and oversee the subject. Modern prisons (unlike the dungeons of the dark) bring its inhabitants to light: the prisoners are seen and overseen and subject to a normalization (through penitence rather than rehabilitation) on the basis of models of medical, economic, and political expertise. The human sciences legitimize and contribute to disciplinary power. Discipline is both discourse and practice.

Summarizing the characteristics of discipline, it is a form of power that is productive and useful. Punishment should benefit both the offender and society. It should be useful economically, politically, and socially. Discipline does not come in the typical form of power that excludes and is negatively enforced (as a prohibition). Discipline does not prohibit; instead, it prescribes proper modes of conduct. Disciplinary power is also pervasive throughout society as the Panopticon becomes articulated in multiple institutions outside of the prison as a generalized function of panopticism. The panoptical formula of power through transparency permeates the social body. The relations of disciplinary power cannot be captured in a dichotomy of dominators and dominated: discipline is a machine in which everyone is caught. And power is always related to knowledge that justifies power. Yet, discipline is not the one master-concept of power in the modern age. Instead, power relations today are multiple, of various kinds.

The procedures of power today are more diverse than only the disciplinary type, and there remains a trace of torture. Finally, there is also always resistance against power. Disciplinary power is omnipresent but not omnipotent; modern society is disciplinary but not disciplined.

Also developed by Foucault and especially widely applied by contemporary post-Foucauldian scholars, the concept of governmentality broadens the perspective of discipline to focus on the objectives of modern power (Foucault, 1991). Governmentality is defined as "the way in which the conduct of a whole of individuals is found implicated, in an ever more marked fashion, in the exercise of sovereign power" (Foucault, 1991, p. 101). Central to Foucault's notion is that power does not exclude people but that, on the contrary, governmental power centers on the population and its truth by presupposing, measuring, and evaluating individuals in their conduct as living subjects. Especially in 19th-century Europe, Foucault explains, instead of a justification of power in terms of a centered state, power was conceived in terms of an efficient economy directed at furthering the fertility of territories and the health and movements of the population. Governmentality thus broke with any form of state-sanctioned legalism.

According to Foucault, governmental power relies in its effectuation on a triple alliance of criminology, statistics, and police (Foucault, 1980, 1984, 1991; see also Deflem, 1997). Indeed, in order to concretize the governmental form of political technology, it was critical to know the population. With respect to criminality, it was criminology which, as the science of the criminal species, provided this knowledge, while criminal statistics uncovered the relevant regularities in the population. Police is understood, not in the contemporary sense as law enforcement, but as "a program of government rationality ... to create a system of regulation of the general conduct of individuals whereby everything would be controlled to the point of self-sustenance, without the need for intervention" (Foucault, 1984, p. 241). Corresponding to the objectification of the lives of delinquents in systems of criminological knowledge, governmental policing is targeted at a society of living beings outside and beyond the context of law. This extra-legality does not imply that the practice of police would not be influenced by political and economic developments. However, Foucault maintains, "the type of power that it exercises, the mechanisms it operates and the elements to which it applies them are specific" (Foucault, 1977, p. 213).

Clarifying and extending the concepts of discipline and governmentality, the burgeoning scholarly move towards the study of surveillance and governance can be conceptualized, at its most general level, as referring to the instrumental and goal-directed components of modern manifestations of social control, respectively. Importantly, the concept of social control has thereby come to be understood in an increasingly broadened meaning that is no longer tied up exclusively with crime and deviance, but that is applied in a more general sense to a nation, a world, a society of beings. Sometimes, even, scholars have in this direction altogether abandoned the very idea of social control in order to move away from an implied functionality in surveillance and governance towards an observing attitude in terms of risk and suspicion. Not surprisingly, a tendency of postmodernism, implied or explicit, can often be detected in contemporary surveillance studies.

AN OVERVIEW OF THE CHAPTERS

Traditionally, sociologists have contemplated power in terms of the institutions of politics and its modern apex, the state. Yet, because of many of the contemporary changes affecting the institutions and practices of surveillance and governance, it can be argued that social control today is less a domain of the nation-state alone. At the same time, perhaps, never before has the state been involved with social control as much as today. Surveillance and control also are no longer an exclusively local or regional affair but extend beyond national boundaries to take up the sphere of the global order. Yet, at once, much control is localized and continues to go "down to the finest grain of the social body" (Foucault, 1977, p. 80). Also, many of the new technologies that a decade ago led to analytical reflections of the highest order today have become banal in their everyday application and routine diffusion. What can sociologists intelligently say about these developments in both empirical and theoretical respects? It is from this perspective that the authors of the present volume were invited to contribute a chapter in which they could freely explore any facet of the broader constellation of contemporary surveillance and governance strategies with respect to both crime control and related developments that push social control processes beyond the concerns of crime and deviance. Based on their research efforts, the contributors were encouraged to offer provocative and thoughtful reflections that can stimulate our theoretical thinking about relevant issues. As this review will make clear, the authors yield the very rich variety that exists in contemporary sociological thinking about surveillance and governance.

Part I of this volume brings together contributions that primarily focus on the boundaries that modern surveillance practices attempt to break and the spaces they are applied to. In a study of the Minuteman Project at the southern border of the United States, James Walsh offers a penetrating analysis of the history, ideology, and practices of a peculiar form of citizen surveillance. Walsh argues that such non-state projects, in fact, represent an effort by citizens to align themselves with the surveillance apparatus of the state. In the Canadian context, Kevin Haggerty, Laura Huey, and Richard Ericson analyze the political contests that waged about the installing of closed circuit television (CCTV) surveillance systems in the city of Vancouver. On the basis of interviews, the authors show that the application of such camera systems is not always embraced despite the often-propagated attractiveness of such systems. Turning to the city of New York, Kirsten Christiansen examines the intrusion of systems of surveillance and control in large public spaces. She concludes that public urban space greatly impacts our understanding of rights to free speech and assembly, which affects the health of the contemporary democratic process. Taking us on a journey to France, Fabien Jobard and Dominique Linhardt scrutinize the control systems at the international airport of Orly, south of Paris, and in the housing projects in the town of Dammarie-lès-Lys. They draw illuminating comparisons between these very different spaces of surveillance that particularly indicate a strong difference in the intensity of surveillance. Extending the analysis to the international level, Thomas Mathiesen scrutinizes various transnational systems of surveillance, especially in the context of the European Union. This global order, Mathiesen argues, presents a system of control without a state.

The chapters in Part II focus on the technological and strategic elements of surveillance and governance. William Staples and Stephanie Decker examine the techniques of house arrest as they are applied in a midwestern town in the United States. The authors use ethnographic interview data to contemplate on the implications of house arrest for the person's sense of self in view of the objectives of docility. Scott White also takes on a formal means of social control by investigating the practices enacted by the Federal Bureau of Investigation (FBI) to control activities in the academia. What White's analysis brings out is the centrality of the extraction of information in modern systems of surveillance. FBI activities are also at the heart of the chapter by David Cunningham and John Noakes in their analysis of counter-intelligence programs. What the authors particularly focus on are the implications of such activities for the course and outcome of social movement activities and, by extension, the lessons thereof for social movements theorists. Focusing on the much-discussed CCTV systems, Michael McCahill contemplates on the formation of a plural policing system, extending beyond the confines of formal policing activities. Plural policing, McCahill shows, is blurring the traditional divides between public and private systems of control in a multitude of respects.

Part III contains chapters that contemplate on the objectives as well as the counter-objectives of surveillance. Janet Chan examines what she calls the new lateral surveillance, especially as it took place since the events of September 11, that centers on the involvement of citizens in reporting suspicious behavior and people. In this new constellation of state and public collaboration, Chan argues, a new culture of suspicion is formed that, like the high policing efforts of old, is both dangerous and political. Karen Glover centers her analysis on racial profiling strategies that are aimed at racial minorities in the United States. Glover argues that such systems of hyper-surveillance instill in minorities a sense of double-consciousness that separates them from the dominant groups in society. Turning to counter-objectives, Benoît Dupont notes without irony that modern technological systems can also be turned against surveillance strategies. Specifically focused on the use of the internet, Dupont argues that a democratization of surveillance takes place whereby the categories of those who watch and those who are being watched can interchange and shift. Kevin Stenson contemplates on the changes including, but also extending beyond, the internet to argue that surveillance scholars ought not to make claims that the nation-state would be disappearing. Instead, Stenson argues, many advanced practices of surveillance are to be seen as extensions of the powers of the sovereign state.

In Part IV, finally, a group of chapters examines systems of surveillance that are not primarily focused on matters of crime and deviance. John Gilliom discusses the surveillance aspects of recent educational programs in the United States that have been designed to test the progress of children. Gilliom argues that these educational policies will particularly affect lower-income schools and their students, who are subjected to sanctions and shaming as expressions of power oriented at shaping institutions and those who inhabit them. Nathan Harris and Jennifer Wood also focus on the younger members of society by discussing child-protection programs. Theoretically, the authors raise important matters related to responsive regulation, regulatory theory, and nodal governance. Finally, Minas Samatas examines the curious development in Greece, since the Olympic games were held there in 2004, that implied a movement against the installation of CCTV in matters of traffic control and other surveillance systems. Even though the use of CCTV to secure traffic control and prevent car accidents is widely accepted, the people of Greece are generally opposed to systems because of the burdens posed by the authoritarian political past of the country. Covering analyses that discuss surveillance and governance from a plurality of perspectives and centered on a multitude of important components, the chapters in this book collectively show the vibrancy of serious scholarship on the nexus of surveillance and governance.

ACKNOWLEDGMENTS

I am grateful to the authors who contributed to this volume in such exemplary fashion. All those who are interested in important theoretical and empirical puzzles surrounding surveillance and governance can learn much from their efforts. I also thank Shannon McDonough for her kind assistance in preparation of this volume.

REFERENCES

Contemporary Sociology. (2007). A symposium on surveillance studies. *Contemporary Sociology*, *36*, 107–130.

Deflem, M. (1997). Surveillance and criminal statistics: Historical foundations of governmentality. In: A. Sarat & S. Silbey (Eds), *Studies in law, politics and society* (Vol. 17, pp. 149–184). Greenwich, CT: JAI Press.

Forrest, J. D. (1896). Anti-monopoly legislation in the United States. *American Journal of Sociology*, *1*, 411–425.

Foucault, M. (1977). *Discipline and punish: The birth of the prison*. New York: Vintage Books.

Foucault, M. (1980). In: C. Gordon (Ed.), *Power/knowledge: Selected interviews and other writings 1972–1977*. New York: Pantheon Books.

Foucault, M. (1984). In: P. Rabinow (Ed.), *The Foucault reader*. New York: Pantheon Books.

Foucault, M. (1991). Governmentality. In: G. Burchell, C. Gordon & P. Miller (Eds), *The Foucault effect* (pp. 87–104). Chicago: University of Chicago Press.

Haggerty, K. D., & Ericson, R. V. (Eds). (2006). *The new politics of surveillance and visibility*. Toronto: University of Toronto Press.

Langmuir, A. D. (1965). Developing concepts in surveillance. *The Milbank Memorial Fund Quarterly*, *43*, 369–372.

Lyon, D. (Ed.). (2006). *Theorizing surveillance: The Panopticon and beyond*. Devon, UK: Willan Publishing.

Zurawski, N. (Ed.). (2007). *Surveillance studies: Perspektiven eines Forschungsfeldes*. Opladen, Germany: Budrich.

PART I:
BOUNDARIES AND SPACES

COMMUNITY, SURVEILLANCE AND BORDER CONTROL: THE CASE OF THE MINUTEMAN PROJECT

James Walsh

ABSTRACT

As political interfaces, national borders are subject to extensive surveillance and policing within the interstate system. But what happens when the state's gatekeepers emerge from within the social body? How do such instances impact scholarly understandings of governance and surveillance? This chapter investigates these questions empirically, analyzing the Minuteman Project, a grassroots vigilante movement dedicated to directly policing the nation's borders. Situating the movement within the existing literature on "governmentality" and "community policing", I analyze its history, ideology, practices and interactions with authorities, arguing that, despite their status as non-state actors, its members appropriate, enforce and extend many of the principles of governance and statecraft; whether, surveillance, policing, security or territoriality. Like community policing, the Minutemen highlight the pervasive and decentralized nature of government, social control and surveillance. In occupying and monitoring the border, the group serves as the state's "eyes and ears" without impinging upon its juridical or

Surveillance and Governance: Crime Control and Beyond
Sociology of Crime, Law and Deviance, Volume 10, 11–34

ISSN: 1521-6136/doi:10.1016/S1521-6136(07)00201-1

coercive capacities. However, in contrast to community policing, the Minutemen are not an instance of the state or police engaging or reaching down into the public, but represent a distinct segment of the public reaching up and aligning itself with the "arms" of the state.

INTRODUCTION

Nation-states are fundamentally invested in securing and governing their territorial borders and posses the legitimate authority – backed by the threat of force – to do so. As stated by Joseph Carens "Borders have guards and the guards have guns" (Carens, 1987, p. 251). His succinct observation remains true to this day. As interfaces between the foreign/domestic divide, national boundaries – while often taken as axiomatic and uncontested – are social constructs formed through extensive institutional surveillance, coordination and policing. In demarcating, regulating and protecting the liminal spaces of the social order, policies governing citizenship and "border control" enhance the nation-state's authority over a bounded territory and population and provide mechanisms of social closure and control (Brubaker, 1992). Through such controls, nation-states exercise their sovereignty and communal rights to self-determination, establishing their status as administrative units, political identities and sociocultural complexes. Additionally, when deciding who can formally enter and join the receiving society, states provide normative evaluations of what their members *should* look like (Joppke, 1999; Walzer, 1983).

Although seemingly paradoxical, the above observations are especially salient given the present phase of intensified globalization. With new levels of planetary interaction and interdependence advanced capitalist states have generally agreed to open their borders to transnational flows of capital, goods and services, but have yet to cede their sovereign authority in regulating migratory flows – even as globalization has expanded the scale and scope of human mobility. Rather than "losing control" (Sassen, 1996), many countries have actually recalibrated their migration policies to obstruct flows of "unwanted" and undocumented populations. As a result, access to citizenship and the "legitimate means of movement" (Torpey, 2000, p. 4) have emerged as the "major new axis of inequality" within the late-modern period (Robinson, 2006, p. 82). Nowhere are such trends more evident than along the United States' southern border.

In the US context, undocumented border-crossings are not new. The socio-legal categories of "alien" and "illegality" can be traced back to the Southwest's conquest and annexation in the 19th century and the Border Patrol's foundation in the early 20th century (Ngai, 2004). What is new is the present "escalation" of border policing and surveillance following NAFTA and the economic liberalization of North America. Since the early 1990s immigration has been elevated from a socioeconomic to law enforcement and national security issue governed by the logic of "prevention through deterrence" (Andreas, 2000). This regulatory shift has included, inter alia, new restrictive legislation, rising law-enforcement budgets, the utilization of high-tech surveillance and information technologies and the adoption of high-profile paramilitary enforcement initiatives ranging from Operation Gate-Keeper to Operation Hardline (Andreas, 2000; Nevins, 2001). Since September 11, 2001, border securitization strategies have further intensified. Media demagogues, government officials and political figures have all invoked the border as an ungoverned frontier prone to terrorist infiltration. According to republican congressman Duncan Hunter (California): "You have to be able to enforce your borders. It's no longer ... an immigration issue. It's now a national security issue" (Zellen, 2006, p. 3). Such claims have culminated in increased personnel and resources dedicated to "border security" and the establishment of the Department of Homeland Security as the central policing agency (Salter, 2004). Reinscriptions of border control and national spatiality in the wake of the global "war on terror" challenge claims of the "end of geography" and the emergence of a "borderless" and "deterritorialized" world (Albrow, 1996; Ohmae, 1990). As the US case demonstrates, they are perhaps less relevant as economic regulators and cultural containers, but, as noted by Michel Trouillot (2001), "National states produce countries and countries remain fundamentally spatial" (p. 133). In light of such developments, it may be more accurate to say that the significance and functions of territoriality are being transformed rather than retreating in total.

In the foreground of this "re-bordering" (Andreas, 2000) within the United States, are significant and emergent alignments across the state–society spectrum. This chapter empirically investigates these novel dynamics by analyzing the Minutemen, a US-based vigilante movement dedicated to securing the nation's borders via "border watches" which, in many ways, embody the spirit and rationale of community policing. Border vigilantes have attracted significant attention from the press, civil-society groups and policymakers, but have received little, if any, scholarly attention. This investigation fills this gap by answering the following questions: What happens when the nation's gatekeepers are no longer exclusively state-sanctioned agents but emerge voluntarily from within the social body? How do such instances render the concepts of governance and surveillance intelligible? Scholars have already noted that many of the mechanisms of border control have been shifted "up" to supranational institutions and "out" to the market through carrier sanctions and other forms of "remote control", where travel agents and airline personnel are expected to deter unauthorized entries (Guiraudon & Lahav, 2000; Walters, 2006). I want to shift analysis downwards toward the social body and interrogate the role of organized citizens in directly monitoring and governing the limits of the political field. Despite their status as non-statal actors, by appropriating, enforcing and extending many of the administrative principles of governance and statecraft – whether surveillance, discipline, territory, identity or security – the Minutemen force scholars to reevaluate the enterprises of government, social control and "bordering".

The remainder of this chapter is divided into four parts. First, I survey the role and function of collective surveillance within modern territorial states. Second, the pre-existing literature on community policing, surveillance and government is reviewed. This assessment is informed by Michel Foucault's work on discipline and "governmentality" (Foucault, 1977, 1991) and uncovers how the Minutemen, like community policing, highlight the pervasive and decentralized nature of government, social control and surveillance. Third, the Minutemen's history, ideology, practices and interactions with state authorities are analyzed. As will be demonstrated, unlike community policing, the Minutemen are emergent and represent a unique attempt to forge grassroots partnerships with the "arms" of the state. Fourth, I conclude by briefly summarizing the group's theoretical and conceptual implications as well as identifying issues for further research. As a final caveat, while striving to be comprehensive, space constraints have left many questions unanswered, dilemmas unresolved and have precluded a full discussion of the issues at hand. Despite these limits, this chapter will hopefully articulate the previously unobserved and provide an analytic lens for orienting future work. The propositions advanced here are therefore exploratory and are intended to challenge accepted orthodoxies, generate debates and uncover new avenues of investigation.

COLLECTIVE SURVEILLANCE AND THE MODERN TERRITORIAL STATE

Although often ignored in traditional analysis, recent scholarship has stressed the importance of the nation-state's ocular capacities in the successful institutionalization and reproduction of national societies. Centralized and "heightened" surveillance, or the ability to systematically observe, categorize, evaluate and influence the actions and behaviors of individuals and groups, is a significant cause and consequence of both modernity and the nation-state (Dandeker, 1990; Higgs, 2001; Lyon, 1994; Starr, 1987); "reaching an intensity quite unmatched in previous types of societal order" (Giddens, 1987, p. 312). Collective surveillance has come in two predominant forms: (1) the collection and storage of official information, whether censuses, national accounts or tax-lists, meant to document, enumerate and serialize the attributes of the social body and (2) direct forms of hierarchical supervision embedded in multifarious state institutions, whether courts, the police, welfare state, social services, or educational systems, invested in monitoring, directing and normalizing the behavior of collectivities (Giddens, 1987; Gorski, 2003). Carried out by experts, bureaucrats and other state functionaries, the instrumental benefits of surveillance are primarily "infrastructural" and "logistical" (Mann, 1986); they expand the depth of political knowledge and intervention by allowing the various organs of government to efficiently measure, predict, coordinate and organize social life through a panoply of technical controls. Collective surveillance ensures that the state is able to "embrace" (Torpey, 2000), "penetrate" (Mann, 1986), "reflexively monitor" (Giddens, 1987) and render visible or "legible" (Scott, 1998) even the smallest details of daily life. Alongside their administrative power, seemingly quotidian practices of state surveillance, whether maintaining "files", drawing maps, undertaking censuses or creating passports, are also profoundly symbolic in their implications (Anderson, 1991; Torpey, 2000). In particular these insidious gazes naturalize state power and national identity by constituting that which they claim to merely represent (Loveman, 2005). As noted by James Scott (1998), "The builders of the modern nation-state do not merely describe, observe and map; they strive to shape a people and a landscape that will fit their techniques of observation" (p. 82).

Discussions of surveillance's role in institutionalizing state power have almost exclusively adopted an internalist perspective, focusing on strategies of observation and policing within the body politic rather than at its boundaries (see Torpey, 2000 as an exception). This is puzzling given that collective surveillance has played a significant role in transforming the spatial principles of modern political power. With the interstate system's gradual expansion permeable frontiers became sharply demarcated socio-political borders (Giddens, 1987; Scott, 1998). "Bordering" has subsequently emerged as a powerful administrative practice intended to "produce" nationals, fuse collectivities into institutionally and symbolically bounded containers and endow the social order with and inevitable, necessary and desirable quality (Balibar, 1991; Newman, 2006). Drawing attention to these features, Timothy Mitchell (1999) has argued that: "establishing a territorial boundary to enclose a population and exercising absolute control over movement across it ... help manufacture an almost transcendental entity, the nation-state" (p. 90). As a disturbance to these state-building efforts *international* migration emerged as a uniquely modern political problem. Since the late 19th century national borders have been constructed as institutional zones of regulation, whether through law-making, the founding of administrative offices, the centralized collection and storage of immigration statistics or the establishment of a matrix of surveillance, sequestration and documentation that includes border patrols, barbwire fencing, military checkpoints, medical inspections, X-ray machines, surveillance cameras, integrated biometric databases, visas, passports and other official forms of identification (Brubaker, 1992; Sassen, 1999; Torpey, 2000; Zolberg, 2000). Border controls thus demonstrate that while the state's eye is found throughout the social field, nowhere is its gaze more encroaching than at its jurisdictional boundaries. Additionally, borders demonstrate the individuating and totalizing elements of collective surveillance (Giddens, 1987; Walters, 2006). When states govern their boundaries they are simultaneously interested in both controlling individual movement and dividing human collectivities into segmented and reproducible cultural and governmental units.

SURVEILLANCE AND GOVERNANCE: FROM AN INSTITUTIONAL TO PRACTICAL ASSESSMENT

The expansionary and centralizing nature of state surveillance is, however, part of a larger double movement as the mechanisms and techniques it embodies, whether administration, observation, documentation or discipline, have been horizontally dispersed throughout society, surfacing in multiple institutional sites, whether churches, clinics, factories, families or schools (Foucault, 1977; Lyon, 1994; Staples, 2000). While traditional structural and institutional definitions of government and statecraft would exclude such agents and locales from analysis (see Poggi, 1990; Skocpol, 1979), a potentially more fruitful method of interrogation is found within Foucault's concept of "governmentality" (Foucault, 1991).

Expanding upon his earlier attempts to articulate a "micro-physics of power" that was diffused throughout society, operated on the body and permeated subjectivities through pedagogical and corrective regulations (Foucault, 1977, 1980a, 1980b), Foucault's later years were dedicated, in part, to reconstituting definitions of political power to include a web of decentered and localized practices. In place of the traditional fixation with the state and rulers, Foucault defined "government" as any number of regulatory schemes implicated in the "conduct of conduct" (Foucault, 1982, p. 221). More specifically, the term was meant to capture calculated and systematic techniques "both internal and external to the state" (Foucault, 1991, p. 103) intended to affect, mold and manage individual subjects, social groups or entire populations in the interests of directing their actions and behaviors toward specific ends. In parallel with disciplinary power, government was conceptualized *practically* rather than institutionally, representing a "modality", "instrument" and "technology" of power (Foucault, 1990; Rose, 1999) irreducible to a central system of institutional coordination (Rose & Miller, 1992; Mitchell, 1999).

Foucault's insights allow the researcher to venture beyond the empirically obvious and avoid reifying statecraft as a unified process carried out exclusively by a country's formal agents and institutions. Rather than stable and coherent, in practice, the state's boundaries with society are permeable and elastic. While projected as legally distinct domains, through the processes of legitimation, consent and incorporation, the political and the social are in actuality interdependent and interlocking (Abrams, 1988; Gramsci, 1992). As noted by Timothy Mitchell the state is best conceived of as the "network of institutional mechanisms through which a certain social and political order is maintained" (Mitchell, 1999, p. 83). While derived from the modern liberal state and conceptualizations of "civil society" as an independent realm of self-government and free exchange, the indeterminate location of the state is increasingly evident given the rise of new "advanced" or neoliberal models of government and "state restraint" (Jessop, 1994; Rose, 1999). Within the last 30 years extensive socioeconomic intervention and planning intended to manage societal risks, has gradually given way to privatized and individualized strategies of government defined by communal autonomy, self-management and the "hollowing out" of national regulatory measures (Jessop, 1994; Rose, 1999). Rather than signaling a wholesale retreat of the state, present models of government provide examples of its "deplacement" – or reworking and realignment – as state "functions and effects" (Trouillot, 2001) are obtained by citizens and social groups, whether businesses, religious groups, communities and other private organizations (Rose & Miller, 1992; Rose, 1999). These new governmental configurations transcend "the binary and all encompassing opposition between rulers and ruled" (Foucault, 1990, p. 94) and represent a

> pluralization of the centre, [where] the problems of the state to rebound back onto society, so that society is implicated in the task of resolving them. What ... seems to be taking place ... is a ... hybridization of the private and the public, the state and the civil – which has been the very principle of the social for the past century or more. (Donzelot, 1991, p. 178)

In undertaking traditionally statal roles, "non-governmental" actors contribute to the maintenance and reproduction of the social order, demonstrating they are not merely the passive subjects of state power, but are critical constituents in its exercise and realization (Rose & Miller, 1992).

One decisive example of the heterogeneous and multivalent nature of government and surveillance is contemporary strategies of community policing and social control. These practices rely on the recruitment and active participation of community members in police activities, whether through volunteer work (ride-alongs, traffic control) or the initiation and formalization of community-based systems of surveillance, ranging from, volunteer-led patrols, neighborhood watches and other "crime-stoppers" groups (Cohen, 1985; Fielding, 2002; Marx, 1989). Many authors have noted the profound import of community policing as it impacts the core elements of modern political authority. Such initiatives represent the development of alternative and intermediate forms of order enforcement that are non-bureaucratic, locally situated and straddle the state–society boundary (Merry, 1992). Since the police are interested in maintaining a single, cohesive notion of order and represent the state's "heart" as the "domestic specialists in the exercise of legitimate force" (Reiner, 1992, p. 762), new partnerships with private citizens highlight shifts in government responsibility and the emergence of new sites of social reproduction. The domain of government is also widened as citizens are incorporated into practices of surveillance, order enforcement and more generally statecraft. Community policing and other attempts by the government to enroll and craft watchful, responsible and empowered individuals and organizations to legally regulate human activity, enhance the observational and enforcement capacities of the police, allowing them to establish a pervasive and omniscient presence (Darian-Smith, 1993; Saunders, 1999). Citizens and community members serve as the police's "eyes and ears" thereby extending their gaze and establishing new lines of communication and informational exchange (Ericson, 1994). Such programs also serve as mechanisms of communal affirmation that seek to recuperate and reinvent the neighborhood and community as stable and enduring moral entities defined by common goals, norms and perceptions of the larger social order (Crank, 1994). While demonstrating its re-articulation and redistribution, community policing ultimately acknowledges, endorses and legitimizes state power; first, such programs are generally initiated and funded by local police forces and national foundations, and second, by falling on the "right side of the law" and cooperating with government authorities, citizen-based initiatives enforce and uphold the pre-existing socio-legal order (Cohen, 1985; Marx, 1989).

"SEEING LIKE A STATE": THE MINUTEMEN, SURVEILLANCE AND BOUNDARY MAINTENANCE

The preceding sections have outlined (1) the centrality of surveillance in the exigencies of statecraft, nation-building and border control and (2) an expanded, *practical* conception of government in which surveillance, policing and social control are obtained by a plurality of societal agents external to the central state apparatus. The current section situates these observations empirically, analyzing the Minuteman Project – a self-proclaimed "national citizens neighborhood watch" dedicated to "securing the American border" (Simcox, 2006a). The group's history, ideology, practices and relationship with the state, all reveal how citizen involvement in policing, surveillance and, more generally, government, has spread to new domains of state power and political authority. As a result sovereignty, statecraft and boundary maintenance are all defined by a widening net of bordering mechanisms and disciplinary schemes initiated and obtained within the social field.

History

Vigilante patrols along the United States' borders have existed in a number of different forms over the last 150 years. However, within the last 15 years citizen-based organizations attempting to police the border and "take the law into their own hands" have proliferated (Andreas, 2006; Walker, 2007). These groups, ranging from the American Border Patrol, Civilian Homeland Defense, Ranch Rescue and the Arizona Guard have generally existed as spontaneous, local and ad-hoc groupings of citizens (Vina, Nunez-Neto, & Weir, 2006). However, in April 2005 the Minuteman Project – the largest and most politically active vigilante group – drew international attention as hundreds of volunteers converged upon a narrow 23-mile stretch of the Arizona–Mexico border to monitor the movements of suspected undocumented migrants and "assist" the US border patrol. Unlike prior movements, the group engages in extensive media campaigns, actively recruits new members and maintains a highly institutionalized organizational structure (Vina et al., 2006). Additionally, whereas other civilian patrols – most notably the now defunct Ranch Rescue – have used paramilitary tactics to track and physically apprehend suspected undocumented border crossers, the Minutemen's website, www.MinutemanProject.com, stresses that the group is "non-confrontational" and "not a military-oriented organization".

Jim Gilchrist, a retired Californian businessman, founded the Minuteman Project in October 2004. Much of the planning and organizing behind the group's foundation was also carried out by Chris Simcox who has since split with the organization and founded the closely related Minutemen Civil Defense Corps (Vina et al., 2006).[1] The ultimate objective of the group is the implementation of a series of draconian measures that would ensure the end of all undocumented migration and the permanent fortification of America's borders. "Adequate" state measures are envisioned to include both the complete militarization of the United States' borders and the criminalization of "illegal" populations through stricter societal regulations, whether rigorous enforcement of employment laws, the denial of social services or mass roundups and deportations (Gilchrist & Corsi, 2006; Simcox, n.d.). However, embracing the potential of communal forms of social control and viewing the government as "hamstrung" and "paralyzed", the organization proclaims it their "civic duty" to "secure" the country's borders and deter "illegal" immigration – a force that is simultaneously labeled as an economic, cultural and security threat (Feldman, 2006). Since April 2005 the group has mobilized several "border watches" to "observe and report" undocumented border-crossings. These actions serve a dual purpose; they offer direct assistance to state agents and are a symbolic publicity-stunt meant to demonstrate that "border security" is attainable by drawing attention to what is viewed as a failed set of policy choices. Since the group's foundation civilian patrols have been extended to other border states in the southwest and along the Canadian border (although actions and resources remain clearly targeted at the former) and local chapters have emerged in nearly every state. In late 2005, the group claimed to have over 4,000 official national members and several more sources of financial and political support (Turnball & Tu, 2005).

Ideology

Underlying the Minutemen movement is a series assumptions and beliefs about the nature of the social order and the necessity of their actions. Like other social movements, the ideational themes and rhetorical maneuvers disseminated by the group serve as a complex set of "cultural frames" that involve "collective processes of interpretation, attribution and social construction that mediate between opportunity and action" (McAdam, McCarthy, & Zald, 1996, p. 2). The framing process is particularly important given the movement's highly symbolic mission of drawing public attention to immigration and border security. Ultimately, however, it is employed to cultivate a subterranean form of statecraft and border control. As a result most of the visual and verbal emblems associated with the group and its objectives are derived from the language and principles of the nation and state. Four discursive foci are especially central in constructing migration, the border and the group's actions: (1) legality, (2) nativism/nationalism, (3) security and (4) patriotism/citizenship. Additionally, the group has pursued an extensive media campaign to amplify these frames.

Central to the group is the language of legality. Compared to other border vigilante movements, the Minutemen's official message appears less openly racist as the perceived problem is defined in formal terms: "illegal" migrants, and not migrants per se, are the official targets of the group's acrimony. Legality is therefore mobilized to give the group's claims a heightened level of normative and moral authority. In transgressing the nation's jurisdictional boundaries, migrants are constructed first and foremost as threats to its sovereignty and legal authority. Rigid proceduralism is also used to justify the group's actions. The organization's founders have gone to great lengths to project its members as nothing more than a collection of responsible citizens "operating within the law, to support [the] enforcement of the law" (Simcox, 2006a). Rather than renegades that circumvent and usurp the nation-state's juridical authority, group members voluntarily fill gaps in the state's administrative power.[2]

Although appearing less xenophobic than other border vigilantes, upon closer scrutiny the Minutemen clearly rest upon nativist and nationalist principles. In continuity with prior anti-immigrant movements, migrants provide convenient scapegoats for a host of societal ills (Higham, 1988; Sanchez, 1997). The Minutemen have portrayed immigrants as corruptive: upsetting the economic, cultural and social integrity of American society.[3] In spite of a wealth of countervailing evidence, the group asserts that migrants depress wages and are a drain on social services and government largess. Additionally, the group provides a series of reactionary cultural arguments as its leaders view migration as eroding cultural values, dividing allegiances, attenuating social bonds and creating a fragmented society composed of disparate and conflictive parts. Despite their disavowal of ties to white supremacist organizations and appeals to America's immigrant past (Gilchrist & Corsi, 2006; Simcox, 2006b), the organization articulates an insular and ethnically absolutist American identity founded upon the removal and absorption of linguistic and cultural difference. The group's leadership has proclaimed undocumented migration to be tantamount to "colonization" and "balkanization" (Simcox, n.d.), ensuring that "future generations will inherit a tangle of rancorous, unassimilated and squabbling cultures with no common bond to hold them together" and "political, economic and social mayhem" (Turnball & Tu, 2005, p. 1A). According to one group member, the border is both a political and communal necessity: "A border is like your skin. There's got to be a place where you stop and something else begins. A nation can't exist without a border for long." (Hendricks, 2005, p. 1A). Despite allusions to collective identity and belonging, the ethnonationalist ideology of the group transforms citizenship into an exclusive and closed category of cultural identity by embracing a fictive homogenous and monolingual past.

Further indicating the group's nativist position, "illegal" migration is also projected as a vital threat to the biological and physical security of the social body. Employing racialized characterizations and neo-eugenicist arguments, undocumented migrants are regularly referenced as spreading diseases, leading to overpopulation and introducing criminals, ranging from "drug smugglers, murders and rapists" (Simcox, n.d.) into the United States (Nash, 2005). Additionally, the present state of emergency or "exception" (Agamben, 2005) induced by the global "war on terror" has allowed the Minutemen to go even further in constructing the border as "under siege". In the wake of September 11, the group has argued that the country's borders are "hemorrhaging" and have been breached by a "human flood" of terrorists and "predators" (Simcox, 2006a, 2006b). Tropes of terror, insecurity and invasion have served as expedient symbolic currency: conflating two drastically different phenomena under the common rubric of national security and converting the immigrant into an automatic security risk and potential terrorist. These perceptions, encouraged by the mass media and Bush administration, have pervaded mainstream public discourse and have added a new appeal and legitimacy to restrictionary groups and policies. While the Minutemen have been met with great contention for their historical amnesia, myopic view of border security and racism and virulent nationalism, currently their actions are more readily accepted due to the public's perception of border control and immigration as security rather than socioeconomic issues.[4]

Finally, idioms of patriotism and citizenship are engaged. In addition to viewing transnational migrant flows as dislocating the core rights, entitlements and sentiments associated with national citizenship and belonging, the group also proclaims it their "civic duty" to secure the country's cultural and territorial borders. In the months following September 11, President Bush reminded the nation that "our citizens have new responsibilities, we must be vigilant" (CNN, 2001). Reflecting these calls and embodying the voluntary, enterprising spirit favored by neoliberalism, the group's name and logo are derived from the civilian-led militias of the Revolutionary War. Alluding to citizenship as participation in ruling and being ruled, the group views itself as more than simply an interest or advocacy group, proclaiming that "we, as citizens, are the government" and that "eternal vigilance [is] the price of freedom" (Simcox, 2006a, n.d.). Their self-acknowledged responsibility for border security highlights the Minutemen's ambiguous position between the state and the market and their willingness to adhere to the principles of personal responsibility and civic empowerment. Additionally, the imagery of the nation's foundation and the virtues of the warrior–citizen both engage the merits of preparedness, loyalty, self-sacrifice and the local provision of security for family, community and nation. Like forms of community policing, the Minutemen also summon a mythical imagined past that their actions are meant to reinstate. However, rather than the tight-knit local community it is the nation, and the purported linkage of soil, people and identity this encompasses, that is meant to be recuperated.

In addition to appropriating strategic symbolism and powerful political imagery, the Minutemen have relied extensively on the electronic media and mainstream press in promulgating their goals, image and ideology. According to Jim Gilchrist, a central aspect of the organization is "bring[ing] the message of patriotism ... 'by any media necessary'"

(Gilchrist, 2006). The group maintains state-of-the-art websites, issues press releases, has a regular staff of media liaisons and has created blog posts, mass emails and multiple radio and newspaper advertisements. In the group's early months, Gilchrist and Simcox both made good use of the internet in mobilizing potential volunteers; whether veterans or former Border Patrol and law enforcement agents (Center for New Community, 2005). However, starting with their high-profile "deployment" in April 2005, the group has been widely successful in garnering coverage and within the popular press. During this initial "border watch", the group was covered by every national news outfit and outnumbered by reporters. This spectacle has been described as "one of the great media simulacra of recent times" (Center for New Community, 2005, p. 6).

Practices

The Minuteman project has pursued its objective of "secure[ing] the sovereign territory of the United States" (Simcox, 2006a) through: (1) "border watch" campaigns and (2) "internal vigilance" measures, whether monitoring day labor sites or lobbying at the local, state and federal levels. These practices demonstrate how the group appropriates, enhances and achieves the logics and effects of the territorial state. With the exception of lobbying, these actions reflect efforts to bolster and provide the state's surveillance and gatekeeping functions. Although unsolicited and unauthorized, by casting their gaze on the border and upholding, reclaiming and renationalizing the political principles of territory, order and identity, the Minutemen have added a new node to the broad archipelago of observational practices that exceed the boundaries of state bureaucracy and other formal institutions.

The "border watch" or "civil defense" arm of the organization embodies two measures of gatekeeping and boundary maintenance: direct surveillance and sequestration. The first is linked to self-organized border patrols, while the second involves the erection of high-tech fencing on private land near the border.

During the height of their nationally orchestrated operations in April 2005, the Minutemen patrolled 23 miles of the Arizona–Mexico border, and drew from 857 participants, 200 of which were generally active at any given time (Sterngold & Martin, 2005). Proclaiming themselves to be popular gatekeepers and "force multipliers", these civilian-led patrols were indented to supplement the state by setting up surveillance outposts, documenting irregular entries and relaying information to the Border Patrol (Simcox, 2006a, 2006b). Since these initial actions, several other border watch campaigns – with names ranging from "operation sovereignty" to "operation spotlight" – have been undertaken. During these campaigns volunteers are assigned roles as either minutemen, those physically standing watch on the border, or signalmen, those responsible to transmitting communications to other group members and the authorities when necessary (Gilchrist, 2006). Minutemen are often given binoculars, night vision goggles, Global Positioning System receivers, cell phones and walkie talkies to assist in monitoring the border and communicating with either the Border Patrol or signalmen. Additionally, private helicopters and airplanes are frequently used to enhance the gaze of those stationed on the ground (Gilchrist, 2006; Simcox, 2006a). The primary objective of these patrols is to discipline the border and "alien" bodies by tightly controlling the conduct, location and movements of the later. Mirroring the observational nature of disciplinary power, the group embodies an "anti-nomadic technique" that "arrests or regulates movements ... dissipates compact groupings of individuals wandering about ... [and] establishes calculated distributions" (Foucault, 1977, p. 219). Signaling an extended web of governmentality on the border, migrants are frequently referenced as "targets" to be "intercepted" and contained (Gilchrist, 2006; Simcox, 2006a, 2006b).

Although many carry firearms – estimates say around 40% – volunteers are explicitly instructed to avoid all interaction with suspected border crossers and follow the group's standard operating procedures of "no contact, no engagement" and "observe, report, record and direct" (Gilchrist, 2006; Simcox, 2006a, 2006b; Vina et al., 2006). Participants are conceived of as assisting the state through their collective gaze and vigilant presence rather than coercive or extra-legal measures. According to one group member, more than anything the group expands the state's ocular capacities by providing an "extra pair of eyes" during these "extraordinary times" (in Juffer, 2006, p. 664). Nonetheless, in attempting to enhance the surveillance and bordering capacities of the United States, the group ensures that territoriality continues to function as a mechanism of social control and exclusion.

Since the closing months of 2006, the group has also repaired damage to current border fencing and has begun construction on new fencing on private land near the border. These "border fence" operations reflect attempts to create more permanent methods for strengthening border architecture, enforcing sedentarization and documenting the movement of irregular migrants. Currently, the group is offering landowners near the border free perimeter fencing in exchange for permission to build on their land. The fences constructed by the group consist of two 10-ft high "no climb" steel-plated panels similar to those constructed by the Israeli military in Gaza (Zellen, 2006). Additionally, once completed, the fencing is to be equipped with barbwire, fiber-optic sensors and surveillance cameras that can be accessed online. Using real-time video streaming, these cyber-panopticons will allow volunteers throughout the world to remotely "scan the border, look into Mexico ... and track illegal aliens that cross" (Simcox, n.d.). Citing the neoliberal principles of voluntarism, market rationality and communal autonomy, Simcox, the brainchild of the operation, explains it is meant to show Washington "how easy it is to build these fences ... by private people and free enterprise" (Zellen, 2006, p. 3). As testaments to the group's investment in societal-based securitization, communal responsibility and market-based solutions, all of the funding, engineering and labor for the project have been obtained through private channels. In addition to surveyors and contractors who provided assistance in the fence's planning and design, FOMGuard, a Washington-based private security firm, donated over $7 million in fencing and surveillance materials (MCDC, 2006).

Going beyond the border, the group's members have also spread their gaze throughout the social field by monitoring day labor centers. "Internal vigilance" campaigns have become a major focus of the movement and are now organized in a number of states including Colorado, Texas and Virginia. These campaigns involve staking out meeting places, videotaping the interactions of contractors and suspected undocumented migrants, recording the former's license plate numbers, and threatening to inform the local authorities or IRS (Feldman, 2006).

All of these practices demonstrate how the "objectifying" gazes of private citizens have been employed to exclude foreign "others" and "aliens" from the national territory and society. However, the group is equally invested in self-discipline as its members are taught to employ a number of "subjectifying" or internal gazes. Even more so than citizens engaged in community policing, Minutemen are constantly reminded to police their own actions and those of others (Gilchrist, 2006; Simcox, 2006a, 2006b). According to the group's code of conduct, volunteers are to be courteous and must operate with "civilized bearing" at all times; "by exhibiting a courteous and professional demeanor, we will continue to gain the trust and support not only of our fellow citizens but of the nation as a whole" (Gilchrist, 2006, p. 5). Group members are forbidden to wear military uniforms, carry handcuffs, openly display firearms and are required to document all external contact and communication through either audio or video recordings (Gilchrist, 2006; Simcox, 2006a). The movement's participants are therefore envisioned – much like the agents of the state – to be restrained, rational and disciplined subjects.

Interactions with State Authorities

The ideological underpinnings and practices of the Minutemen uncover a clear symmetry with programs of community policing. In both instances participants are mobilized by fears of "outsiders" and perceptions of insecurity. Both also represent attempts to reclaim the neighborhood, community, border or nation by embedding private citizens within the process of order enforcement and policing. By participating in collective surveillance civilians serve as co-producers of local and national security. However, in contrast to community policing, the Minutemen's relationship with the state apparatus is unique and warrants an expanded conceptualization of citizen participation in the legal regulation of human activity. Despite their investment in the maintenance and enforcement of territorial boundaries, the Minutemen occupy a hybrid *boundary-spanning* position on the sociopolitical spectrum. In particular, the group occupies a political borderland between the governmental and social or state and society. In place of extensive interfacing with state authorities, security experts and other elites, the Minutemen are emergent and self-organized. Their alignment with the state is largely organic occurring through grassroots mobilization.

While unofficial, the group's actions have not been completely rejected by political figures and state bureaucrats. Many congressional members have expressed sympathy and support for the group's cause. In May 2005, the Minutemen were invited to meet with members of the Congressional Immigration Reform Caucus by its ultra-conservative chair Tom Tancredo (Juffer, 2006).[5] Governor Schwarzenegger of California has also come out in support of the group, indicating he would support the deputizing of civilian border patrols (Sterngold & Martin, 2005). Additionally, referencing the Minutemen, Jarrod Agen, a spokesman for the Department of Homeland Security, noted that "homeland security is a shared responsibility and the department believes … the American public plays a critical role in defending the homeland" (Mansfield, 2005, p. 1A). However, despite garnering some level of political support, endorsement, coordination and assistance from the relevant federal agents has yet to materialize. Retracting upon previous statements, the Department of Homeland Security currently maintains it "does not encourage members of the public to conduct law enforcement activities on their own" (Vina et al., 2006, p. 14) and has no future plans to incorporate civilian volunteers into its border control efforts. Additionally, the Bush administration has dismissed the Minutemen as vigilantes and continues to adopt a stance of effective neutrality, neither endorsing nor obstructing the group's actions (Vina et al., 2006). While endorsing the role of societal actors in the provision of social welfare and human security, when it comes to national security and the country's borders, the Bush administration has made it clear it prefers its own agents and institutions.

The Minutemen's exteriority to the state apparatus – although by no means politics – raises a number of interesting issues. Neighborhood watches and community patrols are testaments to the decentering and expansion of social control, surveillance and government; however, these are instances where government initiation and intervention is intimately connected to the maintenance and cultivation of communal autonomy and local control. Community policing therefore exists as a state-based strategy that recognizes the value of police community partnerships. The Minutemen, in contrast, operate entirely within what has traditionally been labeled the domain of civil society, all the while undertaking tasks and functions that, in symbolic and administrative terms, are patently matters of state. While state agents eventually play an important role in their practices, this does not diminish the fact that the networks of surveillance, communication and assistance provided by the group have been initiated entirely at the societal level. Rather than the state and police engaging and reaching down into society to incite, enable and manage the capacities of self-government, the Minutemen are an example of the reverse: a distinct segment of the public "reaching up" and aligning itself with the "arms" of the state. In doing so, the group indicates that sovereignty and territoriality are potentially articulated and achieved at multiple levels.

Despite their subterranean nature, the group exists as a form of popular mobilization and collective action that supplants and reinforces – rather than contests – state power. Engaging the rhetoric of nation, order and law to secure the country's territorial borders, the motivations and activities of the Minutemen are largely governmental. While tainted by their clearly private agenda of capturing public attention, transforming policy and eliminating immigration, the movement cannot be reduced to an instance of oppositional politics. The group's leadership is in full support of the nation's gatekeepers and notes "we're here to support, not hinder, local, state and federal law enforcement officers and agents ... our relationship with law enforcement is just as important as our relationship with the local community" (Gilchrist, 2006, p. 8). Even as the group finds fault with the state qua bureaucrats and political leaders, their ultimate goal is not to abandon or suspend the law but to heighten its enforcement and strengthen and safeguard state sovereignty and the social order it purportedly encapsulates. However, in complementing the observational capacities of the state, the movement also reveals an underlying instability in the concept of the state itself: processes and locales that exceed its formally projected boundaries inevitably constitute it.

REINVISIONING SOCIAL CONTROL: THE SOCIETAL BASES OF ORDER, BORDERS AND GOVERNMENT

By way of conclusion, the Minutemen force us as scholars to reevaluate and modify the existing concepts and assumptions through which we interrogate the social world. As this chapter has acknowledged, comprehensive systems of order enforcement, bordering and surveillance were all central to the emergence of the modern nation-state as an administrative, political and social unit. However, assumptions that the state is the exclusive agent in these endeavors ignore a host of societal processes and agents that, while failing to coalesce around national institutions, provide similar roles and functions. Advancing a practical definition of government, I have used the examples of community policing and the Minutemen to demonstrate how private citizens assist in enforcing and upholding the nomos of the state and social order. In doing so, this chapter has expanded analysis of social control and surveillance to sites, scales and agents that have been neglected by received theories.

Many authors have already discussed the political significance of community policing and diffuse practices of surveillance within modern societies. Nevertheless, the Minutemen and communal forms of boundary maintenance have yet to receive substantive commentary. In overcoming this silence I have analyzed the historical, ideological, organizational and political dimensions of the group, arguing that the Minutemen are taking part in an enterprise exclusive to states since the 19th century, telling people where they should be, enforcing sedentarization, and documenting their movements. The contemporary resurgence of border controls and political territoriality throughout the advanced capitalist world reminds us, as Ralph Miliband (1969) has stated, that we "continue to live in the shadow of the state" (p. 1). The Minutemen require this statement to be qualified.

The human consequences of the state's "shadow" are nowhere stronger than at its borders, however, vigilante patrols have ensured that this shadow is created by multiple dispersed sources. In proclaiming it their duty to uphold the spatial, legal and security principles of the social order though their vigilant gaze, the group pursues a neoliberal model of surveillance, policing and border control. In place of depending upon the state to manage the purported risks and insecurities of social life, the group – exercising voluntary associations and the "virtues" of self-help and personal responsibility – have taken it upon themselves to initiate, enhance and extend many of the central properties of statecraft. Border vigilantism, thus, points to a larger trend identified by Hardt and Negri (2000), the emergence of "societies of control" where the

> mechanisms of command become ever more 'democratic,' ever more immanent to the social field, distributed throughout the brains and bodies of the citizens. The behaviors of social integration and exclusion proper to rule are thus increasingly interiorized within ... subjects themselves. (Hardt & Negri, 2000, p. 23)

The Minutemen indicate that border surveillance and policing are now, in certain respects, tasks framed as everyone's "civic duty" and not just the responsibility of leaders, experts or the state.

This case study has argued that scholarly discussions of surveillance and government must remain open to locales and agents that exceed the accepted and more obvious boundaries of analysis, however, many issues have fallen beyond its scope of inquiry. Further research is both desirable and necessary. In place of societal groups invested in recuperating, reinventing and intensifying the exclusive psychic, cultural and territorial boundaries of the nation-state, subsequent investigations could focus on actors invested in contesting the Minutemen's actions and the insular and unjust nature of borders themselves. Groups ranging from the Border Action Network, the Anti-Defamation League, Amnesty International, No More Deaths and The Human Rights Coalition/Indigenous Alliance Without Borders, have all challenged border securitization and the mistreatment of migrants by both vigilante movements and the Border Patrol (Center For New Community, 2005; Vina et al., 2006). Through cross-border mobilizations, the sharing of resources with the public and organized actions – ranging from protests to leaving water and maps throughout the harsh terrain of the Sonora desert – these groups represent attempts to construct alternative and more humane cartographies in which narrow visions of national security and border control give way to broader ones of human security and community.

NOTES

1. Since the ideology, organizational structure, participants and practices of both groups are largely identical (e.g., both espouse a right-wing populist message, maintain a "no contact, no engagement" policy and use the same training manuals and handbooks) and since both organizations were thoroughly integrated for most of their history, for the purposes of this chapter they will be referenced synonymously as the Minutemen or Minuteman Project.

2. The Minutemen also seem to implicitly acknowledge the flexibility of the law when faced with exceptional circumstances (Agamben, 2005; see also Gilchrist & Corsi, 2006): if their practices seem legally ambiguous, this is no more so than the federal government's recent attempts – through the Patriot Act, torture and domestic spying – to suspend the law under the guise of supporting the "rule of law" and national security.

3. In contrast to classical nativist movements at the turn of the 20th century, the Minutemen's rhetoric is based predominantly on cultural, rather than biological, understandings of race and ethnicity.

4. Recent public opinion polls indicate that 77% of American's believe the government is not doing enough to control the borders and screen entrants; 85% feel that as a result the country is more vulnerable to terrorist activity; and 79% support the militarization of the border (Walker, 2007, p. 152).

5. Tancredo went on to proclaim the group to be "good citizens" merely trying to enforce the law.

ACKNOWLEDGMENT

The author would like to thank Howard Winant for his very helpful comments.

REFERENCES

Abrams, P. (1988). Notes on the difficulty of studying the state. *Journal of Historical Sociology*, *1*(1), 58–89.

Agamben, G. (2005). *State of exception*. Chicago: University of Chicago Press.

Albrow, M. (1996). *The global age*. Cambridge: Polity Press.

Anderson, B. (1991). *Imagined communities*. New York: Verso.

Andreas, P. (2000). *Border games: Policing the U.S.–Mexico divide*. Ithaca: Cornell University Press.

Andreas, P. (2006). Politics on edge: Managing the US–Mexico border. *Current History*, *105*(688), 64–68.

Balibar, E. (1991). Nation form: History and ideology. In: E. Balibar & I. Wallerstein (Eds), *Race, nation, class: Ambiguous identities*. New York: Verso.

Brubaker, R. (1992). *Citizenship and nationhood in France and Germany*. Cambridge: Harvard University Press.

CNN. (2001). Transcript of Bush Speech in Atlanta, November 8.
Carens, J. (1987). Aliens and citizens: The case for open borders. *Review of Politics*, *49*(3), 251–273.
Center for new community. (2005). Shell games: The 'minutemen' and vigilante anti-immigrant politics. Retrieved on October 12, 2006 from http://www.buildingdemocracy.org/shellgames.pdf
Cohen, S. (1985). *Visions of social control*. Cambridge: Polity Press.
Crank, J. P. (1994). Watchman and community: Myth and institutionalization in policing. *Law and Society Review*, *28*(2), 325–352.
Dandeker, C. (1990). *Surveillance, power, and modernity*. New York: St. Martin's Press.
Darian-Smith, E. (1993). Neighborhood watch – who watches whom? Reinterpreting the concept of neighborhood. *Human Organization*, *52*(1), 83–88.
Donzelot, J. (1991). The mobilization of society. In: M. Foucault, G. Burchell, C. Gordon & P. Miller (Eds), *The Foucault effect*. Chicago: University of Chicago Press.
Ericson, R. (1994). The division of expert knowledge in policing and security. *The British Journal of Sociology*, *45*(2), 149–175.
Feldman, M. (2006). The hunted. Minutemen train their sights on a new target: Hispanic day laborers. *Dallas Observer* (August 23), p. 1A.
Fielding, N. (2002). Theorizing community policing. *The British Journal of Criminology*, *42*(1), 147.
Foucault, M. (1977). *Discipline and punish*. New York: Pantheon Books.
Foucault, M. (1980a). Two lectures. In: M. Foucault & C. Gordon (Eds), *Power/knowledge*. New York: Pantheon Books.
Foucault, M. (1980b). Truth and power. In: M. Foucault & C. Gordon (Eds), *Power/knowledge*. New York: Pantheon Books.
Foucault, M. (1982). The subject and power. In: H. Dreyfus & P. Rabinow (Eds), *Michel Foucault: Beyond structuralism and hermeneutics*. Chicago: University of Chicago Press.
Foucault, M. (1990). *The history of sexuality*. New York: Vintage Books.
Foucault, M. (1991). Governmentality. In: M. Foucault, G. Burchell, C. Gordon & P. Miller (Eds), *The Foucault effect*. Chicago: University of Chicago Press.
Giddens, A. (1987). *The nation-state and violence*. Berkeley: University of California Press.
Gilchrist, J. (2006). Minuteman project 'Operation Sovereignty' border watch operations manual. Retrieved on October 3, 2006 from http://www.minutemanproject.com/pdf_files/MMP-BorderWatch-OperationsManual.pdf
Gilchrist, J., & Corsi, J. (2006). *Minutemen: The battle to secure America's borders*. Los Angeles: World Ahead Publishers.
Gorski, P. (2003). *The disciplinary revolution: Calvinism and the rise of the state in early modern Europe*. Chicago: University of Chicago Press.
Gramsci, A. (1992). *Prison notebooks*. New York: Columbia University Press.
Guiraudon, V., & Lahav, G. (2000). A Reappraisal of the state sovereignty debate: The case of migration control. *Comparative Political Studies*, *33*(2), 163–195.
Hardt, M., & Negri, A. (2000). *Empire*. Cambridge: Harvard University Press.
Hendricks, T. (2005). On the border. *San Francisco Chronicle* (December 5), p. 1A.
Higgs, E. (2001). The rise of the information state. *Journal of Historical Sociology*, *14*(2), 175–197.
Higham, J. (1988). *Strangers in the land: Patterns of American nativism*. New Brunswick, NJ: Rutgers University Press.

Jessop, B. (1994). Post-Fordism and the state. In: A. Amin (Ed.), *Post-Fordism: A reader*. Oxford: Blackwell.

Joppke, C. (1999). *Immigration and the nation-state: The United States, Germany, and Great Britain*. Oxford: Oxford University Press.

Juffer, J. (2006). Introduction. *South Atlantic Quarterly*, *105*(4), 663–680.

Loveman, M. (2005). The modern state and the primitive accumulation of symbolic power. *The American Journal of Sociology*, *110*(6), 1651–1684.

Lyon, D. (1994). *The electronic eye: The rise of surveillance society*. Oxford: Blackwell Publishers.

Mann, M. (1986). *The sources of social power* (Vol. 1). Cambridge: Cambridge University Press.

Mansfield, D. (2005). Minutemen groups grow. *Albuquerque Journal* (July 18), p. 1A.

Marx, G. (1989). Commentary: Some trends and issues in citizen involvement in the law enforcement process. *Crime and Delinquency*, *35*(3), 500–519.

McAdam, D., McCarthy, J., & Zald, M. (1996). Introduction. In: D. McAdam, J. McCarthy & M. Zald (Eds), *Comparative perspectives on social movements*. Cambridge: Cambridge University Press.

Merry, S. (1992). Popular justice and the ideology of social transformation. *Social and Legal Studies*, *1*(2), 161–176.

Miliband, R. (1969). *The state in capitalist society*. London: Weidenfeld & Nicolson.

Minutemen civil defense corps. (2006). Multimillion dollar donation of fiber-optic mesh for minutemen border fence. Retrieved on April 21, 2007 from http://www.minutemanhq.com/hq/article.php

Mitchell, T. (1999). Society, economy and the state effect. In: G. Steinmetz (Ed.), *State/culture*. Ithaca: Cornell University Press.

Nash, K. (2005). Watchful eyes. *Albequerque Tribune* (October 3), p. A1.

Nevins, J. (2001). Searching for security: Boundary and immigration enforcement in an age of intensifying globalization. *Social Justice*, *28*(2), 132–148.

Newman, D. (2006). The lines that continue to separate US: Borders in our 'borderless world'. *Progress in Human Geography*, *30*(2), 143–161.

Ngai, M. (2004). *Impossible subjects: Illegal aliens and the making of modern America*. Princeton: Princeton University Press.

Ohmae, K. (1990). *The borderless world*. New York: Harper Business.

Poggi, G. (1990). *The state: Its nature, development, and prospects*. Stanford: Stanford University Press.

Reiner, R. (1992). Policing a postmodern society. *The Modern Law Review*, *55*(6), 761–781.

Robinson, W. (2006). 'Aqui estamos y no nos vamos!' global capital and immigrant rights. *Race and Class*, *48*(2), 77–91.

Rose, N. (1999). *Powers of freedom*. Cambridge: Cambridge University Press.

Rose, N., & Miller, P. (1992). Political power beyond the state. *The British Journal of Sociology*, *43*(2), 173–205.

Salter, M. (2004). Passports, mobility, and security: How smart can the border be? *International Studies Perspectives*, *5*(1), 71–91.

Sanchez, G. (1997). Face the nation: Race, immigration, and the rise of nativism in late twentieth century America. *International Migration Review*, *31*(4), 1009–1030.

Sassen, S. (1996). *Losing control?: Sovereignty in an age of globalization*. New York: Columbia University Press.

Sassen, S. (1999). *Guests and aliens*. New York: W.W. Norton.

Saunders, R. (1999). The space community policing makes and the body that makes it. *Professional Geographer*, *51*(1), 135–146.
Scott, J. (1998). *Seeing like a state*. New Haven: Yale University Press.
Simcox, C. (2006a). The official minutemen civil defense corps training manual. Retrieved on October 12, 2006 from http://www.minutemanhq.com/pdf_files/training_manual2.pdf.com
Simcox, C. (2006b). Standard operating procedure for the minutemen civil defense corps. Retrieved on October 12, 2006 from http://www.minutemanhq.com/hq/borderops_sop.php
Simcox, C. (n.d.). About minuteman border fence. Retrieved on April 21, 2007 from http://www.minutemanhq.com/bf/about.php
Skocpol, T. (1979). *States and social revolutions*. Cambridge: Cambridge University Press.
Staples, W. (2000). *Everyday surveillance: Vigilance and visibility in postmodern life*. Lanham, MD: Rowan & Littlefield Publishers.
Starr, P. (1987). The sociology of official statistics. In: W. Alonso & P. Starr (Eds), *The politics of numbers*. New York: Russell Sage Foundation.
Sterngold, J., & Martin, M. (2005). Governor signals he'd welcome minutemen on the California border. *San Francisco Chronicle* (April 30), p. A1.
Torpey, J. (2000). *The invention of the passport*. Cambridge: Cambridge University Press.
Trouillot, M. (2001). The anthropology of the state in the age of globalization. *Current Anthropology*, *42*(1), 125–138.
Turnball, L., & Tu, J. (2005). Minutemen watch Canada border. *The Seattle Times* (October 4), p. A1.
Vina, S., Nunez-Neto, B., & Weir, A. (2006). *Civilian patrols along the border: Legal and policy issues*. Congressional Research Service Report. Library of Congress, September 28.
Walker, C. (2007). Border vigilantism and comprehensive immigration reform. *Harvard Latino Law Review*, *10*, 135–174.
Walters, W. (2006). Border/control. *European Journal of Social Theory*, *9*(2), 187–203.
Walzer, M. (1983). *Spheres of justice*. New York: Basic Books.
Zellen, B. (2006). America's southern front: Immigration, homeland security, and the border fencing debate. *Strategic Insights*, *5*(5).
Zolberg, A. (2000). Matters of state: Theorizing immigration policy. In: C. Hirschman, P. Kasinitz & J. DeWind (Eds), *The handbook of international migration*. New York: Russell Sage.

THE POLITICS OF SIGHT/SITE: LOCATING CAMERAS IN VANCOUVER'S PUBLIC SPACES

Kevin D. Haggerty, Laura Huey and Richard V. Ericson†

ABSTRACT

This chapter is about the politics of surveillance and more specifically about the politics of siting public closed circuit television (CCTV) systems within urban neighborhoods. Through an exploration of political contests waged around attempts by local police to install public surveillance systems in the City of Vancouver's Downtown Eastside and Granville Mall districts, we argue that the success of public surveillance proposals is hardly inevitable. Instead, a combination of local factors play vital roles in variously supporting or constraining such attempts. Although this present chapter can be read as providing a useful counterpoint to the dominance of accounts about such developments in Great Britain, where public CCTV is a routine fact of daily urban life, we conclude on a cautionary note: with the current proliferation of public and private forms of surveillance throughout urban spaces, surveillance analysts risk missing the forest for the trees if we only concentrate on the fate of one surveillance tool or tactic.

Surveillance and Governance: Crime Control and Beyond
Sociology of Crime, Law and Deviance, Volume 10, 35–55

ISSN: 1521-6136/doi:10.1016/S1521-6136(07)00202-3

INTRODUCTION

The widespread introduction of closed circuit television (CCTV) cameras in Britain has been one of the most high-profile surveillance-related developments of the 21st century. For the uninitiated, it can be startling to witness the remarkable infrastucture of public cameras that now exist in the United Kingdom. Visitors to Britain now routinely add tales about the ubiquity of cameras to the familiar tourist narratives of queuing at museums and being delayed at Heathrow.

Cameras in the United Kingdom have been justified as a means to catch criminals, reduce disorder and revitalize neglected downtowns. Enthusiastic local authorities have installed systems at such a pace that it is difficult to acquire accurate data on the number of cameras. All of this makes for a compelling and often-recounted story about a remarkable intensification of surveillance capacity. More problematically, the surveillance studies literature also occasionally takes Britain as the paradigmatic example of CCTV-related developments. This is unfortunate because most countries continue to lag behind Britain in terms of the number and sophistication of camera systems. Britain remains an extreme case, arguably *the* extreme case. Even the United States, still traumatized by the 9/11 terrorist attacks, has not (yet) seen a comparable installation of public camera systems.

This chapter examines the local politics of introducing CCTV cameras, accentuating the contestation that can surround such proposals. As such, we contribute to a burgeoning literature on the 'politics of surveillance' (Haggerty & Ericson, 2006; Huey, 2007; Huey, Walby, & Doyle, 2006; Monahan, 2006). This literature typically has a dual focus, with the first concentrating on the institutional and political factors that encourage the introduction of surveillance initiatives, and the second concerned with the broad sociopolitical implications of such measures. Such an orientation provides suggestive avenues to explore the dynamics of surveillance; however, despite a few notable exceptions, there remains a dearth of empirically grounded analyses of the factors that condition the introduction of public surveillance systems. This chapter takes steps towards remedying this lacuna by analyzing efforts to implement CCTV systems in the City of Vancouver, Canada.

Although what follows is an analysis of developments in one city, it provides a useful counterpoint to the dominance of accounts about developments in Great Britain. Given the specific history of CCTV in

England, analysts might be excused for embracing a sense that CCTV cameras are inevitably part of our collective future. The sheer number of cameras and the pace with which they have been installed can nurture a sense that political efforts to thwart such cameras have failed, and that it is only a matter of time until cameras are everywhere.

While the British example makes such fatalism understandable, it is not an accurate representations of wider international developments. The case of Vancouver suggests that while CCTV has an undeniable momentum, the global expansion of public CCTV remains an open question. The embrace of CCTV in Britain was shaped by a series of local factors, some of which are apparent in other locations, but others of which were comparatively unique to that country. Some of the unique factors include the way that the IRA bombings in London fundamentally transformed the downtown, leading to the introduction of an abundance of surveillance cameras in the financial district. At one time this area had claim to being the most monitored place on earth. Other iconic developments include the 1993 abduction and killing of two-year-old Jamie Bulger. The security camera footage of Jamie's 10-year-old abductors leading him by the hand to his eventual murder remains one of the most famous, heart-wrenching images ever captured on CCTV. While the cameras could not prevent the Bulger murder, the image – both what could be seen and what could not – sensitized the British people to the prospect that public cameras might be used to confront crime (Young, 1996). Finally, and arguably most importantly, the British Home Office emerged as a principal proponent of CCTV cameras as a preferred crime-prevention technology. This support gave the cameras a powerful centralized advocate eager to help fund the installation of such tools.

Our research into developments in Vancouver highlights the operation of a host of other local and highly contingent political factors that can come into play in the siting of a CCTV system. It provides a glimpse into an evolving series of contestations over the promise and dangers of surveillance as played out in a self-consciously 'global city.' The heterogeneous series of factors that became germane to this politics include traditions of community activism, ill-advised police initiatives, municipal elections, urban planning decisions and, finally, a global sporting event. We detail some of the ways that these diverse factors shaped and continue to shape the prospect for introducing public CCTV in two prominent Vancouver neighborhoods. Our conclusion positions such initiatives in the context of the more general proliferation of cameras in urban spaces.

METHODOLOGY

We began our research in early 2000 in response to the Vancouver Police Department's (VPDs) announcement that they wanted to install CCTV cameras in the city's poorest neighborhood, the Downtown Eastside (DTES). We interviewed individuals who were publicly identified as being involved in surveillance-related issues. Other respondents were contacted through a process of 'snowball sampling' where interviewees suggested other individuals who they thought might contribute to the research. The result was a dataset of 25 interviews with members of the police and various community groups. When the police proposal was shelved in 2001, we also temporarily abandoned our research plans. We did, however, maintain our transcribed interviews and other materials we had collected, including policy reports and news clippings. We draw upon this data here as it provides insights into both the history of these developments and community reaction to the plans.

Fortuitously, in 2000 we were also engaged in a separate ethnographic study of private policing in Vancouver's Granville Mall district (Huey, Ericson, & Haggerty, 2005). For this second study, 48 interviews were conducted with 68 relevant community stakeholders (police, the business improvement association, business owners, city planners, local community groups and area residents). This chapter also draws upon these interviews as they contain insights into the area, its history and redevelopment, crime issues, public and private policing of the site and social divisions within the community.

In 2006, the VPD announced that it intended to install CCTV cameras in the Granville Mall entertainment zone, a prime downtown tourism and leisure destination. We immediately resurrected our initial study. We collected additional documents on the proposal and in the spring and summer of 2007 conducted further interviews with police, local business representatives and community groups. One of the authors returned to Vancouver in 2007 to follow-up on our initial research by mapping, photographing and observing activities within the Granville Entertainment District.

THE FIRST SITE: THE DOWNTOWN EASTSIDE

On the other side of the railway tracks at the foot of Main Street is Crab Park. There is an almost jarring contrast between this grassy park overlooking the water and distant

> mountains, and the open drug trade only three blocks away. Traces of the problems in the Downtown Eastside, however, are still apparent. Just yards away from the water sits a boulder surrounded by bushes. On it is carved: 'The Heart Has its Own Memory. Dedicated July 29, 1997. In honor of the spirit of the people murdered in the Downtown Eastside. Many were women, and many were native aboriginal women. Many of these cases remain unsolved. All my relations.' As I read this inscription into my hand-held tape recorder a man who has been exercising his dog on the beach walks over and tosses a used syringe into the bushes that encircle the monument. (Field notes)

Advertising campaigns extol Vancouver's natural beauty, accentuating its proximity to the Pacific Ocean and to British Columbia's forests and Rocky Mountains. These adverts are part of a concerted effort by Vancouver's planners and politicians to position the city amongst the pantheon of 'global cities.' Undoubtedly, the key moment in this continuing process was Expo 86, which was simultaneously an international tourist event and a global advertising platform for the city itself. Expo 86 was held in the heart of the downtown on lands which officials promised to subsequently develop with community interests and social housing in mind. The government reneged on this promise, privatized the land and sold it to billionaire developer Li Ka-shing who built luxury condominiums on the site. Ka-shing's centrality in Vancouver's urban development also contributed to the city's increasing orientation towards the Pacific Rim. Vancouver experienced an influx of capital and migrants from Hong Kong during the 1980s and 1990s, a process which continues to the present. Today, Vancouver's skyscrapers, inflated housing market and thriving movie industry are testament to the triumphant reimagining of the city (Tinic, 2005). Its successful bid for the 2010 Winter Olympics put an exclamation point on this urban transformation. Not everyone, however, has reaped the rewards of an invigorated city.

Vancouver's DTES, a community of approximately 16,000 residents with the lowest socio-economic status of any urban area in Canada, lies immediately adjacent to the central business and retail districts (City of Vancouver, 2001; Lees, 1998; MacPherson, 2001). Its history is also intimately tied to the legacy of Expo 86. Anticipating a financial windfall in the lead-up to Expo, landlords expelled scores of poor residents from long-term housing to transform those units into tourist accommodations, exacerbating the existing poverty and homelessness problems.

While much of the public discourse on the DTES focuses on the site as a 'skid row' district (Huey, 2007), it is also home to lower working class individuals and families, and is populated by groups of varying ethnic backgrounds. This population also includes local artisans and writers, the mentally ill, community activists and a variety of other social groups.

Its streets are also a home for the homeless. What many of the members of these groups share is chronic poverty. As Hasson and Ley (1994, p. 175) point out, here 'struggle is far more than an intellectual abstraction; it is a central preoccupation of everyday life.' The median income for this neighborhood is $12,084, which is less than one-third of that for the rest of Vancouver (City of Vancouver, 2006). High unemployment rates mean that 40% of its residents receive some form of income assistance (*ibid.*). Although the community is ethnically diverse and contains a mix of young and old, most residents (62%) are male (Trade and Convention Centre Community Impact Team, 1999). It is also home to a disproportionate number of Vancouver's drug addicts and mentally ill. Those individuals, such as the poor families and itinerate workers also found here, are drawn by the 'affordable' housing. Most of the city's low-income housing is located here, mainly in the form of single-room occupancy (SRO) hotels. In 2005, 82% of the City's SROs were located in these 15 blocks (*ibid.*).

The site's reputation as a 'skid row' district began in the 1940s and 1950s, when it became associated with alcoholism (Huey, 2007). By 1960, the DTES housed a core population of approximately 600 chronic alcoholics (Obe, 1960). In the 1960s and 1970s, residents turned to heroin as the drug of choice, and much of the city's drug trade became contained within this neighborhood (Huey, 2007). Dealers introduced crack cocaine in the 1980s, with devastating effects. By the 1990s, there was a flourishing open-air drug market around the intersection of Main and Hastings (*ibid.*). By 2000, heroin users treated alleyways as shooting galleries, and lanes were littered with syringes and condom wrappers. The approximately 6,000 intravenous drug users in the area and the active sex trade have also contributed to a massive problem with Hepatitis C, and the highest per capita HIV infection rate in North America (Health Canada, 1995), prompting the City of Vancouver to declare its first-ever medical emergency in 1997. To further compound the area's problems, alcohol abuse is rampant, exacerbated by the massively disproportionate number of pubs and liquor license seats in the neighborhood (City of Vancouver, 1998).

Long viewed as a policing problem, the DTES is home to about 3% of the city's population, but statistics for 2002 show that approximately 19% of all recorded serious violent crimes occur here (City of Vancouver, 2006). In that same year, 36% of all recorded drug offences took place in the DTES (City of Vancouver, 1999).[1] Recent increases in the number of police officers in the area have done little to reduce these problems (Huey, 2007).

Few legitimate businesses remain and entire blocks of storefront property on Hastings Street are boarded up. The police and drug users alike acknowledge that corner stores are often fronts for the illegal drug trade (Huey, 2007). The pawnshops catering to the petty thief's need to obtain cash for stolen goods are the only thriving commercial interest in the area (Matas, 2001). In 1996, 32 pawnshops were located within three blocks in the DTES, representing an average of four per square block (Huey, 2007).

This intense concentration of social problems and the neighborhood's proximity to upscale commercial districts has contributed to the DTES becoming a popular site for informal urban sightseeing. Middle-class Vancouverites often drive their out-of-town friends through 'the skids' to show-off some of the post-apocalyptic scenes to be found in the exoticism of the dispossessed. These sights can include open drug dealing, prostitutes working the corners, drunks stumbling blindly into traffic and ambulance crews working to revive addicts from the grips of cocaine psychosis, all paced by the frenetic tempo of emergency vehicle lights.

THE FIRST CCTV PROPOSAL

Given the sheer number of problems associated with the DTES and their seeming intractability, it might appear natural that CCTV cameras would be introduced in this neighborhood. In 1999, the VPD took up this cause and started to champion the use of cameras to curb the 'open-air drug market' that prevailed at the corners of Main and Hastings. They hoped to install cameras in the public spaces within and immediately adjacent to this intersection.

While this proposal sought to confront a local problem, its proponents were well aware of its significance in the wider politics of CCTV in Canada. At that time the country had very few public surveillance camera systems. As the DTES represented the 'toughest case' for any Canadian urban policing initiative, everyone knew that its success or failure would be closely monitored. The author of Vancouver's proposal accentuated how this system 'has the potential to become the showcase for all future public video surveillance systems on this continent' (Fredericks, 1999, p. 11).

The police were key advocates for such measures, a factor that differentiates the Canadian situation from that found in the United Kingdom. While the police are undeniably important supporters of CCTV in Britain, they operate with the vital aid of the Home Office, which embraced CCTV as their premier crime-fighting strategy. Canada does not have a comparable powerful centralized organization championing and funding the introduction of cameras. Neither Canada's federal Department of Justice nor the provincial Solicitor General came close to providing this type of unreserved support for cameras, meaning that local authorities must serve as the advocate for such proposals. In Vancouver, the city police assumed this role.

The police developed a two-pronged approach to introducing cameras in the DTES. The first strategy involved attending local community meetings to make the case for the potential benefits of the cameras in terms of their ability to reduce crime and disorder. Local residents got their first inkling that the VPD might pursue such a plan at a monthly meeting hosted by a community service group. An individual who attended the meeting described the response

> It was really mixed. There were people opposed, and there were people for it. And, generally, like I say, it was generally the older residents that said 'Let's give it a try.' The older, longer-term residents that also supported the increased policing in the DTES. They said, 'we need something.' And some of the younger, more active people said, 'No damn way. This is an infringement of my freedoms.'

The second prong of the police strategy focused on the technological challenges of getting the CCTV system to work. Towards that end they installed a preliminary 'test camera' on the roof of the Royal Bank at the corner of Main and Hastings. Erected unannounced, it proved to be a public relations disaster that helped undermine the police's efforts. As word of the camera rapidly circulated throughout the community, people in the drug trade realized that they were being filmed. Also being filmed, however, were the community activists running the needle exchange program, commuters waiting for the bus, visitors to the public library and so on. As a local activist explained, 'That was ridiculous. That lost them a lot of favour.' One of the most ill-conceived attributes of the camera was that members of a community action group that frequently clashed with the police could look directly out of their second-story office windows at the police camera which was positioned slightly above them on the opposite side of the street. This ostensibly meant that the police could film them while they worked, held meetings and planned events. It was a situation almost designed to foster distrust.

A wave of criticism followed in both the local and national media. The B.C. Civil Liberties Association (BCCLA) and other stakeholders tried to clarify the benefits and dangers of the CCTV cameras. Local activists printed and circulated an 'anti-camera' pamphlet suggesting that the cameras would merely displace crime to adjoining neighborhoods. In interviews, they revealed that they had contemplated resorting to more direct forms of resistance, including a fanciful scenario to attach helium balloons to a bed sheet, which would be levitated in front of the camera.

In the context of this increasingly vocal opposition, the VPD submitted their proposal to install 16 surveillance cameras in the DTES to the City's Police Board. Operators would be able to tilt, pan, zoom and rotate the cameras a full 360 degrees. Smoked plastic domes would cover the cameras to prevent tampering and prohibit anyone from knowing where the cameras were pointing. Rather than respond to the potential privacy concerns, the VPD report instead chose to emphasize the potential 'security benefits.' The author of the report claimed that

> the advantages of CCTV, properly managed, speak for themselves: Crime prevention, the deterrent effect of knowing that there is observation, the alerting of police at an early stage to stop dangerous situations from escalating, the operational assistance to police in evaluating a situation, the safer conditions that can be obtained, the enormous savings in court time, and above all, renewed public confidence, which has led to many town centres being revitalized. (Fredericks, 1999, p. 76)

Local activists and service providers were unconvinced. One service provider who studied the report described its treatment of potential citizen concerns as 'pretty bogus.' He added that its author 'sort of assumed that the only opposition to it was from the civil liberty types ... that communities would have no problem [with its recommendations].' His objections, and those of other like-minded community members, were not based on 'mere abstractions,' as some camera proponents implied, but on the lived reality of working and living in one of the most heavily policed neighborhoods in North America. As one community organizer worried, 'video surveillance by the police sort of constitutes creating a police state.'

A major failure of the police effort was that they underestimated the concerns, power and organizational abilities of local activists and service providers. The concentration of social problems in the DTES also means that the area has a disproportionate number of community activists and committed service providers catering to the needs of local residents. These agencies have a long history of clashing with officials, heightened sensibilities about police practices and considerable experience mobilizing residents to secure media attention and oppose police practices.

In the late summer of 1999, the Vancouver Police Board effectively shelved the proposal by returning it for 'further study.' Beyond the street-level opposition and critical media coverage that the cameras had generated, it also appears that the timing of this proposal was extremely inopportune. Politicians were concerned that the issue could present a potential stumbling block in forthcoming municipal elections. Thus, as a member of Vancouver's City Hall advised us, there was informal pressure to 'keep this [proposal] until after the election.' He added, 'That was the message they got from the Chief [of Police], but the Mayor is the Head of the Police Board, so that is what I heard informally.'

Throughout the remainder of 1999 and 2000, DTES community activists and local privacy advocates braced for the possibility that the CCTV proposal would once again be placed on the public agenda. They proved to be quite prescient. A new proposal was introduced in the summer of 2001 with a call for seven additional cameras. Again, this prospect met with significant local opposition and by 2003 it appeared to have been abandoned. At an 'in camera' meeting of the Vancouver Police Board on May 18, 2005, a senior Vancouver police commander publicly recommended withdrawing the proposal. However, instead of scrapping the project altogether, as privacy advocates had hoped, the inspector suggested that the VPD should explore the possibility of installing CCTV cameras in another area of the city where police surveillance would provide 'maximum benefit[s].' The Police Board voted in favor of both recommendations. The police's ambitions to install cameras went dormant, as officials waited for an opportunity to reintroduce such measures in less hostile environs.

THE SECOND SITE: THE GRANVILLE STREET ENTERTAINMENT DISTRICT

Throughout the 1940s and 1950s, Vancouver's Granville Street was a popular entertainment destination, containing several thriving movie theaters, ballrooms and a variety of nightclubs. Mimicking the plight of most North American urban centers throughout the 1970s and 1980s, the area fell into disrepair and disrepute. Retail and entertainment enterprises fled, and the street became populated with sex stores, panhandlers, street youth[2] and other symbols commonly associated with 'urban blight.'

In the 1990s, two forces of redevelopment combined to redesign the Granville Mall pedestrian zone. First, in 1991, the City rezoned the area surrounding Granville to facilitate the building of residential high-rises. City officials also adjusted zoning regulations along the Mall to reduce the dense cluster of sex shops in the area. Property owners, seeing signs of increased tourism in Vancouver as a whole, also began to redevelop their businesses and the surrounding storefronts in order to attract tourist dollars. Local businesses, through their business improvement association, worked with the City on projects to update street furniture and reduce the amount of graffiti and litter in the area (Huey et al., 2005).

While all of these developments had some impact on the Granville Mall, it was Vancouver City Council's approval of plans to recreate the Mall as a district catering to nighttime entertainment, which most significantly altered the Mall's future. According to a city planner we interviewed in 2000, the plan sought 'to inject some life back into Granville Street.' To facilitate this process, regulators approved an additional 1,000 liquor license seats for the area in 1998 (Gordon & Barth, 1998). This was followed in 2003 by City Council approving an increase in the operating hours of liquor establishments in the Entertainment District: the 57 bars, pubs and cabarets located here would be able to extend their hours from 2 to 4 a.m. (Hume, 2006). Planners justified these extended hours on the assumption that staggering closing times would reduce overcrowding on the street and thus the potential for violence and disorder. This strategy of increasing consumption times in order to reduce street crime has been pursued in other jurisdictions, but as Deborah Talbot (2006) reports, it has also tended to be unsuccessful. By 2006, 5,000 liquor-licensed seats were concentrated along the downtown portion of Granville Street, with several of these establishments operating until 4 a.m. (Bohn, 2006). In a survey we conducted in the spring of 2007, we found 12 liquor-licensed venues in a two-block section of the Mall, including pubs, bars, cabarets, nightclubs, eateries and the lounge of a bowling alley. Two further licensed establishments (a pub and a nightclub) are now operating in the 1,000 block and there is a license application for a thirteenth location in the 900 block of Granville.

A city planner advised that the idea behind locating Vancouver's new 'fun zone' along the downtown portion of Granville Street was to 'concentrate the bars in three or four blocks of Granville street where people aren't living,' and thus presumably limit complaints from residents about noise from late-night revelers. It proved to be a faulty assumption. As a city official interviewed in 2000 noted, nearby residents complained vociferously about disorder problems associated with intoxicated partiers visiting the Entertainment District:

> Now they are seen as a problem because of people coming out of bars at two o'clock in the morning, creating noise, other problems that could jeopardize the marketing success of the residential area that was being planned for the area. You can see it in the press. There have been instances of people complaining about late night noise from those bars.

Noise complaints from nearby residents are, however, the least of the area's problems. The plan to 'inject life' back into the new Entertainment

District by rebranding it as a nighttime adult entertainment center had another entirely foreseeable undesirable result. As Jayne, Holloway, and Valentine (2006, p. 461) note, one of the problems associated with nighttime entertainment districts is that 'drinkers are consumers who are often badly behaved and do not conform to discourses of polite, civilized and cosmopolitan urbanity.' Predictably, the numbers of assaults and other crimes that are prone to occur in a small space awash in inebriated people began to increase significantly. Between April and October 2006 alone, the police issued nearly 900 liquor-related tickets within the Entertainment District, including 310 for drunkenness, 246 for intoxication in a public place, 194 for consuming alcohol in a public place and 113 for public urination (CBC News online edition, 2006). A senior VPD official lamented that the Granville Street Entertainment District is now a site where 'open public disorder, fighting in and around the bars and assaulting police officers has become all too common' (Hume, 2006, p. A1). Another commander complained that, 'We are going to fight after fight after fight ... Things are getting out of control' (*ibid.*).

Various attempts have been made to control what one interviewee described as 'the carnage' in this area. This included an agreement by club owners in 2004 to push back their closing time to 3 a.m. When other club operators failed to do the same, the experiment was deemed a failure. A local business community representative felt that even if all the clubs had rolled back their hours the problem would still remain because 'the culture had changed.' As he explained, 'rolling it back to 2 a.m. will do nothing to change the culture. [Granville] is now seen as the place to come and party.' A spate of radio ads aired in 2007 to convince patrons to refrain from disorderly behavior in the Entertainment District. They appear to have had little impact on the crime and disorder generated by the local nighttime economy.

RESURRECTING THE CCTV PROPOSAL

It was in this context that the Vancouver police's dormant CCTV-related ambitions reemerged. In November 2006, the chief constable of the VPD returned to the Police Board to recommend that the city conduct a feasibility study for a CCTV system in the downtown core, concentrating particularly on the Granville Entertainment District. According to Chief Graham, 'limited use of CCTV could effectively reduce crime in D1 [district one], the entertainment district' (Vancouver Police Board, 2006).

The police had considerable support for this proposal from local businesses. As early as 2000, we were informed by key proponents of the DTES camera proposal that business interests along the Granville corridor had been pushing for a similar plan for the Entertainment District. In a 2007 interview with a business representative, we were advised that members of the area's business improvement association supported the police plan, and that they expect to 'see some returns with a properly applied CCTV' in the form of reductions in local crime and disorder.

> Q: So you guys are not only in favour of the proposal, but you believe it's long overdue?
>
> A: We'd even fund some of this.

Police media briefs suggest that the police rationale for supporting a network of cameras in the Entertainment District was a lack of available police resources and increasing demands for police service generated by the site (Hume, 2006). According to representatives of the VPD's Planning and Research Section, a CCTV system along Granville would help them address a 'high call load for violent offences' and a 'high call load regarding quality of life offences (mischief, theft, aggressive panhandling)' (Vancouver Police Department, 2006). Other putative benefits include improvements to police crime detection and response, additional investigational support, increased ability to conduct crowd management, target-hardening, counter-terrorism, reductions in violent crime and improvements in local safety (*ibid.*). The city's chief constable, conscious of the earlier failed attempt to install such cameras in the DTES, and the lingering suspicions among local activists, 'assured the delegations' of the Vancouver Police Board 'that the cameras would not be focused on the DTES' (Vancouver Police Board, 2006). The board approved the study.[3]

Civic groups immediately responded when they heard that the VPD was once again resurrecting their plans to monitor public spaces with CCTV, this time in the Granville Entertainment District. Representatives of the BCCLA spoke against the proposal in a variety of fora, including print media editorials, public meetings, on the Association's website and at a meeting of the City's Police Board. They also wrote to the Vancouver Police Board to seek assurances that the Board would hold extensive public consultations, and that the public would be provided full access to information as to where cameras would be placed, how they would be used and all associated costs (BC Civil Liberties Association, 2006). The Vancouver Public Space Network (VPSN), a loose coalition of local activists concerned with the use of Vancouver's public spaces, joined the

BCCLA's efforts. This time, however, activists could not generate significant public outcry against the proposal.

In published interviews, civil liberties advocates speculated on the strategic nature of the police's shift from trying to install cameras in the DTES with its distinctive mix of highly mobilized activists to a site where public CCTV would likely engender significantly less local opposition (Bucher, 2006). Indeed, one activist characterized this move in the following terms: 'It appears now that the VPD is shopping for a neighborhood where [the cameras] can get a foothold' (*ibid.*). In our interviews, privacy and public space advocates expressed similar sentiments. For example, a privacy advocate wryly noted that club patrons – who would be likely targets of CCTV surveillance – would be unlikely to participate in public debates about the increased use of cameras in nightlife zones. As she explained, 'There's no population affected who is going to be vocalizing on this ... that's where it starts to make strategic sense.'

CONCLUDING REMARKS ON URBAN VISIBILITY

> The Granville bus jerks along Hastings Street, slicing through the heart of Vancouver's Downtown Eastside. It is standing room only when I board after my second day of interviews. We turn left on trendy Granville Street, now officially deemed Vancouver's 'entertainment' district. An elderly man is seated near the rear exit. Obviously mentally unstable, he mutters mildly to himself and the woman beside him. She does not know him and conveys the signs of discomfort that can accompany public encounters with the mad. Smelling beer, I notice that he is drinking a can of Wildcat and has clutched on his lap what remains of a twelve-pack. As we approach the Granville Street Bridge, a small commotion ensues as a stench starts to emanate from the back of the bus. We all crush to the front in an effort to escape the odor. The old man has shit himself, risen from a pool of his own diarrhea and exited onto Granville. A mother on my left urges her two adolescent boys to breath through their mouths. On my right, a woman repeatedly gags, making me fear that she might vomit directly in front of me. Finally reaching the other side of the bridge, the passengers pour out. Some ruminate on what the poor bus driver is supposed to do with his polluted vehicle. Nobody mentions the man now lumbering, beer in-hand, through Downtown Vancouver in fouled trousers. (Field notes)

Our condensed account of developments in Vancouver provides a glimpse of the dynamics pertaining to the local politics of surveillance. While these struggles might not qualify as a wholesale grassroots 'revolt against the gaze,' they provide an antidote to the sense of inevitability that can come from watching the seemingly inexorable rise of CCTV cameras in other parts of the world. History may still prove that the wholesale introduction of such cameras is a foregone conclusion, but even so there remain important local stories to tell about how this was accomplished and resisted, and about how the specific sites of cameras and the populations on which they set their sights, were established.

A heterogeneous mix of actors, processes and technologies operate in Vancouver's ongoing politics of surveillance. For example, crime, alcohol and drug problems play a vital role in fostering a sense that there is a need for such technologies. The public perception of the DTES has long been shaped by a perpetual series of media images of addicts swaying in dirty back alleys, usually coupled with stories of fatal stabbings in the area's bars or of shots fired in skirmishes over local drug turf. The belief that the DTES is an out-of-control urban jungle is such that Vancouver residents are often heard to tell stories of driving through Main and Hastings with their car doors locked. Today, these images of a space awash in crime and disorder are supplemented by new stories of Granville Mall as a site similarly out of control. Replacing the stock shots of emaciated addicts convulsing on sidewalks in the DTES, we see images of well-fed and well-dressed suburban youth, belligerent and drunk, creating disorder as they stumble through the Entertainment District.

Paradoxically, within one of the neighborhoods studied, the problems attendant on drug and alcohol use have helped to populate the site with a small army of dedicated activists predisposed to oppose measures that would bring greater police scrutiny of DTES residents. In their fight against the implementation of public CCTV systems within their community, these activists were supported by a vocal citizenry, who attended public meetings and spoke out against police proposals in the media and to policy-makers. Conversely, despite attempts by two community groups to mount opposition to a similar set of proposals for the Granville Mall corridor, support for the project by the local business community, a high volume of complaints about public disorder by outraged area residents and an inability to generate resistance from those who such cameras would target – club goers from other communities – cause activists to rightly worry that the battle over public CCTV may be lost.

City Hall also figures prominently in this account. Concerns about the potential vote-losing implications of the CCTV cameras helped shelve the original proposal. Equally important, City Hall was willfully blind to the urban disorder problems they would inevitably create when establishing a downtown 'entertainment zone,' essentially nurturing the problems the police then sought to solve through introducing CCTV cameras.

The role of local businesses and area residents, and their willingness to support public CCTV initiatives, must also be taken into account when considering the politics of surveillance. In the DTES, residents were vocal opponents, while area businesses, the few that exist here, remained largely silent on the issue. Thus, public police did not have an identifiable 'community' from which they could claim support when presenting their proposals to civic policy-makers. With respect to the Granville Mall, no such impediment to police plans exists. Indeed, the local business association has even offered to pay for some of the costs of implementing a public CCTV system.

If we look to the future, it appears that the odds are stacked against Vancouver's opponents of surveillance cameras. When the city was awarded the bid for the Winter Olympics in 2010 the police were handed what in other locations has proved to be an unassailable justification for greater CCTV surveillance. It is now almost inevitable that the Olympics' host city becomes wired with enhanced electronic monitoring, and that such increased scrutiny remains as one of the urban 'legacies' of the games (Boyle & Haggerty, n.d.; Samatas, 2006). The Vancouver Chief of Police has already started to pave the way for such a development, suggesting that cameras have become more pressing with the Olympics on the horizon, as Vancouver is 'on the sights of many individuals and groups who intend to do Canada harm' (Gibson, 2007).

While these developments suggest that we are likely on the cusp of a widespread introduction of cameras into Vancouver's urban places, it is also worth accentuating, by way of conclusion, that fixating on CCTV cameras risks drawing attention away from the wider phenomena of urban visibility. Even in the absence of a public surveillance camera system, one would be hard-pressed to argue that downtown Vancouver is free from photographic scrutiny. Simply studying these neighborhoods necessitates that you subject yourself to surveillance. It commences as you travel downtown by car, where you are apt to be photographed by the photo-radar system or the 'red-light cameras' mounted at intersections. There are 24 traffic cameras on Hastings Street, 12 on the Lions Gate bridge and 48 on the Deas Island tunnel. More than 400 cameras monitor the 28 km route of the commuter Sky Train. If you travel by taxi you will likely have your picture taken by a 'taxi-cam,' a miniature camera mounted on a taxi's dashboard to take a customer's picture upon entry and every few seconds thereafter. Dispatchers can also monitor the taxi's physical location on a geographic positioning system.

Moreover, it is not as if the lack of public CCTV cameras in the Downtown Eastside means that the police cannot monitor the local crime scene. Such surveillance occurs, but it is more clandestine (Marx, 1988). As noted, the Downtown Eastside is notorious in policing circles as Canada's extreme 'hot spot' of crime. Consequently, it is already under constant human and technological scrutiny. In addition to locating their downtown precinct directly adjacent to the DTES, the police monitor the area with undercover officers and surreptitiously video-record the goings-on of the public streets. As one resident observed, 'There are already cameras on Main and Hastings all the time, but they are called "observation posts."' Another resident made the same point, noting that 'A lot of surveillance you aren't going to see, because they will go and get a room in the Regent [hotel] and stand in the f—ing windows and put their cameras on the sidewalks. There are tones of hotels where they do this.'

Many of the DTEs remaining businesses have also installed their own cameras. On the outside wall of a local pub there is a sign written in English and Chinese warning that the surrounding area is videotaped, and that 'All recordings of criminal activity will be given to the police for prosecution. Information will also be given to Canadian Immigration.' Adjoining neighborhoods have installed surveillance cameras in bank machines, retail stores and on private premises, many of which are trained on the public streets. An upscale high-rise apartment building at the edge of the DTES is wired with cameras on all the exits, the parking garage and in the hallways, leaving only a small apartment-sized private realm free from photographic scrutiny. A similar pattern is apparent in some of the SRO hotels which now video record their entrance and immediate environs.

Private CCTV systems are also found throughout the City's central business and retail districts. One such system is operated by the Downtown Vancouver Business Improvement Association (DVBIA), which has cameras installed in places frequently, but not exclusively, targeted by graffiti artists. The DVBIA is hardly the only business group to use CCTV to monitor behavior in public space. Indeed, an organization of local bar owners, Vancouver Bar Watch, has made it a mandatory requirement of its 22 participating members that CCTV cameras be installed at their business entrances to capture patrons entering and exiting. Signage near the entrance to a club on Granville Street advises patrons that they are being taped and the police department can subpoena any evidence of a criminal matter caught on tape. Further, a survey of CCTV cameras along the Granville Mall conducted in 2006 by the VPSN, suggests that there are at least 100 privately owned cameras in this district alone. These cameras are variously located at ATM machines, in bars, retail stores and restaurants (VPSN, 2007). While the vast majority of these cameras likely capture images from action occurring within private businesses, many others are positioned so as to also capture street-front activity occurring in public space.

Not all of the surveillance that takes place throughout Vancouver's city streets utilizes CCTV technology. Indeed, the crime, disorder and vagrancy problems in the DTES have also contributed to an increased presence of private security guards. In addition to using their own surveillance cameras, guards frequently take Polaroid photographs of troublesome individuals who they have arrested or banned from the premises they are guarding (Huey et al., 2005). Some residents have criticized the guards for surpassing their legal authority in 'moving-along' panhandlers and buskers, which has, in turn, led to an instance of citizen-initiated video monitoring (*ibid.*). A resident of the DTES produced a half-hour video documentary entitled *Private Security, Public Spaces* which details some of these policing tactics. Much of his most damning footage was shot surreptitiously and shows guards hustling vagrants (and actors pretending to be vagrants) from the public streets outside of private establishments (*ibid.*).

A particular irony of calls to install a CCTV surveillance network to monitor public space in the Granville Street Entertainment District is that the site is already one of the most heavily surveilled areas in the City of Vancouver. Police officers are assigned to a 17-person squad that operates patrols in and around the bars of Granville Street and the nightclubs of the neighboring Gastown district. This police presence is complemented by regular car and foot patrols and civilian volunteers who operate from the Granville Community Police office located in the heart of the Entertainment District. Further, the public and private spaces of the Entertainment District are already subject to a dizzying array of forms of private surveillance, including two different patrol-based private security programs funded by the area business improvement association (Huey et al., 2005).

Ironically, local activists have also resorted to the use of surveillance techniques and technologies in their attempts to resist 'Big Brother.' For example, an organization in the DTES known as Copwatch, uses video cameras to police the police (Huey et al., 2006). In an interview for this study, one Copwatch member noted how they have filmed police presence at local community meetings, often to the chagrin of the police: 'We just stand back and film them. One cop actually told us that we could be charged with the obstruction of justice.' In similar fashion, a mode of resistance to surveillance adopted by members of the VPSN is to uncover and document the volume of public and privately owned surveillance devices trained on public spaces. To this end, in 2006 members of the VPSN conducted a 'surveillance mapping project' throughout the Granville Street corridor, using cameras to record the volume of cameras trained on public space.

The proliferation of inexpensive cameras means that the range of uses, targets and agents of surveillance will continue to expand. And while the focus of this chapter is on attempts to implement public CCTV systems in downtown Vancouver, it is also true that analysts risk missing the forest for the trees if we only concentrate on the fate of one surveillance tool or tactic. What is important is not so much the success or failure of any particular high-profile surveillance initiative, but the general trend towards ever-greater visibility. Installing surveillance cameras on Vancouver's public streets would not be a radical unprecedented development in the history of surveillance in these neighborhoods. Instead, it would introduce yet another refinement in an existing complex, interweaving and overlapping set of urban optics.

NOTES

1. In previous years, the number of drug arrests were significantly higher. In 1998 and 1999, the DTES accounted for 75% of all drug arrests, this figure decreased to 65% in 2001 (City of Vancouver, 2006).

2. From the 1960s onwards Granville Mall has been 'home' to successive generations of street youth. A survey by the City's Social Planning Department in 2000 found that, on an average, there were 205 street youth subsisting in and around the Mall (Verdant, 2000).

3. On receiving the endorsement of the Police Board, police personnel were tasked with conducting a literature review, completing a public consultation and exploring partnership opportunities, among other tasks. In December 2006, VPD researchers anticipated having a final report to the Police Board in May 2007, making a budget request to City Council in 2007 for a 2008 implementation (VPD, 2006). The May deadline came and went.

REFERENCES

BC Civil Liberties Association. (2006). http://www.bccla.org. Last accessed December 28, 2007.

Bohn, G. (2006). High-tech solution pitched to end Granville Mall violence. *The Vancouver Sun, online edition*, Accessed January 21, 2007.

Boyle, P., & Haggerty, K. D. (n.d.). Spectacular Security: Mega-events and the security complex. Unpublished manuscript.

Bucher, J. (2006). Big brother eyes Granville Mall. *UBC Thunderbird online*, http://typo3.arts.ubc.ca. Retrieved July 5, 2007.

CBC News online edition. (2006). Police take aim at Vancouver's entertainment district, http://www.cbc.ca. Retrieved accessed January 16, 2007.

City of Vancouver. (1998). *Program of strategic actions for the Downtown Eastside*. Vancouver: City of Vancouver.

City of Vancouver. (1999). *Downtown Eastside community monitoring report*. Vancouver: Planning Department.

City of Vancouver. (2001). Downtown Eastside Community Monitoring Report. Vancouver.

City of Vancouver. (2006). Downtown Eastside Community Monitoring Report, online. Vancouver.

Fredericks, G. (1999). *CCTV: A community policing option for the Downtown Eastside*. Vancouver: Vancouver Police Department.

Gibson, E. (2007). CCTV brawl over Granville Street heats up. *The Thunderbird*.

Gordon, M., & Barth, E. (1998). Downtown Granville Street and Mall study.

Haggerty, K. D., & Ericson, R. V. (Eds). (2006). *The new politics of surveillance and visibility*. Toronto: University of Toronto Press.

Hasson, S., & Ley, D. (1994). The Downtown Eastside: One hundred years of struggle. In: *Neighbourhood organizations and the welfare state* (pp. 172–204). Toronto: University of Toronto Press.

Health Canada. (1995). Federal government responds to the HIV/AIDS crisis in Vancouver. Retrieved August 9, 2002.

Huey, L. (2007). *Negotiating demands: The politics of skid row policing in Edinburgh, San Francisco, and Vancouver*. Toronto: University of Toronto Press.

Huey, L., Ericson, R., & Haggerty, K. D. (2005). Policing fantasy city. In: D. Cooley (Ed.), *Re-Imagining policing in Canada* (pp. 140–208). Toronto: University of Toronto Press.

Huey, L., Walby, K., & Doyle, A. (2006). Cop watching in the Downtown Eastside: Exploring the use of (counter) surveillance as a tool of resistance. In: T. Monihan (Ed.), *Surveillance and security: Technological politics and power in everyday life* (pp. 149–165). New York: Routledge.

Hume, M. (2006). Vancouver's 4 a.m. Bars create havoc for police, *The Globe and Mail* (November 8), p. A1.

Jayne, M., Holloway, S. L., & Valentine, G. (2006). Drunk and disorderly: Alcohol, urban life and public space. *Progress in Human Geography*, *30*(4), 451–468.

Lees, L. (1998). Urban renaissance and the street: Spaces of control and contestation. In: N. R. Fyfe (Ed.), *Images of the street: Planning, identity, and control in public space*. London: Routledge.

MacPherson, D. (2001). *A framework for action: A four pillar approach to drug problems in Vancouver*. Vancouver: City of Vancouver.

Marx, G. T. (1988). *Undercover: Police surveillance in America*. Berkeley: University of California Press.

Matas, R. (2001). Vancouver takes aim at fenced goods. *Globe and Mail* (December 6), p. A15.

Monahan, T. (Ed.) (2006). *Surveillance and security: Technological politics and power in everyday life*. London: Routledge.

Obe, D. (1960). Skid road aid picks pockets of taxpayers, Expert says. *The Vancouver Sun* (December 2), p. 1.

Samatas, M. (2007). Security and surveillance in the Athens 2004 Olympics: Some lessons from a troubled story. *International Criminal Justice Review*, *17*(3), 220–238.

Talbot, D. (2006). The licensing act 2003 and the problematization of the night-time economy: Planning, licensing and subcultural closure in the UK. *International Journal of Urban and Regional Research, 30*(1), 159–171.

Tinic, S. (2005). *On location: Canada's television industry in a global market.* Toronto: University of Toronto Press.

Trade and Convention Centre Community Impact Team. (1999). *Which legacy? Vancouver's downtown communities and the expansion of the VCEC.* Vancouver: Ministry of Employment and Investment.

Vancouver Police Board. (2006). *Minutes of the Vancouver Police Board, May 17, 2006.* Vancouver: Vancouver Police Board.

Vancouver Police Department. (2006). *Closed circuit television project powerpoint presentation.* Vancouver: Planning and Research Section.

Vancouver Public Space Network (VPSN). (2006). http://www.vancouverpublicspace.ca. Last accessed December 28, 2007.

Verdant. (2000). *Homeless street youth on downtown south: A snapshot study.* City of Vancouver Social Planning Department Report.

Young, A. (1996). The Bulger case and the trauma of the visible. In: *Imagining crime* (pp. 111–145). London: Sage.

THE CONQUEST OF SPACE: NEW YORK CITY'S NEW FRONTIER OF SOCIAL CONTROL

Kirsten Christiansen

ABSTRACT

Public space has traditionally been space of and for the people, a place where the community could congregate and engage in public expression, including expression of a political nature. In the current environment of increasing insecurity, police departments are encouraged to join the fight against terrorism by increasing surveillance and control over ever-larger areas, including public space, and over ever-greater segments of the community. This chapter examines two recent initiatives by the New York Police Department that are changing the nature of public space in the city.

BEING IN PUBLIC

Most people have seen worse things in private than they pretend to be shocked at in public.

– Edgar Watson Howe (1853–1937)

The Oxford English Dictionary (1989, 2nd edition) defines *public* as "[o]f or pertaining to the people as a whole; that belongs to, affects, or concerns the community or nation ..." Common understanding of what is public is

Surveillance and Governance: Crime Control and Beyond
Sociology of Crime, Law and Deviance, Volume 10, 57–74

ISSN: 1521-6136/doi:10.1016/S1521-6136(07)00203-5

intertwined with its presumed opposite – what is private. Public space as a physical and metaphorical concept exists alongside our notions of community and ownership. In the ideal sense, public space is space for all, while private space is restricted; public space is owned by the people collectively, while private space is owned by people or small groups individually; public space is visible, while private space is hidden; public space is open, while private space is closed.

The reality of public space is rarely this simple. First, the conceptual meaning attached to any space is socially constructed – it is continuously created by an interaction between nodes of power, general cultural norms, and the day-to-day lived experiences of the people who use it (Burkitt, 2004). Second, what we think of as public is often not the true opposite of private in practical terms. Public space is not unrestricted and is not necessarily freely accessible at all times and to all people. Rules exist that govern the behavior of people in open space – when it can be used, how it can be used, and by whom it can be used. The understanding of what it means when space is designated public or private is therefore not just semantic but affects the daily movements and activities of residents and visitors in any community. Public space is also political space. The democratic process requires that citizens have access and exposure to information and ideas – Habermas's "public sphere" (1964) – so that they can make informed decisions about the use of collective power and resources and in keeping with the model of the "agora" of Athens, physical space that is open to all citizens is often the site of this exposure and thus a site of the necessary precursors to political debate (Crawford, 1995). So an understanding of the restrictions and exclusions inherent in the rules and designs of public spaces is crucial to understanding the health of a democracy.

Further confusing the concepts of public and private, the space made available for public use is increasingly owned by private entities. New York City's (NYC) 1961 Zoning Resolution created incentives for developers to include public plazas in their building plans. In a city that was becoming a forest of skyscrapers, city leaders wanted to prevent the streets from becoming dark, airless, and closed-in and wanted to ensure the creation of "usable open space".[1] By the year 2000, NYC had 503 privately owned, publicly used spaces, comprising 82 total acres, approximately one-tenth the area of Central Park (Kayden, 2000). While use of "the commons" has always been subject to some control and exclusivity (Kohn, 2004; McNamara, 1997), public use of space under private ownership and the policing of privately owned space are not subject to the same governmental oversight, and, in fact, the 1961 Zoning Resolution does not spell out specific rules for the management of public use of these spaces, leaving the status of these spaces as forums for democratic participation open to question.

Susan Bickford (2000) argues that the built environment itself can and often does discourage democratic participation. She points out that cities are places where people are most likely to encounter strangers, people with different backgrounds and viewpoints. This exposure to difference can be invigorating and a stimulus to the public debate necessary for political participation in a democracy. It can also be disturbing and scary. While most Americans would probably agree that diversity is a hallmark of the American political and cultural experiment, it often seems like many people would be happier if that diversity occurred somewhere else and not anywhere they need to be. This contemporary segregationism can be seen in the proliferation of gated and planned communities, in the increase of privately owned publicly used space, and in the workings of Business Improvement Districts, which create a controlled and planned version of reality to replace the uncontrolled and therefore dangerous "real world".[2]

This fear of uncontrolled space can also be seen in the increased focus on the detection and control of suspicious or potentially dangerous behavior in public. Societies have probably always struggled with the question of how to manage public places. Democratic space is not anarchical space. Some rules of conduct and use are necessary. As human beings become increasingly interconnected and increasingly insecure, control over public space and particularly the behavior of people in public space becomes a greater concern. This concern pre-dates the current War on Terror and in the contemporary United States can be traced directly back to the crime waves of the 1980s and 1990s. In NYC, police responded to high levels of violent crime in the 1980s in part by focusing on public order offenses which directed police resources towards controlling an "inappropriate public" (Mitchell, 2003a) – the homeless, drug dealers and users, and loitering teenagers.[3] This "broken windows policing" (Wilson & Kelling, 1999) emphasized the reduction of everyday disorder in order to ultimately reduce more serious crime. Places characterized by litter, graffiti, buildings in chronic states of disrepair, and large numbers of people hanging around on the streets with no apparent purpose or destination all signal to evildoers that no one is guarding the fort. This vulnerability attracts criminals and conversely frightens decent people away.

The restriction on the use of public space to only those who are "appropriate" is therefore somewhat ironically designed to *preserve* its public nature and to keep it from being co-opted by undesirable members of the population. But the designation of "appropriate" and "inappropriate" is hardly scientific. Often, appropriateness is determined by the way people look and by the way they make others feel and not by anything specifically dangerous or illegal about their behavior. The homeless are rarely dangerous but they are disturbing. So communities move the homeless along out of the "nice" parts of town and away from "appropriate" people, by creating rules that criminalize spending too much time in one spot or by removing public bathroom facilities and then enforcing laws against public urination, or by any number of regulations designed to address "quality-of-life" issues. Those with access to private space, those who own property and have other places they can be, are welcome almost everywhere. Those with no access to private space are denied access to public space as well – they have no place to be (Mitchell, 2003a).

And while communities seek to make certain members of the population invisible, there is also a heightened concern with making other types of people more visible than they already are. Concerns over security have led to greater surveillance over all members of the population. Surveillance, of course, is nothing new – societies have long sought to keep track of their members and non-members (Parenti, 2003). Most citizens are probably not consciously aware of how watched they are – and many are not that concerned. From tracking devices in cell phones and vehicles to swipe cards used to enter buildings or board public transportation, vast numbers of a modern society's citizens can be located within minutes and tracked as they move across space. Public space is not anonymous space.

What "public" means in the United States has particular importance when looked at in the context of the First and Fourth Amendments of the United States Constitution, the former being key to the democratic process and the latter the subject of increasing scrutiny in recent years, especially in light of the information-gathering capabilities of a digital world.

YOU GOTTA FIGHT FOR YOUR RIGHT (TO SPEAK, ASSEMBLE, AND BE LEFT ALONE)

Congress shall make no law respecting an establishment of religion, or prohibiting the free exercise thereof; or abridging the freedom of speech, or of the press; or the right of the people peaceably to assemble, and to petition the Government for a redress of grievances.

– Amendments to the United States Constitution, Article 1.

> *The right of the people to be secure in their persons, houses, papers, and effects, against unreasonable searches and seizures, shall not be violated, and no Warrants shall issue, but upon probable cause, supported by Oath or affirmation, and particularly describing the place to be searched, and the persons or things to be seized.*
>
> – Amendments to the United States Constitution, Article 4.

While the First Amendment's guarantee of the freedom of assembly more obviously touches upon the issue of the existence and availability of space, the right to free speech is no less a question of geographical definition. Speech cannot exist in a vacuum[4] and political speech even less so. For an act to have an expressive component it requires both a speaker and an audience; this dialogic relationship requires a forum in which those speakers can appear before those audiences. And all audiences are not the same. For political speech to retain its democratic function, the audience present must be the appropriate targets of the expressive activity. The United States Supreme Court has upheld the use of time, place, and manner restrictions on speech subject to the findings that the restrictions are content-neutral, that there is a significant government interest that requires the restriction, and that alternatives exist for the communication of the particular speech in question, which includes the existence of alternative space for expression. But all space is not equal – simply because protestors have the option of conducting their protest in an empty parking lot miles away from the intended target of the protest does not erase the constitutional question. Speech requires a forum and political speech requires a forum that includes a meaningful audience (Mitchell, 2003b).

Public forum doctrine, the set of formal guidelines for regulating speech developed by the United States Supreme Court, separates public property into three types: traditional public forums, designated public forums, and everything else (Mitchell, 2003a; Kaye, 2006). Restrictions of speech in traditional public forums must be content-neutral and as non-restrictive as possible (subject to the guidelines for time, place, and manner restrictions listed above). Designated public forums are afforded the same protections as traditional public forums but can be removed from public use altogether by the state at its discretion. Non-public forums are not completely unprotected but the standard by which any restriction is judged becomes much more relaxed – for example, restrictions must be viewpoint-neutral but can be content-based (Kaye, 2006).

The Court has taken two general approaches to assessing what kind of forum any given space represents and therefore how much restriction of speech is allowable. In part, the Court has looked at whether the space has traditionally been used as a space for speech. Those that have are treated as traditional public forums, regardless of their current use or ownership. Of course, this approach ignores the need for the continual creation of new public forums as the built environment and social habits change.[5] Another approach that has been used by the Court is to look at whether the speech activity is compatible with the way in which the space is normally used – does the activity disrupt the function for which the space exists? This latter approach has the advantage of weighing the First Amendment's interest against the government's interest before deciding what kind of forum a particular space represents but it increases the subjectivity and potential for bias involved in the decision (*ibid.*).

The creation, definition, and maintenance of public forums are key to the ability of citizens to engage in the democratic process, both as disseminators of political ideas and as receivers of alternative opinions. Whether we use the Holmesian metaphor of a "marketplace of ideas" (*Abrams v. United States,* 1919) or the Madisonian ideal of a "government by discussion" (Sunstein, 1995) in interpreting the First Amendment, whether we define democracy as the will of the majority or use a more pluralistic vision that encompasses the acknowledgment of minority opinions, "[i]n a democracy it must be possible for the minority to become the majority" (Kohn, 2004, p. 43). For any minority to have that chance, that minority must have the space to become visible – to educate and inform, and to convince others to join them. It is not that democracy can only occur in public. Many of the key elements of citizenship can and must occur in private. But citizens can only be fully represented politically in public because it is only in public that they can be seen.

Of course, the visibility of people in public space and the ways in which state and local authorities seek to make people both more and less visible in these spaces do not just reflect our understanding of freedom of speech and assembly but have also drawn increased attention to the ever-shifting concept of "privacy rights". The experience of the United Kingdom with domestic and foreign terrorist activity within its borders has led to the development of a surveillance blanket in many locations, particularly the city of London. American cities are increasingly taking a page from the British playbook and installing cameras in greater and greater numbers to monitor open and public spaces (Blitz, 2004). Although the expectation of privacy is traditionally considered lower in public places, the increasing use of surveillance equipment to monitor public areas and questions regarding the use of surveillance footage has prompted a re-examination of what constitutes "unreasonable", "search", and "seizure" under the Fourth Amendment.

Insecurity in public places encourages increasingly tighter control over citizen behavior. In NYC, two recent initiatives by the city and the police department illustrate the increasing desire for control and surveillance of public space – the change of parade permit rules and the Lower Manhattan Security Initiative.

NYC DOES (NOT) LOVE A PARADE

> Parades should be classed as a nuisance and participants should be subject to a term in prison.
>
> – Will Rogers (1879 – 1935)

In the Summer of 2006, the New York City Police Department (NYPD) sought to amend the existing rules governing marches and parades in the city to require permits for

> (1) any procession or race which consists of a group of two or more pedestrians, vehicles, bicycles or other devices moved by human power, or ridden or herded animals proceeding together on any public street or roadway in a manner that does not comply with all applicable traffic laws, rules and regulations; (2) any procession or race which consists of a group of 20 or more vehicles, bicycles or other devices moved by human power, or ridden or herded animals proceeding together on any public street or roadway; or (3) any processions which consists of 35 or more pedestrians proceeding together along a sidewalk on foot or in any other manner authorized by law. (New York Police Department, 2006)

Prior to a public hearing on these proposed changes scheduled for August 2006, and presumably in response to a very vocal opposition, the NYPD announced it was backing off of the proposed changes and would be reworking the definition of parade (Gendar, 2006). In February 2007, the Department published revised changes to the rules, now requiring groups of 50 or more – in vehicles, on bicycles, or on foot – to apply for a permit or face up to 10 days in jail or a fine of $25. Bicycling groups have sued the city, seeking to overturn the new regulations.[6]

The new rules took effect in February 2007 and represent a significant expansion of the authority of police to control expressive activity in public spaces. The NYPD already exercises tremendous control over public demonstrations in the city. Organizers of demonstrations must get permits from their local precinct (or police headquarters for larger events) for everything from the use of bullhorns or other equipment to amplify sound to the holding of marches and rallies in public places. After a permit application is submitted, demonstration organizers are required to meet with police officials to describe their demonstration and identify speakers and topics. Frequently, organizers will receive a follow-up phone call from the Intelligence Division of the NYPD, requesting additional information. And if the organizers make it through these regulatory barriers and the demonstration does finally occur, police presence is generally quite impressive – to the point that at some demonstrations, participants and onlookers have remarked that there are more police than protestors. And not just your typical uniformed officers will be present – police may be dressed in riot gear and officers with rifles have been stationed on rooftops (Lieberman & Dunn, 2004).

During marches, police vans and officers on foot both lead the procession and bring up the rear. Police on motorcycles flank the marchers in a line with a minimum of space between each officer, effectively boxing the marchers into a very limited space that ensures the least amount of disruption of vehicle and foot traffic (and ostensibly protects marchers from the traffic as well). During less mobile demonstrations, protestors are herded into pens created by temporary metal barricades – these pens are colloquially referred to as "free speech zones".

This type of strategy for policing collective activity is what McPhail, Schweingruber, and McCarthy (1998) refer to as a public order maintenance system (POMS). While the policing of protest in the 1960s was characterized by the use of "escalated force", McPhail et al. describe the current style as one of "negotiated management", characterized by the goals of protecting First Amendment rights while limiting the amount of disruption caused by protest activity, necessitating extensive communication between police and protest organizers (in part effected through the use of the permit system), negotiated arrests,[7] and the use of cordoned-off areas as a way to avoid the use of force. However, every bureaucratic step added to the process of engaging in constitutionally protected political activity limits the freedom of citizens to be part of the democratic process – first, by creating a cumbersome and time-consuming procedure that can discourage participation and second, by installing guardians to the process who can approve, deny, or force modifications of requests at their discretion.

New York is unique among American cities in the use of what the New York Civil Liberties Union calls a "suffocating police presence" (Lieberman & Dunn, 2004) at protest events. This overwhelming display of state power and the use of open surveillance create an impression that protestors are deviant, unpredictable, and dangerous, and discourage people from participation; in essence, these police tactics discourage people from participating in the exercise of a constitutionally guaranteed right. Though the NYPD does not publicly discuss its strategic plan for handling rallies, marches, and demonstrations, they state that they make sure to have an "adequate number of officers" at protest events.[8]

Subsequent to the 2004 Republican National Convention (RNC), articles in the *Boston Globe* (NYPD Blues, 2004) and *Boston Herald* (Wedge, 2004)[9] compared the RNC to the 1968 Democratic National Convention in Chicago, during the height of the protests against the U.S. war in Vietnam. During that convention, military snipers were positioned on the top of the Museum of Natural History, thousands of troops surrounded the city, and Democratic National Committeeman Donald O. Peterson called for a two-week adjournment to and relocation of the Convention because "young people are being beaten in the streets" (*The War at Home*, 1979). The Herald also noted that more people were arrested in New York in 2004 than in Chicago in 1968.

It has been suggested that the new rules are primarily directed at a particular group, Time's Up! (Chung, 2007), which seeks to promote sustainable environmental objectives – including the organization of group bicycle rides to encourage New Yorkers to adopt a less polluting form of transportation and to encourage the city to provide or improve the infrastructure for cyclists. Among the group rides promoted by Time's Up! is the monthly Critical Mass ride which occurs in conjunction with similar events in several cities around the world. These monthly events invariably disrupt traffic flow in Manhattan, which has a difficult traffic situation in the best of conditions, and flout existing traffic laws (Critical Common Sense, 2007). The first Critical Mass ride after the adoption of the new parade permit rules was held on March 30, 2007. Although three people were arrested, none were charged under the new rules (Farnam, 2007).[10] In fact, the new rules' first enforcement did not occur until the May 25 Critical Mass ride at which 18 cyclists received tickets for parading without a permit (Siegel, 2007).

ALL EYES ON YOU

> Unwarranted police surveillance will destroy our capacity to tolerate – and even encourage – dissent and non-conformity; it promotes a climate of fear; it intimidates, demoralizes and frightens the community into silence.
>
> – New York State Supreme Court Justice Peter J. McQuillan, *People v. Collier* (1973)

In July 2007, the New York Police Department began implementation of a comprehensive surveillance system, much like the one in London, designed to "detect, track and deter terrorists" (Buckley, 2007). The Lower Manhattan Security Initiative includes the installation of 3,000 new cameras, license plate readers, and state of the art roadblocking equipment[11] by the end of 2008. In 1998, NYC had approximately 2,400 publicly and privately owned cameras pointed at public-used space.[12] In 2005, a local activist and performance group called the Surveillance Camera Players estimated that there were 15,000 cameras in use in Manhattan.[13] While most of these cameras are owned and operated by private businesses and many of the remaining cameras are operated by the Department of Transportation, law enforcement authorities can attain access to surveillance footage from any of these cameras under specific circumstances – and when the concern is that of threats to security, those circumstances may be easy to obtain.

The NYPD has a checkered history when it comes to using surveillance, particularly in the policing of political activity. In NYC, the first official unit created to "surveil and repress radicals" (Donner, 1990, p. 30), called the Anarchist Squad, was formed in 1909 and engaged in "harassing and arresting anarchists on pretexts, intimidating hall owners, and blocking the distribution of the anarchist journal *Mother Earth*" (*ibid.*). In 1914, a new unit with the same function was created, called the Bomb Squad. During World War I, many urban police departments collaborated with the American Protective League (APL) to keep an eye on radicals around the country. The APL was a civilian watchdog organization created to keep an eye on the neighbors – ordinary citizens on the lookout for enemy spies and subverters of the American way of life. In several major cities, including New York, APL members were given the power to arrest by police departments.

The Great Depression of the 1930s precipitated a resurgence of radicalism. The 1930s also saw the stirrings of a social consciousness regarding criminal justice abuses. Local red squads took on a mainly intelligence-gathering function, turning over the information they collected to federal law enforcement agencies. New York's Bomb Squad became the New York Radical Bureau and then the Bureau of Special Services (BOSS) in the 1950s. As the threat from communism then began to fade and the federal agencies took over more and more of the investigation of radical and subversive elements in American society, local red squads became nearly dormant. The 1960s would change all that.

NYC is one of the focal points of political activity in the United States because of its high profile both domestically and internationally. Although much of what is popularly known about anti-war activities during the 1960s and 1970s is centered around California and Washington DC, NYC's experience with protest and police action provides a picture of the conflict between political expression and social control. BOSS conducted surveillance and intelligence-gathering activities of a wide-ranging cast of political organizations (some more political than others). In contrast to the police activities of other major urban centers that focused on raids of groups deemed subversive, BOSS often used the weapon of prosecution to disrupt the activities of radicals – particularly in regard to the Black Power Movement (of which the Black Panther Party was the key group targeted by law enforcement). This ultimately proved to be both a successful and costly strategy. While prosecutions, even when unsuccessful, often succeeded in dismantling political organizations, they also exposed the tactics of the unit to an increasingly disgusted public. However, nothing much was done until the mid-1980s with the settlement of *Handschu v. Special Services Division.*[14]

The *Handschu* case was a class-action lawsuit filed by a group of activists in 1971 against the activities of the New York red squad (now called the Special Services Division) towards political protest activity. The lawsuit took over a decade to reach a settlement but in 1985, Federal Judge Charles Haight signed the agreement. The agreement created the Handschu Authority, a committee charged with overseeing police surveillance of legal political activity. The three-person committee is made up of two police officials and one member appointed by the mayor's office – hardly a group likely to be hostile to the police.

After September 11, 2001, the NYPD began calls to eliminate the restrictions of the *Handschu* agreement, saying it needed "broader authority to hunt terrorists who may masquerade as regular law-abiding folks until the moment they strike" (Lee, 2002). In 2003, Judge Haight granted this request and relaxed the requirements of the agreement to allow the department to develop new guidelines along the lines of those developed for the Federal Bureau of Investigation.[15]

It is an open question, though, how effective the agreement ever was in curtailing the potentially abusive surveillance activities of the NYPD, much less how much of a change this relaxation ultimately represents. Past members of the committee state that during their tenure requests by the police to surveil public activities were rarely denied (*ibid.*). And anti-war protestors report the continued videotaping of participants by the NYPD's Technical Assistance Response Unit (TARU) – the current name for NYC's red squad.[16]

According to the official City of New York website, TARU "provides investigative technical equipment and tactical support to all bureaus within the department, [*sic*] in addition they also provide assistance to other City, State and Federal agencies".[17] What this description does not say is what exactly any of that means, particularly as regards public protest. In practice, TARU assists with investigations that utilize video footage, and telephonic and email communications.[18] It also provides and utilizes surveillance equipment, including electronic newsgathering vehicles, helicopters, and cameras affixed to surrounding buildings, as well as videotaping at public events.[19]

On March 19, 2005, members of several different anti-war organizations, including the War Resisters' League, Troops Out Now, and United for Peace and Justice, held a rally and march to protest the second anniversary of the beginning of the current war in Iraq.[20] The rally was held at the UN and the march ended up at Times Square, where some members of the protest planned to engage in civil disobedience.

Law enforcement presence at the rally and march was impressive. Several small groups of police officers were situated around the perimeter of the crowd gathered to listen to the speakers. As the march began, officers on motorcycles created a moving barricade as protestors walked up the far right lane of the streets from the UN to Times Square, although whether to protect the protestors from traffic or to keep them hemmed in would be difficult to say. These motorcycles, as well as a police van also blocked cross streets as the march moved through intersections. Groups of officers headed the march and brought up the rear.

Among the police officers, but standing off on their own, were two TARU officers. Both officers were carrying video cameras; however, no videotaping visibly occurred during the rally and march (Fig. 1).

When the march arrived at Times Square, participants held a demonstration in front of the entrance to the Armed Services Recruiting Station. Both TARU officers climbed on top of police vehicles parked alongside the demonstration area and began videotaping. A woman standing near one of the vehicles turned to the TARU officer behind her and said, "I thought you weren't allowed to do that anymore". He did not respond (Fig. 2).

Among the guidelines of the *Handschu* agreement is the restriction on conducting investigations on the political activities of individuals unless they are committing or *about to commit* a crime.[21] Although none of the participants was engaged in any criminal activity at the time the videotaping began, the participants had publicly stated their intention to commit an act of civil disobedience after the demonstration in front of the Armed Services Recruiting Station. In fact, not only had this intention been stated (and

Fig. 1. TARU Officers at the March 19, 2005, UN rally.

Fig. 2. TARU Officer Filming Demonstration at the Armed Forces Recruiting Station, March 19, 2005.

printed on flyers and in online documents advertising the event), those who intended to participate in this part of the event were given special vests to wear, with pictures of US soldier and Iraqi civilian casualties of the war on the front and back respectively. So while the *Handschu* guidelines technically do allow TARU to videotape the demonstrators due to their stated intention to commit a crime, this does raise the question of whether the guidelines allow TARU to videotape anyone *other* than those demonstrators wearing vests.

TARU also made an appearance at the May 1, 2005, anti-nuke rally in Central Park, this time without video cameras. Police presence was again impressive – by midway through the rally, there seemed to be more police officers than protestors. TARU presence at this event had increased to four officers, all of whom appeared to leave before the rally ended.

Although the rally was peaceful and the police appeared relaxed, the sheer numbers of police present at this and other demonstrations, as well as the barricades routinely used to contain protestors to specific areas (including barricades on May 1 that created an open area down the middle of the crowd which was restricted to police personnel) cannot fail to send a message to people both participating and passing by.

What this history demonstrates is how hazy the line is between acceptable and illegal tactics in policing political activity and how easily and often law enforcement agencies can cross it. The creation of a surveillance net across the city further blurs this line. If all activity in public is watched and recorded, then all political activity is watched and recorded. While the current plan does not include the installation of facial recognition software, it is under discussion. The effectiveness and reliability of this technology is unproven but it is a popular tool for creating the appearance of a secure environment. The privacy issues of the technology are arguably minimal – it is in essence simply a mechanical method of standing in a public place and looking at the faces of the other people there (Blitz, 2004). More problematic is the potential for recording images. The ability to re-visit images constitutes a dramatic expansion of information-gathering ability over that of a police officer merely watching people in a public place. And more disturbing still is the potential future use of automated profiling systems, which do not just compare faces with a database to identify known or suspected criminals but actively use pattern recognition algorithms to identify suspicious behavior (Hildebrandt, 2007). These systems go beyond information gathering into the creation of information by appending labels to people based on their activity in public. And behavior that may be labeled as suspicious can be as simple as remaining in one spot for too long. In 1996, an Australian artist incorporated surveillance cameras into several public performances by choosing a particular camera in a particular public place and for a succession of days standing in front of the camera and staring at it.[22] On a number of occasions, he was told to leave. Simply watching the camera caused his exclusion from a public place (Levin, 2004).

TO BOLDLY GO WHERE NO MAN HAS GONE BEFORE

> There is danger from all men. The only maxim of a free government ought to be to trust no man living with power to endanger the public liberty.
>
> – John Adams (1735–1826), *Journal 1772*

By rewriting the rules on mobile public gatherings and blanketing the city in surveillance equipment, the City of New York and the New York Police Department are changing the way residents and visitors to New York engage in civic and political activity in public spaces. These changes may have dramatic effects on the ability of people to be a part of the democratic process. And these changes are not unique to NYC. Cities across the United States and across the world are facing new questions of global and national security and whether there should be limits to freedom and are responding by increasing control over larger and larger sections of their populations.

The shifting definition of public space in American society has a powerful impact on any contemporary discussion of the First Amendment's guarantees of freedom of speech and assembly and the Fourth Amendment's protection of privacy. Space that can be clearly described as "public" is shrinking in most major cities as private companies and neighborhood organizations come to control more and more property. But even space that is publicly owned is subject to increasing surveillance and stricter control as cities attempt to prevent crime and acts of terrorism and make themselves attractive to outside investment.

Ultimately the health of a democracy can be seen in the ability of its people to be part of the democratic process. Restrictions on expressive political activity in the guise of protecting public order and the knowledge that all activity is being watched by someone somewhere can have a chilling effect on all but the most extroverted citizens' willingness to engage in the public debate that is a cornerstone of American governance. It is a social science truism that the observation of behavior causes behavior to change. There is also conflicting evidence as to whether greater control and greater surveillance actually make societies safer. London's "ring of steel" did not prevent the train bombings in 2005. The Lower Manhattan Security Initiative would not have prevented the events of September 11, 2001. So, the question is not really whether control is necessary for security but rather what kind of control and what kind of security, and whether security at the expense of democracy does not ultimately make us more vulnerable than we are already.

NOTES

1. Private ownership of publicly used space also benefits the city by shifting the responsibility for care and maintenance away from city coffers and onto the wallets of private entities.

2. In New York City, this can clearly be seen in Times Square, which has become an idealized image of a "Great White Way" that may not have ever even existed (Reichl, 1999).

3. Political activists are also frequently lumped into this "inappropriate public".

4. Even speech in "cyberspace" has a geographical foundation; "virtual" space still requires physical equipment and physical space. Questions of public and private also have resonance here as well – accessibility can determine the ability of electronic forums to have meaningful audiences. For other concerns related to cyberspace as an alternative public sphere, see Koch (2005).

5. For example, shopping malls have become the main site of public life in many communities. These spaces are unlikely to meet the requirement of historical precedent.

6. In a decision denying the groups' request for an injunction barring enforcement of the rules until the suit is decided, Federal District Court Judge Lewis Kaplan stated that the parade permit requirement was not a violation of the First Amendment (Sullivan, 2007).

7. McPhail et al. point out that police using the POMS strategy will provide pre-arrest forms and solicit numbers of people expected to engage in illegal acts of civil disobedience.

8. Personal telephone communication with the Office of the Deputy Commissioner, Public Information of the NYPD.

9. Boston hosted the Democratic National Convention the previous June.

10. Forty-seven tickets were issued at the same event.

11. This equipment consists of moveable gates at intersections that can be employed to close roads for the purpose of stopping traffic, particularly "suspect" vehicles (Buckley, 2007).

12. This number taken from the NYC Surveillance Camera Project website accessed at http://www.mediaeater.com/cameras/overview.html. Importantly, counts of surveillance cameras conducted by community groups or non-governmental organizations are frequently confined to those that are visible, meaning the true level of coverage is difficult to ascertain.

13. Based on increases in counts in selected neighborhoods since the 1998 NYCLU report. Information accessed at http://www.notbored.org/times-square.html
14. *Handschu v. Special Services Division*, 605 F. 1384 (S.D.N.Y. 1985)
15. http://www.nyclu.org/node/731
16. http://scaryny.com/archives/2004/09/nypd_taru.php
17. http://www.nyc.gov/html/nypd/html/chfdept/chfdept-taru.html
18. http://www.nycop.com/Stories/Nov_00/Stalking_the_Stalker/stalking_the_stalker.html
19. http://www.notbored.org/taru.html
20. The information provided regarding the March 19 and May 1 demonstrations is based on personal observations of the author.
21. http://www.nyclu.org/ny_advisory_commission_oncivilrts_052103.html
22. These performances also frequently involved the artist holding signs with various messages up to the camera – such as one asking for a copy of the tape being made and asking the camera to indicate yes or no to the request by either moving from left to right or up and down.

REFERENCES

Bickford, S. (2000). Constructing inequality: City spaces and the architecture of citizenship. *Political Theory*, *28*(3), 355–376.

Blitz, M. J. (2004). Video surveillance and the constitution of public space: Fitting the fourth amendment to a world that tracks image and identity. *Texas Law Review*, *82*(6), 1349–1481.

Buckley, C. (2007). New York plans surveillance veil for downtown. *The New York Times* (July 9), Section A, Column 0, Metropolitan Desk, p. 1.

Burkitt, I. (2004). The time and space of everyday life. *Cultural Studies*, *18*(2/3), 211–227.

Chung, J. (2007). Parade permit rule of 50. *Gothamist*. Available at http://gothamist.com/2007/02/01/parade_permit_r.php. Retrieved June 12, 2007.

Crawford, M. (1995). Contesting the public realm: Struggles over public space in Los Angeles. *Journal of Architectural Education*, *49*(1), 4–9.

Critical common sense (Editorial). (2007). *The New York Sun*, April 20, p. 10.

Donner, F. (1990). *Protectors of privilege: Red squads and police repression in urban America*. Berkeley: University of California Press.

Farnam, T.W. (2007). Cops' Critical Mass crackdown; 3 held, dozens during cyclists' first try after NYPD ban on groups of 50 or more cycling without permit. *Newsday* (March 31), p. A12.

Gendar, A. (2006). Cops step back on parade plan. *The New York Daily News* (August 19), Sports Final Edition, News, p. 10.

Habermas, J. (1964). The public sphere: An encyclopedia article (1964). 49–55(Lennox, S. & Lennox, F., Trans.). *New German Critique*, *74*(3). Original work published 1964.

Hildebrandt, M. (2007). Profiling into the future: An assessment of profiling technologies in the context of Ambient Intelligence. *FIDIS (Future of Identity in the Information Society) Journal*, *January*. Available at http://journal.fidis.net/fileadmin/journal/issues/1-2007/Profiling_into_the_future.pdf. Retrieved September 1, 2007.

Kayden, J. S. (2000). *Privately owned public space: The New York City experience*. New York: Wiley.
Kaye, J. E. (2006). No joy in Mudville for the first amendment: A critical examination of the public forum doctrine in light of United Church of Christ v. Gateway. *Case Western Reserve Law Review, Summer 2006*, *56*(4), 1147–1175.
Koch, A. (2005). Cyber citizen or cyborg citizen: Baudrillard, political agency, and the commons in virtual politics. *Journal of Mass Media Ethics*, *20*(2&3), 159–175.
Kohn, M. (2004). *Brave new neighborhoods: The privatization of public space*. New York: Routledge.
Lee, C. (2002). The NYPD wants to watch you, *Village Voice*. December 18–24. Available at http://www.villagevoice.com/news/0251,lee,40627,1.html. Retrieved September 1, 2007.
Levin, T. Y. (2004). Surveillance and the Ludic reappropriation of public space. *Open, Insecurity* (6), 50–71.
Lieberman, D., & Dunn, C. (2004). Pulling the NYPD out of protest. New York Civil Liberties Union press release, December 9, 2004. Retrieved October 1, 2005 at http://www.nyclu.org/node/777.
McNamara, M. J. (1997). From common land to public space: The frog pond and mall at Newburyport, Massachusetts, 1765–1825. *Perspectives in Vernacular Architecture*, Vol. 6, Shaping Communities, 77–91.
McPhail, C., Schweingruber, D., & McCarthy, J. (1998). Policing protest in the United States: 1960–1995. In: D. della Porta & H. Reiter (Eds), *Policing protest: The control of mass demonstrations in western democracies* (pp. 49–69). Minneapolis: The University of Minnesota Press.
Mitchell, D. (2003a). *The right to the city*. New York: The Guilford Press.
Mitchell, D. (2003b). The liberalization of free speech: Or, how protest in public space is silenced. *Stanford Agora*, Vol. 4. Retrieved March 23, 2007, at http://agora.stanford.edu/agora/volume4/mitchell.shtml
New York Police Department. (2006). Notice of opportunity to comment on proposed rule, July 18.
NYPD blues (Editorial). (2004). *Boston Globe*, September 19, p. D10.
Oxford English Dictionary. (2nd Edition). (1989). Retrieved 8/15/2007 at http://dictionary.oed.com.ez.lib.jjay.cuny.edu/cgi/findword?query_type = word&queryword = public.
Parenti, C. (2003). *The soft cage: Surveillance in America from slavery to the war on terror*. New York: Basic Books.
Reichl, A. J. (1999). *Reconstructing Times Square*. University Press of Kansas.
Siegel, J. (2007). Mass cyclists are critical of new parade permit rule enforcement. *The Villager*, *76*(1). Retrieved June 7, 2007, at http://www.thevillager.com/villager_214/masscyclistsarecritical.html
Sullivan, J. (2007). Bicyclists lose in court. *The New York Times* (April 19), Late Edition, Section B, Column 1, Metropolitan Desk, p. 4.
Sunstein, C. R. (1995). *Democracy and the problem of free speech*. New York: The Free Press.
Wedge, D. (2004). Headline: Republican National Convention. *Boston Herald* (September 3), p. 11.
Wilson, J. Q., & Kelling, G. (1999). Broken windows. In: V. Kappeler (Ed.), *The police and society* (pp. 154–168). Prospect Heights, IL: Waveland.

THE CHECK AND THE GUARDIANSHIP: A COMPARISON OF SURVEILLANCE AT AN AIRPORT AND A HOUSING-ESTATE AREA IN THE PARIS OUTSKIRTS

Fabien Jobard and Dominique Linhardt

ABSTRACT

This chapter approaches the question of government and surveillance through a comparison between the control practices observable in two types of places. First, we focus on international airports, specifically the French international airport of Orly. Airports are maximum security zones where persons perceived as having no legitimate business are expelled and where suspicious objects are destroyed. The second kind of places are the ones labeled as "no-go areas", violent pockets within urban space. Social housing projects located in the bleak suburbs of French cities are such dangerous zones. Both kinds of places – airports and no-go areas – have very different time and space features: people briefly pass through anonymous airports where relationships are kept at an impersonal minimum, whereas the population of a housing estate area is made of "permanent transients" pinned down by a shared fate of which there seems no escape.

Surveillance and Governance: Crime Control and Beyond
Sociology of Crime, Law and Deviance, Volume 10, 75–100

ISSN: 1521-6136/doi:10.1016/S1521-6136(07)00204-7

INTRODUCTION

The term surveillance raises a lexical difficulty which complicates its theoretical implications as well as its empirical specification. Narrowly understood, it refers to the set of processes and measures through which the State is informed of the activities of a person or of a group of persons while avoiding repressive action, either because no offence has actually been identified or because the government prefers, for one reason or another, to be discrete (Fijnaut & Marx, 1995; Sharpe, 2000). Yet, Michel Foucault's seminal work *Surveiller et punir* was translated into English under the title *Discipline and Punish* (Foucault, 1977). In this latter disciplinary understanding, the notion of surveillance has to do with a much broader field (Deleuze, 1988). It ceases to be a mere policing tool among many others in the State's policing arsenal, and instead becomes a regime of "governmentality" combining and articulating different technologies, strategies, and governmental rationalities (Rose & Miller, 1992). Accordingly, surveillance becomes a notion to describe a specific way through which human behavior is apprehended, and hence ensure predictability, calculability, and "governability" (Gandy, 1993; Lyon, 1994; Wood, 2003). Here, of course, we find ourselves following a path opened by Foucault (1988) and followed by many others since (for instance Rose, 2000).

However, whether one understands the notion of surveillance in its narrow sense, as a mere set of disparate means within a governmental apparatus, or, on the contrary, as the basis for the constitution of a governmentality regime which relies on spotting, identification, and control, using number of techniques, devices, and processes, in both cases, the risk is that surveillance becomes an "all-terrain" notion which has less and less to do with the ground realities of its implementation. As David Garland (1997) strongly underlines, the notion of surveillance could then lead directly to a variant of reductionism: applicable to too many situations, it would, at the same time, suppress the empirical specificities of each one. This inclination is all the more detrimental that, in polishing the ruggedness of reality, it contributes to neglecting the uniqueness of the organizational methods and the institutional layouts, the various types of intervention, and of the stocks of knowledge precisely meant to define the "surveillance society." Yet, if there were one systematic observation to report, it would be the multiplicity and variability of the devices of surveillance. Each of these devices adapts to specific constellations, which have characteristic social, spatial, and temporal indicators, and are defined by the nature of the threats and the risks that operate inside of them. The differences from one constellation to another are what deserve particular attention.

Our approach answers the need felt for a return to the scrutiny of the actual modalities in place in the practice of surveillance. To this effect, we chose to anchor the description in space, reminding the reader that Michel Foucault's analyses of surveillance were above all analyses of the spatialization of power (enclosed areas in the treatment of insanity, confined areas in the containment of contagious diseases, areas of panoptical centrality in the disciplinary process, territorialized space in the security process, etc.).[1] We have deliberately stressed the contrast between two unique sites, both of which underwent an in-depth empirical study. Both are located in the southern Paris agglomeration, they only stand a few miles from each other. In both cases, security, control, and surveillance are high-stake and greatly sensitive issues. Yet, everything seems to separate them: when not a blatant antinomy, the surveillance practices at Orly airport on the one side and in the Bas-Moulin housing project of Dammarie-les-Lys on the other diverge entirely, in many respects. As we will establish, in the first case, surveillance takes the form of a checking, in contrast to the second, where it takes the form of guardianship.

An airport, a housing project: are the intrinsic differences between the two not so great that any comparison would only resemble a far-fetched artifact? We find such resistance unnecessary: if indeed surveillance and governance are broad concepts, then they must be able to welcome differences, as extreme as they may seem. But more importantly from an empirical standpoint, one cannot overlook the fact that both sites face the same "problematic of government" (Rose & Miller, 1992): that of security. A technical problem for experts to solve on one level, the question of security nonetheless becomes a public matter when its treatment or non-treatment affects public opinion, and is likely to enter a controversy where lack of security is then seen as the problem and surveillance as its solution.

VARIABLES USED IN THE DESCRIPTION AND THE ANALYSIS

Both the studies of the Orly airport and of the housing project in Dammarie-les-Lys were undertaken using the classical tools of the qualitative method (i.e., empirical observations and interviews). In order to facilitate comparative analysis, however, we entered the collected data into one same analytical grid, in which a number of variables are *tertio comparationis* – their value thus varies according to where they were collected. Table 1 gives an idea of what type of variables we are referring to.

The variables we identified are of two different sorts. The first is purely descriptive: the variables have to do with matters of fact noted from our field observations, and do not require specific interpretation. The first variable (A) corresponds to the institutional arrangement within which the actors of surveillance find themselves; accordingly, it is a description of the various types of cooperation, lines of conflict, distribution of tasks, and share of responsibility in the site under study. The second variable (B) characterizes how the surveillance apparatus defines the targets of surveillance. The third variable (C) allows us to specify the type of interaction existing between the surveillance agents and the surveillance targets, the main point being to identify whether interactions are unique, or whether, on the contrary, there is a principle of reiteration of interaction. The fourth variable (D) allows us to classify surveillance apparatuses according to whether or not they require records and to how collected records are reinvested in the surveillance practices. The fifth variable (E) allows for a better apprehension of the various sources of legitimacy with regard to the authorization and supervision of the interventions in the different spaces, according to the constraints they themselves face.

The second type of variable is analytical. While these may be inferred from the previous sort, they nonetheless require a higher qualification process than does mere observation. The first of these analytical variables (F) seeks to compare the effects of surveillance practices at each site; it distributes the effects on a continuum extending from the "objectification" of the individual targets to their "subjectification." The second analytical variable (G) has to do with the modalities of "we"-formations. For, the deployment of an apparatus affects not only individual subjectivities, but also creates collective subjectivities, or even communities, whose relationship with surveillance services is a relevant matter. The last analytical variable (H), based on a synthesis of the previous variables, seeks to precisely characterize the nature of the intervention at each site. In what follows, and based on the aforementioned variables, we would like to go beyond merely describing the case studies by showing evidence of two contrasting models of surveillance and of governance of behaviors: the check model found on the grounds of the airport, and the guardianship model observable on the grounds of the suburban housing project.[2]

Table 1. Categories of Comparison.

Variables		
Descriptive Variables		**Relevant Questions**
(A)	Institutional arrangement	What is the institutional arrangement between the various security agencies involved in the apparatus? What is the impact of the arrangement on the spatial organization and the responsibility distribution?
(B)	Targets of surveillance	How are the surveillance targets defined? How does this definition affect the surveillance apparatus in itself?
(C)	Type of interaction	What is the nature of the relationship between the surveillance actors and the individuals under surveillance? Are their interactions unique, or reiterated?
(D)	Absence or presence of records	Does the surveillance apparatus rest on recording devices and the production of a type memory? Or is the amnesia of the apparatus intentional?
(E)	Sources of legitimacy	What is the normative justification for the surveillance apparatus? How are the presence and operations of the surveillance actors justified?
Analytical Variables		**Relevant Questions**
(F)	Effects of surveillance on the individual status of its targets	What is the nature of the individual identity produced by surveillance? Does surveillance create subjects, or is it searching for objectification?
(G)	Effects of surveillance on the collective status of its targets	Does surveillance produce a collective identity of individuals under surveillance? How does it infer a sense of collective identity? What are the types of "we" that emerge from it?
(H)	Type of intervention	Everything taken into account, how can the type of surveillance actors intervention be qualified? What does the relationship between surveillance and space and temporality imply?

ORLY AIRPORT: THE CHECK MODEL

Terrorism has a strong relationship with the massive streams of mobility which characterize our societies (Urry, 2000) (E). Indeed, the terrorist enterprise is doubly dependent on the socio-technical systems, which are vital to them. First, mobility infrastructures give terrorists the perfect conditions to achieve their goals. Answering a constant need to conceal themselves, the anonymous streams of movement appear as a perfect hideout for terrorists (Linhardt, 2006). From there, they can attack the infrastructures of mobility themselves and the people who are inside them, since, while offering terrorists a form of protection, the targets of the terrorist enterprise only become more vulnerable. Air transportation was faced with this problem very early on. From this standpoint, the September 11th attacks were only the temporary end of a long-lasting relationship between terrorism and civil aviation (Crenshaw, 1988; Merari, 1998; Lyon, 2003). Throughout the relationship, specific prevention systems have been set up. These systems have quite obviously evolved since the end of the 1960s, in step with the evolution of the threat itself.[3]

Paris Orly airport has faced terrorist threats more than once. It does not, however, show any particular characteristic distinguishing it from other international airports on this level. This is all the more true in light of the fact that air transportation security is a matter for international institutions, which ensure that procedures are normalized beyond national borders and local specificities (Wallis, 1998). Thus, while the site under observation is Orly, the point is not to underline any form of specificity at Orly. Quite the contrary: given the constant normalization of the place, the study shows how Orly's security apparatus more or less fits the framework found in other airports.

Testing and Filtering

In order to understand the mechanisms of airport terrorism prevention, it is appropriate to start from the terrorist enterprise. One way to characterize it is to recognize it as a game with "normal appearance" (Goffman, 1971, p. 256). Terrorists act from an ambush which is not a physical, but, rather, a "social ambush" (Walzer, 1977, p. 176): they blend into the normality of daily coexistence by borrowing a commonplace, negligible or plain physical appearance: nothing looks more like a lambda traveler than a terrorist checking in, or a hijacker going through security.[4] This specificity of terrorists to act in such a way so as to follow through with their goal while concealing it under normality implies that those who seek to stop them must know how to see behind what a person or object is putting on display. Indeed, the real capacity to strike back is measured by the ability to identify and seize terrorists despite their concealment strategies. In the undifferentiated world of streams of travelers, control means a re-discrimination, a sorting between ordinary and ill-intentioned individuals with the implementation of suspicion (B) (Adey, 2003).

The airport example clearly shows that the practice of suspicion is less arbitrary than one might consider it when associating it with a conception of the notion which rests mainly on intuition. Economy of suspicion at the Orly airport rests exclusively on an advanced codification, formalization, and division of labor. The imperative of suspicion is delegated to an apparatus whose quality will depend on the specific lineup of the persons and the objects within the apparatus determined by pre-defined scripts (Akrich, 1992). Hence, suspicion takes form from routine procedures (H). This explains the importance of the organizational aspects of airport security.

Spatial Organization: Zoning

This organization rests above all on a penetration of the area. The airport space undergoes a thorough process of subdivision before the streams of people crossing the zones are actively controlled. Airport security is conceived on a model of concentric circles surrounding the aircraft: the closer one gets, the more limited the access of persons (passengers, employees) and objects (luggage, shipments), which is only authorized after strict inspection, becomes. Accordingly, the space is divided in to various zones of different status; each border between two contiguous zones of different status is either rendered completely hermetic by physical barriers, or offers access opportunities (E).[5]

Zoning and Distribution of Responsibilities

The zoning goes hand in hand with a strict distribution of responsibilities. The Prefect of the department holds ultimate authority in terms of the security of the airport. His main roles are the elaboration, implementation, and monitoring of the "Airport security plan", and the management of crises. More specifically, he decides which zones are open to the public and which are restricted, what the conditions for traffic and parking of persons and vehicles within the restricted zones are, and what the set up ensuring the security of vehicles, equipment, and goods in these zones should be.

In addition, he has ultimate authority over airport coordination bodies such as the Local Security Committee and the Operational Security Committee. As for on-the-ground operations, every zone type is attached to a specific institution. Hence, the national police force is responsible for the public zone, the functional sectors within the airport, and the security sectors for passengers. The air transportation Gendarmerie is responsible for the aircraft security sectors and the functional sectors outside the airport, and for luggage and shipment security on national flights. As for the security sectors for luggage and shipments on international flights, they are taken care of by the customs service. It is important to note that the actual security check operations are not carried out by police officers or customs agents: they are delegated to private companies whose agents are paid for by the airlines and the airport operator. These agents nonetheless answer to a regulatory authority, to which they must immediately turn if they suspect anything is wrong (A).

The Rationale of the Check

The cross-over areas between zones have a particular denotation: they act as cognitive tide gates where vigilance and suspicion are constantly practiced in order to determine whether a person or object may pass or whether their access should be denied. This creates a dilemma: when can one effectively pass judgment as to whether a passenger or an object is "clean" enough? The social ambush strategy of terrorists makes it necessary to take small details into account, as potential clues. The clues agents will be looking for depend on existing available knowledge concerning the terrorist enterprise: a certain number of features are selected, and serve as a basis for control operations. Terrorists need weapons, for instance. These weapons are made of specific, easily identifiable materials. Weapons which can be used to hijack an airplane and could be used in the pilot's cabin, for example, are usually made of metal. Consequently, the clue the agents will be looking for during hand luggage checks is the presence of metallic objects. As for checked luggage, they will look for what is known as the "pyrotechnic chain", that is, the simultaneous presence of three elements: an explosive, a detonator, and a power source.

In order to detect the presence of such elements, agents use sophisticated equipment whose "cognitive artifacts" (Norman, 1991) help see beyond what can be seen with a naked eye. Despite such technical sophistication, however, what vigilance ultimately requires is a sense of normality. This sense of normality is directly integrated into the devices and can be detected independently from the operator, but when the equipment informs an agent of some form of abnormality, he or she will have to assess it according to their own sense of normality. It is rare for an agent to instantly recognize the nature of the abnormality, or what it is due to. His or her degree of certainty will affect his or her interpretation. This is the "investigation" stage. The depth at which the investigation is be pursued will depend on the "importance" the interpreted detail may have, and, more particularly, on the assessment of the risk taken in closing a case when the degree of certainty is still low (Schütz, 1971, p. 77). On the other hand, one must avoid falling into an attitude of constant suspicion and paranoia. Thus, there must be predetermined criteria which allows agents to assess the value of the clue (H). This check rationale implies putting every human being in circulation in the airport through a short test, whose purpose is to verify the absence of pre-defined clues. Once this absence is confirmed, the test is over (Pinch, 1993).

Controlling Customers and Citizens of Law

This test-oriented filtering technology can be understood as a form of political semiology (Linhardt, 2001). Here, actors assess in situ whether or not they are facing danger through the interpretation of small and predetermined details. This semiology, which allows for an appropriate economy of suspicion, is seen as beneficial by all security actors: indeed, the ability to differentiate between dangerous and non-dangerous situations is considered a pledge of efficiency. This does not prevent the security apparatus from being strongly criticized. In fact, two types of criticism are quite common. The first points at the flaws in the apparatus and considers the airport to be completely inefficient in preventing well-prepared terrorists from acting even though it may, at best, succeed in stopping "amateurs." This criticism obviously plays a crucial role in the constant modification of the apparatus. However, we will focus on the second criticism, which points to risks of violation of the rights and freedoms of those individuals who undergo security checks. Potential violations, the argument holds, are: the violation of the freedom of movement, the invasion of privacy, and the violation of the protection against arbitrary treatment by authorities (E). The evocation of such rights and liberties is made particularly relevant by the fact that the subjects of the law also happen to be customers (B).[6] In all three cases, nonetheless, we will see that it is possible to demonstrate that the airport's security apparatus can coexist with individual rights and freedoms.

The Risk of Violation of the Right to Move Freely Within a Given Territory
With regard to mobility rights, the issue was addressed by the introduction of the security apparatus in specific areas of the airport designed to fill the requirements of an efficient air transport system. Zoning came as an additional tool for distinguishing various sectors from each other according to their role within the system. Such differentiations according to functionality show that security is an intrinsic part of the air transport system: the security apparatus is a just another element of the entire process. Rather than being permanently open or closed, the doors remain "half-open." The apparatus allows the airport access zones to act like a binary switch: if the "cleanness" of the passenger is proven, he or she can go through. If not, the doors remain shut (E).

The Risk of Invasion of Privacy
The various forms of security check also limit the risk of an invasion of privacy. Metal detectors only permit agents to see if passengers are carrying metallic objects; besides, the images of the contents of luggage provided by X-rays have more to do with expressionist painting than photography. This indicates that the goal here is not to show everything, but rather to show as little as possible while ensuring that the important elements become visible. It corresponds quite precisely to what Bruno Latour and Emilie Hermant call the "oligoptic" (Latour & Hermant, 1998, pp. 76–80), as opposed to Foucault's "panoptic": to see very little, but to see very clearly. Foucault's "microphysics" do indeed appear rather gigantic, here, in comparison to a kind of "nanophysics" of vanishing clues: the beep of a metal detector, colors on a screen, the glance in a bag. But at the same time as the field of vision diminishes, the precision of what can be seen increases (F).

The Risk of Violation of the Protection Against Arbitrary Treatment by Authorities
It takes a triple operation to calm the feelings of unfair treatment created by misplaced suspicion. The first operation consists in shifting the attention from the terrorist to the "unclean" passerby, for security devices do not, indeed, detect terrorists per se, but only "unclean" individuals or objects.[7] The second operation consists in disconnecting suspicion from subjectivity by making the practice of suspicion "mechanical": vigilance is made a "machine-like action" (Collins, 1990). The point is to make the process uniform, and to ensure the iteration of identical actions at any given point in the mechanism (F). A tempting analogy is that of a "taylorization" of suspicion, where vigilance is no longer individual or subjective, but

collective and objective. The regulation and normalization of the security apparatus is largely comparable to the technical formatting of vigilance by the preliminary definition of "behavioral coordinates of action" to which the components of the apparatus must react regardless of their qualities or flaws, psychological states, moods, mental representations, or ideologies. This taylorization is, in fact, the guarantee of impartiality: just as the terrorist is "objectified", so is suspicion, in that it is under coded restraints which go beyond the agent (F). Finally, the third operation consists in refusing to keep any database or record of the security checks, or to link them to a "center of calculation" (Latour, 1987, p. 235) (D). Each security check operation is closed on itself, and restricted to a certain location; it cannot be moved, and can rarely be expanded (C). For instance, when a passenger leaves a boarding area he has been authorized to access and then wishes to return, he or she will have to go through the security check process again, since the apparatus will not have kept any records of the previous security check (D).

The extent of protection guaranteed by the absence of records and databases can be measured by the recent debates concerning the introduction of measures requiring centralized databases. Following the September 11th attacks, the U.S. Transportation Security Administration (TSA) announced its intention to introduce a new passenger profiling system. Amongst other things, the system would have required every passenger to reveal their name, date of birth, address, and phone number. Security guards would have had to check the information along with other available data before giving a "risk potential" score to each passenger (Adey, 2003; Lyon, 2003; Singel, 2003). The plan was strongly criticized, and raised concern that it would permit for passenger surveillance to become an excuse for the scrutiny of private information such as financial transactions, and the use of biometric databases (D).

It is possible that, in the long term, airport terrorism prevention converges with the guardianship model developed below. Up to the present, however, developments at Orly airport have remained experimental. The general framework is still one of repetition of standardized methods to reducing suspicion through binary-type tests, which everyone needs to undergo so that agents may detect the absence or presence of previously defined clues. Given the rare occasions on which the airport will actually be confronted with terrorists, it is tempting to question how the efficiency of the whole apparatus can be measured. On the other hand, the airport is constantly confronted with millions of passengers passing through it. In a sense, then, the apparatus can be only looked at in the context of preparation for defense against an absent enemy. It must also be understood as a form of alliance between the checking agents and the checked individuals: "*we together* must protect ourselves against *them*." The checking agents and checked individuals form a micro-system where, in order for the political technology to be considered part of the democratic machine (Linhardt, 2000), the former must show respect to the latter. In light of recent developments, it is appropriate to hope that Orly airport does not come to resemble Dammarie-les-Lys in the near future. A heteropia today (Foucault, 2001; Salter, 2007), Orly would then inevitably become a dystopia.

DAMMARIE-LES-LYS: THE GUARDIANSHIP MODEL

Second policing area: Dammarie-les-Lys and the Bas-Moulin housing project. The type of policing practiced there is in direct opposition to the type of policing exercised in Orly. Here, too, there is a specific security apparatus, a characteristic relationship between surveillance agents and individuals under surveillance, between those who hold a monopoly on legitimate violence and the subjects to the law. The relationships and apparatus do not, however, rest on the success of a test system, but on a form of guardianship.

Dammarie-les-Lys was chosen because of a local historical specificity. In the summer of 2002, an unexpected event took place: there was a political mobilization in the housing estate. This mobilization had developed in reaction to two deaths, which had occurred during police intervention, one on May 21st, the other on May 23rd. The event was indeed unexpected, since it was one of conventional kind: no disorders, riots, destructions, street-battles, but demonstrations, public claims made to the local government, press coverage, calls to political organizations, etc. It marked a departure from the contentious repertoires usually resorted to by youths from French deprived urban areas (known as "*banlieues*") in similar cases, such as "coordinated destructions" and "scattered attacks" (Tilly, 2003, p. 15). Typically, in December 1997, when a 17-year-old youth was shot and killed by a police officer, Dammarie's housing project underwent three nights of violent attacks comparable to those which took place on the entire French territory in October and November 2005 (Roy, 2005; Jobard, 2007). The 2002 rallying broke the cycle of the "routinization of rioting" (Campbell, 1983) noted in Dammarie-les-Lys, and other similar places.

Two specific features of the political mobilization are addressed below.[8] First, the interlacing of routine and exception (H), a feature which is characteristic to the place and will be illustrated by an intervention of the new repressive unit, the "GIR" (A); second, the priority given to two overlaying approaches: the familiarity (B) and longstanding nature (C) of the interpersonal relationships between the surveillance agents and individuals under surveillance.

Routinization of Exception

End of May 2002: after a presidential race dominated by crime and disorders issues, President Chirac is reelected with almost 90% of the votes in the second run against Jean Marie Le Pen, the far-right candidate. A new Interior Minister, Nicolas Sarkozy, is appointed, and immediately sets up a new squad, the GIR (*Groupes d'intervention régionaux*). These groups are the product of a superposition of all the existing policing, administrative, and judiciary services (A): Gendarmerie (i.e., military police force), national police (i.e., local civil police force), customs, public prosecutor, fiscal administration, etc. Investigating and administrative agents show up under the protection of a number of forces similar to the paramilitary police units described by Kraska and Kappeller (1997). Their conformity with the general principles of law is fragile, for their authority emanates from elected officials and prefects, when judiciary matters are independent from the government. This is why the use of GIR is exceptional in itself (H), which consist in "crackdown" missions on drugs, gun trade, prostitution networks, or illegal immigration operations.

The housing project of Dammarie-les-Lys, and more specifically the Bas-Moulin block, where the family of the second youth killed in the May 2002 tragedies lives, was surrounded by a GIR on the morning of June 27th, at 6.00 a.m. But the GIR, who entered the local youth center, destroyed it and finally managed to get the court bailiffs to shut it down and seal its doors; the GIR, who proceeded to identity checks on every inhabitant of the building, did not come alone. Indeed, they came accompanied by riot police forces and marksmen positioned on the roof of the high-rise facing the Bas-Moulin building (A).

On July 10th, Interior Minister Sarkozy shared his views on his own security policies implemented in skid-row areas:

> Police forces must regain control in abandoned territories. Let's take an example. There is a housing project, in Dammarie-les-Lys, in which neither police nor gendarmerie couldn't set foot any more. For years, people lived in fear there. A few days ago, the GIR went there. It was disappointing on a penal level, but for the people who live there, and

> the vast majority of them are honest citizens whose only claim is their right to lead a peaceful daily life, well, they suddenly thought 'we're not abandoned anymore.

Two political rationalities were put forward[9]: the first was the consecration of the legitimacy of the State, of the power of public institutions, and of their continuous territorial coverage. The second was the preservation of public order (E). The GIR intervention rests on an economy of fear and of rights. The target areas thus became a platform for the public display of State sovereignty. State's administration spread out in three different ways: a deployment of deterring physical and military forces, identity checks, and the eviction and rampage of the local youth center.

The Show of Power

There is no doubt that the goal of the intervention was to physically close the protest area off from the rest of the city. The number of police officers (around 250), and their lay out (two officers every 10 m, deployment of forces on the roofs, a continuous line of police cars driving around the building, revealing policemen armed with flash-balls and ensuring that the Bas-Moulin was entirely surrounded) allow for no ambiguity concerning their geographic target.

The physical separation between the flashpoint (Waddington, Jones, & Critcher, 1989) and the rest of the city places the presence of the police forces within a purely military semiology, where the display of weapons appears as a deterring sign of the power of the State (E), as opposed to the case of Orly, where police presence can be understood as part of a political semiology.

Identity Checks on Inhabitants

From 6:30 a.m. to 8:00 a.m., every Bas-Moulin inhabitant had to submit to identity checks, under so-called "administrative" (i.e., routine, not judiciary) procedures provided by articles 78-2 and 78-2-2 of the Criminal proceedings code (CPP), allowing the authorities to check the identity of "any person, regardless of his or her behavior … where the authorities have knowledge of repeated offenses, but have not identified their perpetrators." The whole operation led to the arrest of two illegal migrants ("The operation was disappointing on a penal level", M. Sarkozy then said).

The intervention sheds light on another specificity of the type of policing in the area. Contrary to Orly, where the administration is based on the presumption of "cleanness" paired with a technical apparatus allowing for the emergence of suspicion if necessary, here, individuals are always a priori suspected of disrupting public order. Accordingly, the administration demands proof that a person is complying with his or her legal status (through the identity check): the use of this method limits the individual's identity to the one the policeman is verifying, leading to a process of subjectification through police intervention (on control and subjectification, see Rose, 2000, pp. 330–331) (F).

The Eviction and Shutting Down of the Grassroots Organization Center

At the time, the building hosted a grassroots organization ("Bouge qui Bouge"), which was headed by the brother of Mohammed Berrichi, the second youth killed in the May events. The center had become the effective nerve center of the political mobilization: it was the place where journalists, political parties, and other grassroots organizations were invited to meet, where equipment was kept, where decisions were made, information brochures were written, etc. A court bailiff let the organization know that the center would be closed and restituted to the Public Housing Society (headed by the conservative representative of a neighboring city) who some years ago had agreed to lend the center to the organization free of charge. According to the bailiff, the deal had been broken following signs of "behavior which is incompatible with the social purpose announced by the organization, and specified in the contract establishing the free lease." In a decision taken on July 18th, the court of appeal ruled that the center should be given back to the grassroots organization. The keys were to be handed back to the organization on July 29th. On July 27th, however, the center went up in flames, and was entirely destroyed.

What is the social background of the protesters? Low education levels, geographic alienation from city-centers, limited options in terms of vital resources (housing and employment): the scarcity and precariousness of resources available to them stand in stark contrast with the stability and perpetuation over decades of the local conservative political elites, who not only occupy all elected official positions (national assembly, senate, city councils), but also (and, in fact, consequently) disproportionately head local administrations, including the one in charge of social housing. One of the consequences of this asymmetry in the distribution of public resources is the use of police forces by local elites to try to control the expression of public opinion. The signs of undesirable "behavior" referred to by the local court, the eviction demanded by the social housing administration, and the final restitution of the center are all based on a political economy of suspicion. Clearly, this economy of suspicion leads to the strong polarization of collective identities – to the formation of a "we" which is not, as in the case of Orly, directed against a common enemy, an absent terrorist, but, rather, against a particular State's authority: the police (G).[10]

These features of the local administration make Dammarie-les-Lys, and more specifically the Bas-Moulin building, a place which is in a permanent state of exception (H). Contrary to the implications of the generalization of this concept by Giorgio Agamben's (1998), "permanent state of exception" does not make the area a place with no link to politics, no relationship with political forces: it is not a zone of "bare life," which has become an abandoned "camp." The Bas-Moulin is, however, a zone where the formation of individual (F) and collective (G) identities are determined by an administrative body, the police. Traditionally, such identities would normally form out of politicization processes anchored in the usual social spheres (the workplace – politicization through a labor union; the family – the shaping of opinion; income, or capital – the formation of political preference; the grassroots organization – the expression of local or universal concerns). The local administration uses repressive means (with police forces, but also, as we will see, via judiciary means) to hang over individual lives and collective destinies, where, in Orly, its discretionary power is blocked by the prevalence of socio-technical procedures.

A Perennial and Personal Relationship with the Administration

Adding to the consecration of the state of exception (H) in the Bas-Moulin, there is a clear personalization of the relationship between police and targeted subjects, which gradually takes on a perennial character (C). One of the central features leading to this intimacy between police and individuals was the criminalization of verbal assaults on police officers and resistance against police officers (art. 433-5 and -6, French Penal Code, CP). On July 6th, a gathering of protesters was to be held in the city center: the aim was to break the invisible walls confining the protest within one excluded area symbolically defined by the GIR intervention, and to bring the issue onto larger public areas. The Mayor, however, allegedly reacting to the "constant climate of tension and insecurity in the town over the past month," decided to prohibit the gathering. Note that violent protests are common in France, but their prohibition is actually very rare (Fillieule & della Porta, 1998). Here again, the routinization of the state of exception in the area is striking (H).

Charges of Verbal Assault on a Police Officer as a Tool for Political Regulation

On July 5th, the day before the gathering was to take place, Abdelkader Berrichi, president of the grassroots organization "Bouge qui Bouge" and brother of the deceased Mohammed, was taken into police custody. He was accused of verbally assaulting police officers during their intervention in the organization center on June 27th. Shortly before he was taken in, Berrichi had been discussing a way to get around the Mayor's prohibition with one of the Prefect's assistants (while, in France, protests may be previously discussed with the Mayor, it is the Prefect who has authority over the police, and thus over public places – sometimes regardless of what the Mayor says – see Fillieule & Jobard, 1998). Negotiations were then taken up by one of Berrichi's friends,[11] who demanded his release – which the Prefect granted two hours later.

In this case, it clearly appears that police forces can use the criminal justice system in order to serve the local political elites, as shown by the time the decision to put Abdelkader Berrichi in custody had been taken. It is also interesting to note that at times, even the central State (locally embodied by the Prefect) must intervene to restore the balance in the political moves made by the various protagonists (in this case to prompt a de-escalation process, see Edelman, 1969). Thus, the process of politicization of a criminal justice system which allows for local officials to use policing tools in their interests and appears as a constant threat over potential protesters can, in rare instances, be blocked by an administrative act, here illustrated by the Prefect's use of exceptional powers to intervene in judiciary matters (freeing a man from custody and a summoning).

The Personalization of the Relationship between the Targeted Individual and the Administration

Let us return to the charge of verbal assault against a police officer in itself. On July 1st, Berrichi had used "nique ta mere" ("fuck your mother"), an usual insult in deprived suburban areas, against a police officer. Throughout his life, Berrichi was summoned four or five times for verbal assault, the last having been in Paris Court of Appeal in May 2007 for verbal assault and resistance against police officers of Dammarie's neighboring town. What is striking about the June 27th case, though, is that eight police officers sued for damages: eight policemen claimed to have heard the slur, and considered that they were eligible, in a civil lawsuit, for financial reparation.

Usually, this sort of reparation is minimal (about 300 Euros, as stated by Jobard & Zimolag, 2005). However, it is not so much the financial aspect which makes the verbal assault an administrative tool. When a police officer sues for damages, he or she reverses the nature of offense. Originally defined as an offense against a police officer and, as such, against the State's authority (the assault is said to be "against a public authority's deputy", in the terms of art. 433-5 CP), the assault becomes a mere interpersonal offense, since, beyond the assault alleged against the State, the officer asks the judge to recognize his or her own psychological casualty resulting from the offense as a person, not as a State's deputy anymore. The tendency for police officers to sue for psychological damages has increased since the end of the 1980s (at least in Dammarie-les-Lys; see Jobard & Zimolag, 2005). The end of the 1980s is also when the relationship between police and targeted individuals, or, more accurately, between police and police property, as the young protesters in Dammarie-les-Lys were perceived (B), started to crystallize.

Perrenialization of the Relationship

As mentioned above, the reparation for verbal assault is financial (for prison sentences as an alternative, due, for instance, to a lack of financial resources, see Aubusson de Cavarlay, 1985; Hodgson, 2002). Yet, one of the characteristic features of the target individuals here is their constant inability to pay, due to the social background we previously evoked. Police, who generally know this, rarely take the procedure much further. But judges are through these unended civil proceedings provided with an exceptional tool for making their relationship with target individuals perennial: a disciplinary tool, the civil decision to repair the damage always hangs above the head of a target individual when he or she has to appear before the justice system again, or when the justice department itself seeks to close open cases (H).

As we can see, Dammarie-les-Lys is a site of State sovereignty display, which rests on an ancient mode of relationship between the administration and the administered: an inter-individual, immediate, personalized, and asymmetrical relationship, always characterized by an imminent use of violence.[12] Such display of sovereignty is obviously contrary to the neo-liberal governmentality requirements and its three "e"s: efficiency, effectiveness, economy. Its legitimization lies in the government itself. This circular governmental rationality supports D. Garland's hypothesis against the governmentality literature, which takes it as axiomatic that government is a problem-solving activity (1997). Dammarie-les-Lys is, indeed, a place manifesting a "*wertrational*" sovereignty, whose logic is absolutist, not strategic.

TEMPORAL ARRANGEMENTS AND SPATIAL REGIMES OF SURVEILLANCE

Two sites, two antagonistic surveillance practices. It is relevant to assess the consequences of their differences. In the one, the international Orly airport, well-defined authorities control the access modalities onto and in a space divided into units and sub-units. The institutions interact with consumers, strangers, with whom they engage for a strictly defined purpose. During the sorting and checking operations, each interaction is unique and immediate in the sense that it does not allow for any perrenialization of the relationship between surveillance agents and individuals under surveillance. It is not recorded, cannot be used in a different temporality, or even reiterated. Each test is a new test – each time an object or a person passes through the unit or sub-unit, the process starts again. The justification for the apparatus, its political rationality, resides in the combination of the need for security, and the need for the traffic of people, luggage, and shipment to be maintained. Given the extremely low chance of actually being in the presence of the enemy, this norm – a pragmatic norm, combining commercial and moral imperatives – is, *in fine*, the greatest source of constraint but also the best way to assess the efficiency of the apparatus. The apparatus rests on the "iridescence" of the passerby: it does not focus on the intentions, past, or even the being as such of the individuals. Instead, individuals are subject to a series of tests with pre-defined parameters. Every passerby is part of a greater community in that he or she is linked to the surveillance institutions by the invisible presence of a common enemy whose threat must not disrupt the constant mobility within the airport.

From a surveillance standpoint, Dammarie-les-Lys is almost in perfect opposition to the Orly situation. The overlay of institutions is such that they seem to become one, almost in a paramilitary fashion. The intervention targets are familiar targets, and one of the consequences of intervention is the reinforcement of the sorting of the non-familiar faces, and the recognition of those who are already familiar. The relationship is based on a repetition of interactions, which are recorded in several types of memories (individual or collective, codified or otherwise), which can be reinvested in future interactions. The consequences of the surveillance procedure are the subjectification of the surveillance targets and the formation of a community defined by the antagonism felt toward the surveillance institutions. A guardianship relationship follows from it. Indeed, the practice of surveillance in Dammarie-les-Lys creates a space where individuals depend on surveillance institutions, with whom they have a relationship, which is marked in time and highly personal. Throughout the repetition of the interactions, surveillance is reinforced by legal trails and the mutual identification between policemen and targeted individuals that follow, so that the state of dependence is coupled with a form of subjectification, the shaping of an individual identity, a *self*, and of a collective identity, a *we* (Table 2).

What does the comparison between these two entirely different cases tell us about the surveillance/governance couple? First, that it is necessary to keep eventfulness in mind when thinking of governmentality. Both sites also have a peculiar relationship with time. The study focused on the surveillance apparatus of Orly airport on a normal day. The event, or, rather, the emergency, is in virtual reality – and the normalized, standardized, uniform

Table 2. Synoptic Table of the Findings of the Comparison.

Variables		Test		Guardianship	
Descriptive Variables					
(A)	Institutional arrangement	(A1)	Division into sectors	(A2)	Overlay
(B)	Targets of surveillance	(B1)	Strangers (customers)	(B2)	Inhabitants (locals)
(C)	Type of interaction	(C1)	Uniqueness	(C2)	Iteration
(D)	Absence or presence of records	(D1)	No recording	(D2)	Recording
(E)	Sources of legitimacy	(E1)	Mobility	(E2)	Residence
Analytical Variables					
(F)	Effects of surveillance on the individual status of its targets	(F1)	Objectification	(F2)	Subjectification
(G)	Effects of surveillance on the collective status of its targets	(G1)	A common "we" vs. an absent enemy	(G2)	Two antagonistic "we"s
(H)	Type of intervention	(H1)	Routinization of surveillance	(H2)	Permanent state of exception

feature of the various security processes are specifically designed to keep it that way. On the other hand, the study of the Dammarie-les-Lys apparatus was undertaken in the heat of a specific moment, or event (a political mobilization following two deaths). This dimension, the unpredictable feature of an every-day life which is always on the verge of a crisis, is neglected by the bulk of the literature on governmentality. To acknowledge the relevance of the event is to understand how the governmentality agencies both act on the spur of the moment and organize the future while at the same time tightening their grip on society. The observation of the surveillance apparatus during a disruptive event in Dammarie-les-Lys shows how surveillance becomes an element of a form of governance based on the settlement of an extremely unique time: the time of permanent exception.

A close look at history, at the actual interactions between governance agencies and their targets (the citizens, the passengers, the customers ...) also helps understand that, despite the political rationalities in place in advanced liberal democracies (Rose & Miller, 1992, p. 180; Rose, 2000, p. 323) and the subjectification born out of a "new penology" (Simon & Feeley, 2003), some places still bear the marks of ancient forms of governance: personalization and perrenialization of the relationships between agents of governance and the governed, politicization of relationships, permanent imminence of physical contact. Other sites, such as Orly airport, are also places where State agencies are present, but State control does not rely on the display of power or on the threat of violence. Rather, it relies on the introduction of control programs in a machine-like apparatus: control becomes the product of a machine of machines.

In his lectures "Security, territory, population", Michel Foucault had identified this question of eventfulness and of unpredictability. But he had also identified a second question, closely related to the first: the question of space and the necessity of dividing it, creating grid patterns within it – in other words, the necessity of rationalizing space (Foucault, 2004). As different from each other as they may be, both sites under observation show different mechanisms of political rationality in Foucault's sense, that is, manners of "conducting conducts" (Foucault, 1981; Gautier, 1996) closely linked to space-penetration. In both cases, control means a total knowledge and coverage of the geographic space, and the adjustment of every operation to specific spatial constraints. But the way the space is apprehended in each case is entirely different. At the Orly airport, the space is seen as purely transitory. Hence, the division of space into sub-units can be understood as the provision of a set of directions, which allow a better understanding of the space without territorializing it as such. These sub-units are not territories, in that they do not have inhabitants, and that no one has a permanent relationship with them: in fact, the zones are precisely designed for people to pass through them. The resulting organization of the space, almost geometrical, is designed to fit a principle of regulation, which cannot be ignored without affecting the sole purpose of the airport: the preservation of the continuity of the flow of transit and the reduction of the likeliness of friction. An entirely different story in Dammarie-les-Lys. In opposition to the Euclidean regime of Orly airport, the Bas-Moulin project is under a *chôra* regime (Berque, 2000, pp. 20–25): the site cannot be separated from its inhabitants, nor can inhabitants be separated from their place of residency: they "wear" the site, just like site "sticks" to them. Here, the territorialization is at its strongest.

CONCLUSION

The examples we have used are sufficient to reveal the importance of ensuring that analyses of surveillance and of its governance map the sites under surveillance to better identify what singles them out, take the various modes of access across zones into account, and identify the political rationality which emerges from the interlacing of spaces – or on the contrary, from their strict separation. Michel Foucault called this necessity "heterotopology" (Foucault, 2001). But the cartography is also a marked in time: as underlined by Michel Foucault, "more often than not, heterotopies are linked to divisions of time, that is to say that they open up onto what, by pure symmetry, we could call, 'heterochronies.'" If the check and guardianship models put forward a contrast between the time of repetition and the time of the event, it is because all forms of governance of surveillance, in fact, lead to a specific time arrangement. If we keep this in mind during our analyses, we place ourselves in a better position to contribute to the literature on criminology which tends to focus on various forms of "government at a distance" on the one hand, and on withdrawn places of incarceration on the other – and as a result, to neglect the great diversity of modes of control and surveillance, and the way they produce individual and collective identities.

NOTES

1. Right up until his final lectures, M. Foucault paid particular attention to the spatial anchorage of discipline and security devices. See, in particular, his

treatment of urban space in his lectures at the Collège de France in 1977–1978 (Foucault, 2004).

2. In order to make our text easier to follow, our empirical descriptions will include uppercase letters corresponding to the variables where appropriate.

3. To quote James Beniger (1986), the historical feature of terrorism prevention in airports can be interpreted as a sequence of alternations between situations of *control crisis* and *control revolution* – the latter understood as what facilitates the transition from a world which has become uncontrollable because of transformations in the nature of the threat it could face to a world once again under control, thanks to the establishment of an apparatus able to contain and apprehend new risks within a process of rationalization and normalization. The latest crisis to date was triggered by the realization of the existence of liquid explosives concealable in bottles and flasks.

4. Here, one may recall the widely broadcast images after the September 11th attacks showing the hijackers going through security checks at the Portland airport. The outrage produced by the images comes from the double lack of efficiency they disclose: not only did passenger checks not allow security to catch the terrorists, but the images revealing the first breakdown in efficiency come from no other than surveillance cameras themselves.

5. The most important border is that between the "public zone" and the "restricted zone." Its purpose is to ensure that all necessary functions for take off are concentrated within the restricted zone, and that all others are excluded. The restricted zone includes the post-transborder filter area inside the terminal as well as the traffic area, the control tower, and certain technical rooms outside the terminal. Within the restricted zone, there is a subdivision separating "security sectors" from "functional sectors." Given their proximity to the aircraft, security sectors have the strictest access policies.

6. For a description of the relationship between the consumer and the airport, see Rosler (1994).

7. To take a concrete example: at a security checkpoint, an agent signals the presence of a hand-grenade in a piece of hand luggage. We know nothing of the passenger, nor whether he has ill intentions or not. After all, the grenade may be a collector's item (which it turned out to be). Nonetheless, the passenger is not authorized to go through – not because he is considered a terrorist, but because he does not satisfy the criteria of "cleanness." On the other hand, all the passenger had to go through to give his hand-grenade to the security agent – and this would have been the case *even if he had been a terrorist*.

8. A more detailed account of these events was published in Jobard & Zimolag (2005). Numerous documents, archives, and pictures are also available on: http://vacarme.eu.org/rubrique102.html

9. In the narrow sense of Rose and Miller (1992, p. 175): "the moral justification of power." Further on, we will be using this term in the larger sense promoted by Foucault.

10. In a similar vein, see Escobar (1999).

11. The transcripts of the interaction can be found at: http://vacarme.eu.org/article377.html

12. The sovereignty we are referring to is the one described by Foucault (1977), such as the ancient mode of power display which preceded the emergence of discipline.

REFERENCES

Adey, P. (2003). Secured and sorted mobilities: Examples from the airport. *Surveillance and Society*, *1*(4), 500–519.

Agamben, G. (1998). *Homo sacer: Sovereign power and bare life*. Stanford, CA: Stanford University Press.

Akrich, M. (1992). The de-scription of technical objects. In: W. E. Bijker & J. Law (Eds), *Shaping technology, building society* (pp. 205–225). Cambridge, MA: MIT Press.

Aubusson de Cavarlay, B. (1985). Hommes, peines et infractions. La légalité de l'inégalité. *L'Année Sociologique*, *35*, 275–309

Beniger, J. (1986). *The control revolution: Technological and economic origins of information society*. Cambridge, MA: Harvard University Press.

Berque, A. (2000). *Ecoumène. Introduction à l'étude des milieux humains*. Paris: Belin.

Campbell, B. (1983). *Goliath. Britain's dangerous places*. London: Methuen.

Collins, H. M. (1990). *Artificial experts: Social knowledge and intelligent machines*. Cambridge, MA: MIT Press.

Crenshaw, W. A. (1988) Civil aviation: Target for terrorism. *Annals of the American Academy of Social and Political Science*, *498*(1), 60–69.

Deleuze, G. (1988). *Foucault*. Minneapolis, MN: University of Minnesota Press.

Edelman, M. (1969). Escalation and ritualization of political conflict. *American Behavioral Scientist*, *13*(2), 231–246.

Escobar, J. E. (1999). *Race, police, and the making of a political identity: Mexican Americans and the Los Angeles police department, 1900–1945*. Berkeley, CA: University of California Press.

Fijnaut, C., & Marx, G. T. (1995). *Undercover. Police surveillance in comparative perspective*. The Hague: Kluwer International.

Fillieule, O., & Jobard, F. (1998). The Maintenance of order in France. Towards a model of protest policing. In: D. della Porta & H. Reiter (Eds), *The policing of mass demonstrations in contemporary democracies* (pp. 70–90). Minneapolis, MN: University of Minnesota Press.

Fillieule, O., & della Porta, D. (1998). Policing protest in France and Italy. From intimidation to cooperation? In: S. Tarrow & D. Meyer (Eds), *The social movement society. Contentious politics for a new century* (pp. 111–130). Boulder, CO: Rowland and Littlefield.

Foucault, M. (1977). *Discipline and punish. The birth of the prison*. New York: Pantheon.

Foucault, M. (1981). Omnes et singulatim. Towards a criticism of political reason. In: S. McMurrin (Ed.), *The Tanner lectures on human values* (vol. 2, pp. 223–254). Salt Lake City, UT: The University of Utah Press.

Foucault, M. (1988). The political technology of individuals. In: P. H. Hutton, H. Gutman & L. H. Martin (Eds), *Technologies of the self. A seminar with Michel Foucault* (pp. 145–162). Amherst, MA: The University of Massachusetts.

Foucault, M. (2001). Des espaces autres. In: M. Foucault (Ed.), *Dits et Ecrits* (Vol. 2, pp. 1571–1581). Paris: Gallimard.

Foucault, M. (2004). *Sécurité, territoire, population. Cours au Collège de France (1977–1978)*. Paris: Gallimard/Le Seuil.

Gandy, O. H. (1993). *The panoptic sort: A political economy of personal information*. Boulder, CO: Westview.

Garland, D. (1997). "Governmentality" and the problem of crime. Foucault, criminology, sociology. *Theoretical Criminology*, *1*(2), 173–214.

Gautier, C. (1996). À propos du gouvernement des conduites chez Foucault. In: CURAPP (Ed.), *La gouvernabilité* (pp. 19–33). Paris: Presses universitaires de France.

Goffman, E. (1971). *Relations in public: Microstudies of the public order*. New York, NY: Basic Books.

Hodgson, J. (2002). Hierarchy, bureaucracy and ideology in French criminal justice: Some empirical observations. *Journal of Law and Society*, *29*(2), 227–257.

Jobard, F., & Zimolag M. (2005) When the police go to court. A study on obstruction. contempt and assault on a police officer. *Penal Issues*, March, available at http://www.cesdip.org/IMG/pdf/pi_03_2005.pdf

Jobard, F. (2007). Urban riots in France, 2005. Data-based interpretations. Submitted to *Compass Blackwell online*.

Kraska, P. B., & Kappeller, V. E. (1997). Militarizing the American police: The rise and normalization of paramilitary units. *Social Problems*, *44*(1), 1–18.

Latour, B. (1987). *Science in action: How to follow scientists and engineers through society*. Cambridge, MA: Harvard University Press.

Latour, B., & Hermant, E. (1998). *Paris ville invisible*. Paris: Les empêcheurs de penser en rond.

Linhardt, D. (2000). Demokratische Maschinen? Die Vorrichtung zur Terrorismusbekämpfung in einem französischen Großflughafen. *Kriminologisches Journal*, *32*(2), 82–107.

Linhardt, D. (2001). L'économie du soupçon. Une contribution pragmatique à la sociologie de la menace. *Genèses*, *44*, 76–98.

Linhardt, D. (2006). Guerrilla diffusa. Clandestinité, soupçon et provocation dans le conflit entre organisations révolutionnaires subversives et l'État ouest-allemand (années 1970). *Politix*, *19*(74), 75–102.

Lyon, D. (1994). *The electronic eye: The rise of surveillance society*. Minneapolis, MN: University of Minnesota Press.

Lyon, D. (2003). Airports as data filters: Converging surveillance systems after September 11th. *Information, Communication and Ethics in Society*, *1*(1), 13–20.

Merari, A. (1998). Attacks on civil aviation: Trends and lessons. In: P. Wilkinson & B. M. Jenkins (Eds), *Aviation terrorism and security* (pp. 9–27). London: Frank Cass.

Norman, D. (1991). Cognitive artifacts. In: J. M. Caroll (Ed.), *Designing interaction: Psychology at the human–computer interface* (pp. 17–38). New York: Cambridge University Press.

Pinch, T. (1993). "Testing – one, two, three ... testing!" toward a sociology of testing. *Science, Technology, and Human Values*, *18*(1), 25–41.

Rose, N. (2000). Government and control. *British Journal of Criminology*, *40*(2), 321–339.

Rose, N., & Miller, P. (1992). Political power beyond the State. Problematics of government. *British Journal of Sociology*, *43*(2), 173–205.

Rosler, M. (1994). In the place of the public: Observations of a frequent flyer. *Assemblages*, *25*, 44–79.

Roy, O. (2005). The nature of the French riots. *SSRC*. 18th November. Available online at: http://riotsfrance.ssrc.org/Roy/

Salter, M. B. (2007). Governmentalities of an airport: Heterotopia and confession. *International Political Sociology*, *1*(1), 49–66.

Schütz, A. (1971). *Das problem der relevanz*. Frankfurt/M.: Suhrkamp.

Sharpe, S. (2000). *Search and surveillance. The movement from evidence to information.* Aldershot: Ashgate.

Simon, J., & Feeley, M. (2003). The form and limits of the new penology. In: T. Blumberg & S. Cohen (Eds), *Punishment and social control* (Enlarged 2nd ed., pp. 75–116). New York, NY: Aldine de Gruyter.

Singel, B. (2003). JetBlue shared passenger data. *Wired News*. 18th September. Available online at: http://www.wired.com/news/privacy/0,1848,60489,00.html

Tilly, Ch. (2003). *The politics of collective violence*. Cambridge, MA: Cambridge University Press.

Urry, J. (2000). *Sociology beyond societies: Mobilities for the 21st century*. London: Routledge.

Waddington, D., Jones, K., & Critcher, C. (1989). *Flashpoints. Studies in public disorders.* London: Routledge.

Wallis, R. (1998). The role of international aviation organizations in enhancing security. In: P. Wilkinson & B. M. Jenkins (Eds), *Aviation terrorism and security* (pp. 82–100). London: Frank Cass.

Walzer, M. (1977). *Just and unjust wars: A moral argument with historical illustrations.* New York: Basic Books.

LEX VIGILATORIA: GLOBAL CONTROL WITHOUT A STATE?

Thomas Mathiesen

ABSTRACT

The chapter presents and discusses the development of a number of new transnational surveillance and information systems in Europe. It relates their development to the question of whether we here have an example of 'law without a state'. Guenther Teubner's notion of a 'lex mercatoria', *a system of global contract law developed by large law firms and international business lawyers, is paralleled by a* 'lex vigilatoria', *a system of global control in the making. The chapter provisionally concludes that the ties to the nation-state of an integrated surveillance system are increasingly diluted.*

INTRODUCTION

During the past 30 years or so, we have seen a massive and accelerating development of various types of transnational information systems pertaining to law enforcement agencies throughout the world. The information systems actually take the form of control or surveillance systems (control and surveillance here taken as synonyms, leaving the question of compliance open). Economic and political change lies at the

Surveillance and Governance: Crime Control and Beyond
Sociology of Crime, Law and Deviance, Volume 10, 101–127

ISSN: 1521-6136/doi:10.1016/S1521-6136(07)00205-9

base of the development, new technology has made it possible. In Europe alone, new concepts flourish – you have:

(i) the Schengen Information System or the SIS, and SIS II;
(ii) the SIRENE Exchange; Supplementary Information Request at the National Entries;
(iii) the Europol Communication Systems or TECS;
(iv) the European Asylum System or EURODAC;
(v) the ideas and practices concerning storage of communication traffic data, more or less following in the wake of the so-called Enfopol papers; the planned exit-entry system for travellers to and from the European Union (EU); and
(vi) the Echelon System, a US creation aided by Canada, Australia, New Zealand and UK (the UK being a major partner with its listening posts in the UK and Cyprus).

These are only a few of the systems developing in Europe, and if you go further to the United States, Canada and elsewhere in the industrial world, you find still another plethora of information systems. New norms are established within the transnational information and surveillance systems, norms emphasizing a future-oriented monitoring of categories of people, thus creating risk profiles of whole groups.

The various information systems have been described in detail elsewhere (see, among others, Bunyan, 2005, 2006a, 2006b; Mathiesen, 2004a, 2001, 2000a, 2000b, 1999; Hayes, 2004; Karanja, 2006, Wiig, 2007; numerous other contributions by Bunyan and Hayes in Statewatch). Because they are in rapid development and continually expanding, the need for continual updating and revision of descriptions is great.[1] I will here give a quick outline of the main systems, and some recent data on them (for the sake of brevity the Visa Information System (VIS) will be briefly treated as an adjunct to the Schengen Information System (SIS), which it essentially is), and go on to a presentation of a theoretical idea which may possibly be useful when it comes to a deeper understanding of the systems in question.

THE SYSTEMS

The police systems in question are transnational, regional (especially, in this chapter, pertaining to the EU) or even global in character, where vast amounts of information across national borders, within whole world regions, or even globally are entered and/or received. The systems depend heavily on the expanding modern data technology.

The Schengen Information System – SIS and SIS II

The SIS, with close to 895,000 people (in addition to aliases) and over 17 million objects on the central database at the beginning of 2007 (House of Lords, European Union Committee, 2007a, p. 22), is one such system. Briefly about the development of the Schengen accord: In 1985 five European states (Belgium, France, Luxembourg, The Netherlands and Germany) entered an agreement (in the little town of Schengen in Luxembourg) on the principle of abolition of border controls between the states in question, and a parallel fortification of the common borders surrounding the states. A detailed convention, called the Schengen Convention, was signed, also in Schengen, in 1990.[2] In the place of internal control between the states, measures were introduced to strengthen substantially the border controls along the external common borders. For this reason, Schengen is referred to as a 'Fortress Europe'. Originally, the Schengen accord was formally outside the EU, but certainly a part of a long-term EU goal, and at the EU summit in Amsterdam in 1997, Schengen, with its whole *acquis*, was incorporated lock, stock and barrel into the EU (in force by 1999). By 2001, there were 13 full EU member states in Schengen. The UK and Ireland, also EU member states (giving the EU a total of 15 member states at the time) participated in police cooperation, but wish to maintain their own national border control of the islands. In addition, Norway and Iceland, non-members of the EU, fully apply the provisions of the Schengen *acquis*. Also Switzerland, another non-member of the EU, has decided to join the Schengen area. Switzerland will have the same associate status as Norway and Iceland. In 2004 and 2007, a total of 12 Eastern European states were given EU membership, bringing the number of EU member states to 27, and on 21 December 2007, nine of the new member states, including the three Balkan states and Poland, entered the Schengen area.[3]

The general purpose of the Schengen Information System or SIS is stated in Article 93 of the 1990-convention and reads as follows, in extenso:

> The purpose of the Schengen Information System shall be in accordance with this Convention to maintain public policy and public security, including national security, in the territories of the Contracting Parties and to apply the provisions of this Convention relating to the movement of persons in those territories, using information transmitted via the system.

The general purpose quoted here may be considered the 'core goal' of the SIS: Much of Schengen's border control revolves around the issues of public order and security, including State security. The stipulation is very wide and diffuse, and quite open to interpretation. It may pertain to a very wide category of people – on the one hand from foreigners and others who are more or less vaguely suspected of instigating public disorder and insecurity to on the other hand people suspected (often with a low degree of suspicion, however) of transnational crime and even terrorism. The Presidency Programme on Police and Judicial cooperation of 22 December 2006 made the following telling statement concerning asylum seekers and foreigners (quoted in Wiig, 2007, p. 67,[4] *italics* supplied here): '*Frequently*, asylum-seekers and foreigners who are staying in the EU unlawfully are involved in the preparation of terrorist crimes ...'[5] It has been documented that the SIS has been used for political purposes, such as the border control of political demonstrators entering a country hosting a summit (Mathiesen, 2004a, pp. 446–447).

On a single day in 2003, there were 125,000 access terminals (computers) monitoring the SIS (Hayes, 2004). A majority of the 895,000 *persons* on the database at the beginning of 2007, 84.1%, are 'unwanted aliens', entered pursuant to Article 96 in the Schengen Convention. This means, in absolute numbers, *a little over 752,500 persons*. The decision to refuse entry 'may', as stated in Article 96 and in line with the general goal of SIS quoted above, vaguely 'be based on a threat to public policy or public security or national security', which the presence of an alien presumably may pose. This is one point where political demonstrators may come in. Conviction of a custodial sentence of at least 1 year, which may mean entirely different degrees of seriousness in different countries, is a further criterion. '[S]erious grounds for believing' that the alien has committed 'serious criminal offences', is another. '[S]erious grounds for believing' are beliefs based on police hunches. One of the problems with Article 96 is that it appears to be interpreted differently in different Schengen countries, creating variations in degree of restrictiveness as far as entry into Schengen territory as a whole is concerned.

Furthermore, 1.8% of the 895,000 persons were entered for extradition to a Schengen State (Article 95), 4.8% were entered as missing persons (Article 97), 5.7% were entered because they were wanted as witnesses (for the prosecution or for enforcement of judgments; Article 98) and 3.7%–33,000 – for the so-called 'discreet surveillance' (Article 99). This measure is devised as a preventive action. A police prognosis holding that the person in question may commit a crime in the future is sufficient grounds for inclusion in the database (Statewatch analysis, 2007a, p. 2). The number of persons entered for 'discreet surveillance' has almost doubled between 2005 and 2007. Article 99.3 in effect allows for discreet surveillance of political behaviour – it states that 'a report may be made in accordance with national law, at the request of the authorities responsible for State security' – which means the intelligence services of the various states (Mathiesen, 2004a, pp. 438–439).

The *objects* on the database at the beginning of 2007 were typically identity documents (over 13.7 million). Next there were vehicles (over 1.7 million), blank documents (less than 400,000), firearms (less than 300,000) and bank notes (over 250,000; information about persons and objects are taken from House of Lords, European Union Committee, 2007a, p. 22).

Figures on 'hits', indicating the 'usefulness' of the SIS for narrow police purposes, are of course far fewer. Statistics show 'successful' controls in the sense of hits that the police of the Schengen states have had inside the EU on the basis of alerts issued by other Schengen states. In 2005 there were altogether 20,600 hits on persons, the majority of them concerning unwanted aliens (in addition, there were 11,000 hits on objects; information from Statewatch analysis, 2007a).

In other words, the Schengen system is to a large extent, though not exclusively, a system for border control, and risks to 'public order and security, including State security', which Schengen is supposed to limit, is intimately tied to foreigners as presumable risks. A main point is that once you have been denied entry to one Schengen state, you are also denied entry to all other Schengen states. Each state is responsible for control of its own part of the common external border – Norway for the country's over 20,000 kilometres coastal line with deep fjords, rugged mountains, icy waters and scattered fishing communities (admittedly not the best access points for migrants from the third world) – though integration of policing between the various states is developing. The internal borders, between the Schengen countries, have been lowered according to plan: normally there is no passport control (though notably, exceptions may easily be made on short notice pursuant to Article 2.2 in the Schengen Convention,[6] and though you definitely have to bring your passport for a variety of controls by airliners and hotels and hostels in the Schengen states to which you travel).

Up till now, SIS has a central database located in Strasbourg, with a database identical to the one in Strasbourg in each of the participating countries. A string of standardized items of information about individuals may be stored in the Schengen database. Data may only be entered according to national legislation, but by the same token, information may also be taken out according to national legislation.

As the SIS grew larger, it had to be updated to a 'SIS plus'. A so-called SIS II, with a new structure, a far greater capacity and a greater number of items of information included, such as biometric information (Hayes, 2004; Karanja, 2005), was first announced to go online in 2007, but political as well as technical and even legal aspects prolonged the waiting period, and the plan at the time of writing is apparently to go online in 2009.[7] The SIS II 'not only offers new technological functions, but it will also fundamentally change the police practice based on the system' (Statewatch analysis, 2007a, p. 4). The British House of Lords has characterized the system as planned to 'store an enormous volume of sensitive personal data', and goes on to criticize the pending system for lack of transparency in the following words (House of Lords, European Union Committee, 2007a, p. 16):

> A project of this importance and magnitude needs to be developed openly and publicly. It potentially affects not only EU citizens, but also hundreds of thousands on non-EU citizens who may wish to travel to or reside in the EU. Information must be readily available, not just to EU institutions and national experts, but to all those affected.
>
> It is unacceptable for a project with such cost and resource implications to be developed without a prior full impact assessment, and a full legislative explanatory memorandum.

The changes to be expected are, in brief, the following (Statewatch analysis, 2007a):

- The alert categories will be differentiated and extended.
- The data retention period will be extended.
- Alerts can be linked, so that the SIS will be able to carry out true investigative actions.
- More authorities will be able to access SIS data (Europol, Eurojust,[8] national public prosecutors, immigration authorities and the like).
- Alerts relating to persons will contain biometric data – fingerprints, pictures.[9] With the VIS such information will also be collected. The control function of the SIS in relation to non-EU citizens will be greatly enhanced by the parallel creation of the VIS. The two will have a common technical platform. In total, the SIS II will thus lead to a fundamental increase in restrictions and repression of non-EU citizens.

A number of the European detention centres where illegal immigrants and people who are denied refugee status are kept until they are transported out of Schengen territory, have been severely criticized by lawyers and civil rights organizations (e.g. Webber, 1995). Stephen Karanja, who is a lawyer, has in an academic paper demonstrated that in a few ways SIS II is a 'gain' in the sense of improving the legal safety or security of persons, but it mostly represents 'losses' in this respect (Karanja, 2005).[10]

The SIRENE Exchange

So much about the SIS. *The SIRENE exchange* of information is a second system. A so-called 'SIRENE office' is responsible for the administration of the SIS in each participating country. The SIRENE offices store a vast amount of auxiliary information about individuals registered on the SIS. In Norway, information in national police databases is available to the Norwegian SIRENE office, and may be communicated to other SIRENE offices throughout Europe. The information is extremely comprehensive. A single example from *one part of one Norwegian national police base*, concerning police data on foreigners, includes the following (the information is taken from the secret relevant police manual in 1999): registration number; registration date; police office code; identification (birth) number; citizenship; passport number; ethnic group; country of origin; sex; name/alias; address; telephone; height; age; bodily features; date of death; hair; language/dialect; spouse (with name and identification number); occupation; position; employer; information about automobiles; information about close acquaintances; other individuals who are closely tied to the person; the person's history ('... should provide a brief history of the person's escapades; ... supplementary information should be entered as time passes ...'); as well as 'soft info' ('... information we don't wish others to see') and other information concerning where the person comes from (this 'in order for us in the future to take out information about a whole nationality'). The operation of the SIRENE exchange is detailed in a secret manual, which is continuously being updated. The manual, which has been made available to me, places a major emphasis on means of communication of information between the SIRENE offices which are quick and as traceless as possible.[11]

The Europol Computer Systems

Thirdly, there are the Europol Computer Systems (TECS). Europol became operational in 1999; it started operating the information system in October 2005 (initially with only 3 of the 25 Member States – Sweden, France and Germany – involved, no further detailed information on this matter is available, see Statewatch analysis, 2007a). Officially Europol has the aim to combat organized crime, terrorism and the like, while in reality and potentially it controls much more widely and diffusely defined categories, and partly relies on low degrees of suspicion. Europol has three important computer systems:

- A central information system with standardized data about sentenced people and suspects, as well as about possible future offenders within Europol's competence ('persons who there are serious grounds for believing will commit criminal offences for which Europol is competent'). 'The information system thus unites convicts, suspects and not-yet-but-soon-to-be suspects' (Statewatch analysis, 2007, p. 2). We should notice the diffusely future-oriented character of the system.
- Work files for the purposes of analysis. These are special, temporary work files set up for the analysis of specific areas of activity. The work files may contain extensive personal data, not only about persons registered in the central information system, but also about (the following quotes are from the Europol Convention[12]): possible witnesses ('persons who might be called on to testify'); victims or persons whom there is reason to believe could be victims ('... with regard to whom certain facts give reason for believing that they could be victims ...'); 'contacts and associates', and informants ('persons who can provide information on the criminal offences under consideration'); in short, a very wide circle of individuals loosely tied to persons who have been sentenced or are under suspicion. In December 2004, Europol operated 18 or 19 files, altogether holding data on 146,143 persons. Approximately 10,000 people were registered in the work file 'Islamic terrorism', 22,500 were registered in a file on Turkish, 14,000 in a file on Latin American organizations involved in drug trade, 2,200 persons were registered in a file on illegal immigration of Iraqi Kurds (a file created while the US was still bombing Iraq). The largest work file, with tips from financial institutions on financial transactions pointing to money laundering and cross-border cash transfers, contained information on 68,870 persons. In view of the open definitions contained in the Convention, the figures indicate a kind of 'mass suspicion' which is likely to lead to concrete investigation results only in a minority of cases (information from Statewatch analysis, 2007a).
- An index system which enables one to find one's way around the vast amount of information.

A bit more about the second of these levels, the work files. The kinds of personal information which may be stored in the work files are not specified in the convention, only in the so-called implementation rules, given pursuant to the convention. As an example of the kinds of personal information which the work files are designed for, mention should be made of a proposal presented in 1996, concerning supplementary information of a highly personal and intimate kind:[13]

> It shall be forbidden to collect personal data *solely* on the grounds that they relate to racial origin, religious or other beliefs, sexual life, political opinions or membership of movements or organizations that are not prohibited by law. Such data may be collected, stored and processed only if they supplement other personal data stored in the analysis file and only where they are absolutely necessary, taking into account the purpose of the file in question. (*italics* supplied)

The important word here is 'solely'. It will be seen that the proposal in fact *opens* for inclusion of data about 'racial origin, religious or other beliefs, sexual life, political opinions or membership of movements or organizations that are not prohibited by law'.

Later, the proposal went through various new editions, following criticism by the European Parliament amongst others. But the final formulation still allows the inclusion of such intimate personal data. The implementation rules applicable to analysis files states:[14]

> Europol shall also specify in this order whether data related to racial origin, religious or other beliefs, political opinions, sexual life or health may be included in the analysis work file, ... and why such data are considered to be absolutely necessary for the purpose of the analysis work file concerned.

With regard to contacts and associates, victims, possible victims, possible witnesses and informants, such intimate data can only be included after special grounds are given and upon the explicit request from two or more member states. In practice, these limitations are not particularly strict. For other categories of persons, no such limitations are given.

The implementation rules applicable to work files allow the processing of around 70 (at minimum 68) types of 'personal data, including associated administrative data', about persons registered in the central information system. The personal data are grouped in 12 categories, and they are:

> 'personal details' (fourteen types of data); 'physical description' (two types of data); 'identification means' (five types of data; including forensic information such as fingerprints and DNA evaluation results, though 'without information characterising personality'); 'occupation and skills' (five types of data); 'economic and financial information' (eight types of data); 'behavioural data' (eight types of data; including 'life style, such as living above means, and routine', 'danger rating', 'criminal-related traits and profiles' and 'drug abuse'); 'contacts and associates' (subtypes not specified); 'means of communication used' (a wide range of means given as illustrations); 'means of transport used' (a wide range of means given as illustrations); 'information relating to

> criminal activities' under Europol's competence (eight types of data); 'references to other data bases in which information on the person is stored' (six types specified, including 'public bodies' and 'private bodies'); 'information on legal persons associated with the data referred to' under economic and financial information and information relating to criminal activities (ten types of data). (Europol, 1999)

To reiterate, the above-mentioned types of data may not only be included about persons registered in the central information system, but also about possible witnesses, victims or persons whom there is reason to believe could be victims, contacts and associates and informants. People working in Europol have immunity within the EU, and though it is denied by Europol authorities, the Europol police force is said rapidly to approach the status of a European FBI. In 2005, Europol had a staff of 536, and 181,000 operational messages exchanged.

EURODAC

A fourth information system is *EURODAC- a fingerprint system under the Dublin Convention* (1990).[15] The Dublin Convention establishes that only one European state is to be responsible for deciding on an application for asylum, and EURODAC – announced in 1991 with an Agreement which came into force in 1997, to be replaced by a Regulation in 2002, and going online 15 January 2003 – contains fingerprints of all asylum seekers over 14 years of age. During the first year after going online (15.1.03–15.1.04) Member States transferred 246,902 data entries of asylum seekers to the EURODAC's central unit. Because EURODAC started as an empty database, this was considered as a significant result by the European Commission (Statewatch analysis, 2007a, p. 3). During the second activity year (all of 2004), the central unit received 232,205 data entries (same source). Comparisons are made with already existing data. The data are stored for 10 years.[16] There are only two exceptions to this – information about persons who have attained citizenship is to be deleted, and information about persons with formal refugee status, pursuant to the UN Convention relating to the status of refugees, is to be kept for a shorter period and is only to be used for statistical purposes. EURODAC's official goal is, inter alia, to avoid asylum seekers seeking asylum in more than one European country. In reality and potentially it is a vast control system for whole categories of ethnic and immigrant groups.

Telecommunications Traffic Data

Fifth, the storage of telecommunications traffic data (fixed and mobile telephone, fax, e-mail, Internet: traffic data enabling the police to monitor closely who you interact with, but not content data) has been introduced. It began as a collaborative effort between the EU and the FBI in the early 1990s. In 1993, a 'seminar' was held in Quantico, Virginia, where the FBI academy is located, between EU member states and the FBI. An outgrowth of that meeting was the so-called 'International Law Enforcement Telecommunications Seminar', ILETS. To make a long story short, a memorandum on the requirements of law enforcement agencies was signed by EU member states in 1995, a rather secret committee within the EU working further towards the so-called *Enfopol papers*.[17] The Enfopol papers were revealed in the Internet journal *Telepolis* and thus made public, creating quite a stir and debate. The documents in fact became considerably shorter and quite a bit less informative afterwards. Plans for *storage of telecommunications traffic data* are now finalized within the EU. The London onslaught in July 2005 provided the opportunity. In the panic following the onslaught the idea got widespread support. Prime minister Tony Blair quickly made a speech (at the G8 Summit in Scotland, which took place simultaneously) in which he proposed long-term storage of telecommunications traffic data for all British citizens. Critics have pointed out that the masses of data thus stored, will not help, or only help marginally, in catching terrorists.[18] But they will certainly in a massive way threaten civil liberties. This, of course, is an important point, adding to it the more basic question of how far we can go in sacrificing fundamental principles of democracy and the rule of law in our struggle against terrorists. Related proposals were in the making on the EU level, and the final text of a EU directive was issued 3 February 2006.[19] Under mandatory data retention a record will be kept of *everyone's phone calls, e-mails, mobile phone calls (including location) and Internet usage.* The process of decision making has partly been carried out in secret. A week after the EU directive was adopted, US officials raised the possibility of access to the information with the Council (in general, there is a great deal of contact between the EU and US concerning information sharing and the like[20]). The UK government saw to it that the mandatory retention of communications data by telephone service providers was introduced by a Regulation, implying a procedure that the measure is 'laid' before Parliament and passed by 'affirmative vote' in the Commons and Lords – unless a large number of members insist on a debate there will be none. It came into effect in the

UK on 1 October 2007.[21] Later, by March 2009, retention of Internet usage will become mandatory.[22] As Statewatch observes, 'the annual report of the Interception of Communications Commissioner says there were a staggering 439,054 requests to service providers for communication data by law enforcement agencies'.[23]

Along the same line of mass surveillance of everyone concerned goes the controversial EU/US bilateral relations and transfer of Passenger Name Record (PNR), signed on 28 June 2007. Statewatch observes that EU negotiators agreed that PNR data will be held for 7 years, doubling the current 3.5 years, and in addition agree that data can be accessed for a further 8 years (the so-called 'dormant' data). Just after signing the agreement, the US government (30 July 2007) wrote the European Council asking it to agree that all the documents regarding the negotiations leading to the agreement be kept secret for at least 10 years (*source:* statewatch.org/news/sep/eu-pnr-ssecret.htm). A draft reply (same source, EU doc. 12309/07) from the EU Council Presidency in August 2007 says the EU 'shares' this understanding regarding confidentiality. The US is planning to give exemptions for the Department of Home Security and for the 'Arrival and Departure System' (ADIS) from its Privacy Act, thus diminishing citizens' rights to find out what data is held on them and who it is held by (*source:* statewatch.org/news/2007/sep/04eu-usa-pnr-exemptions.htm). The development of mass surveillance systems covering 'everyone' may be viewed as a final surveillance stage. Lucia Zedner (2007) asks, how is the future fixed? She delineates three possibilities, going from prediction of *dangerousness* through the wider assessment of greater or lower *risk* of deviant behaviour, to a final possibility where dangerousness as well as risk are abandoned in favour of *everyone* being targets. There are also other examples of such generalized 'total' surveillance. The new controls at national and international airports are one.

Echelon

A sixth information system is Echelon. This is a regular spy system involving the US, Britain, Canada, New Zealand and Australia. Echelon appears to be able to take down vast amounts of telecommunications data from satellites, finding relevant information by means of a system of code words. It is uncertain how far the technology has come; what is certain is that Echelon exists and that the technology is developing rapidly. A report to the

European Parliament by the journalist and researcher Steve Wright may be quoted to indicate the activities of Echelon:

> A wide range of bugging and tapping devices have been evolved to record conversations and to intercept telecommunications traffic .... However, planting illegal bugs ... is yesterday's technology. ... [T]hese bugs and taps pale into insignificance next to the national and international state run interception networks .... Modern technology is virtually transparent to the advanced interceptions equipment which can be used to listen in .... Within Europe, all email, telephone and fax communications are routinely intercepted by the United States National Security Agency, transferring all target information from the European mainland via the strategic hub of London then by Satellite to Fort Meade in Maryland via the crucial hub in Menwith Hill in the North York Moors of the UK .... The ECHELON system works by indiscriminately intercepting very large quantities of communications and then siphoning out what is valuable using artificial intelligence aids like Memox to find key words. Five nations share the results .... Each of the five centres supply 'dictionaries' to the other four of key words. Phrases, people and places to 'tag' and the tagged intercept is forwarded straight to the requesting country. (Wright, 1998, pp. 18–19)

After 9/11

These are only a few of the systems which are operational or in the making. After 11 September 2001, a large number of new measures and regulations relevant to transnational, regional and global surveillance have been added. A central point has been the rapid development of a broad and diffuse definition of 'terrorism', making it very clear – if it was not clear before – that the various surveillance systems may be, and are, used politically, far beyond any reasonable definition of terrorism (Mathiesen, 2002). In a Statewatch report (2005) Tony Bunyan, the editor, summarizes the situation as of today well:

> The UK and the EU are facing a defining moment in their response to terrorism. Everyone understands placing suspected terrorists under surveillance and bringing them to court to face charges. But to create new offences 'preparatory' acts where no crime has been committed and for apologia, to employ surveillance techniques which could catch the innocent in the net, and to change the normal rule of law so that defendants will not know the evidence against them or its sources and to imprison or put them under house arrest on this basis tips the balance in favour of security over rights. Where the rights and freedoms of the few are curtailed so too are the rights and freedoms of us all.
>
> Since 11 September 2001 governments, ministers and officials at all levels of the EU have maintained that the swathe of new measures introduced have all been 'balanced' as between the needs of security and respect for fundamental rights. Concerned civil society

> groups across Europe know differently as do refugees, those stopped and searched or detained and the communities subject to surveillance.
>
> What has been seen as exceptional and draconian becomes the norm.

COMMON FEATURES OF INFORMATION SYSTEMS

Two features seem to be basic across the various surveillance systems which are developing: (i) the integration of systems; and (ii) the weakening of state ties.

Integration of Systems

The various information systems are established, as well as operated, by the same, or professionally very similar, organizations and agencies. During the early days before 1997, when Schengen was formally outside the EU, planning was organized so that essentially the same people could discuss Schengen issues in one meeting and EU issues in the next. Later, this differentiation of course disappeared completely. It is also possible to be more specific. I select Europol as an example.

The system was clearly planned with a view towards far-reaching integration, inter alia, with the SIS. For one thing, Article 10.4. No. 1–3 in the Europol Convention established a whole range of authorities and bodies within the EU, from whom Europol could request information: the European Communities and bodies within them governed by public law; other bodies governed by public law established in the framework of the EU and bodies based on an agreement between two or more Member States within the EU (also, Article 10.4. No. 5–7 established that information could be requested from international organizations and subordinate bodies governed by public law, other bodies governed by public law based on an agreement between two or more States, and the Interpol). Clearly, this opened the way for integration with the SIS. As a matter of fact, as early as on 9 April 1997, before Europol was operational, the 'High Level Group on Organized Crime' explicitly recommended that Europol should be given access to the information stored in the SIS.[24] This and other recommendations were on the agenda of the Justice and Home Affairs' Council[25] at the meeting on 3–4 December 1998 in connection with the action plan on establishing a so-called area of freedom, security and justice, and the recommendations were also discussed in a report of 26 February 1999.[26]

Furthermore, concrete work directed towards facilitating and easing compatibility between Europol, Schengen and other systems has been going on for a long time. The then Norwegian liaison officer in Interpol, Iver Frigaard, outlined a number of the issues, problems and possible solutions in a lengthy paper as early as 1996 (Frigaard, 1996). Frigaard saw Europol, Schengen and Interpol as three 'mutually interlocking' and 'overlapping' policing initiatives. He discussed their relationship on a systems level and on the level of exchange of information in concrete cases. On the systems level he pointed to the fact that, by 1996, 'only' 10 of the 45 states linked to Interpol's information system were also linked to Schengen and Europol. The number of states linking up to all three systems should and could, he argued, be increased. In connection with the exchange of information in concrete cases, he pointed to a lack of harmonization of the various data systems, and discussed what he viewed as the great need for compatibility between them as well as how compatibility might be attained technologically. The vigorous tenor of the paper clearly suggested that this was a matter of high priority.

It may be added that at its meeting on 19 March 1998, the Justice and Home Affairs Council agreed, without debate and as an 'A' point, on rules allowing Europol to request and accept information from non-EU sources (pursuant to Article 10.4. No. 4 of the Europol Convention). The report covered the receipt of data from 'third States and third bodies' (a relevant country was Turkey), and included only the most minimal safeguards on data protection. The plans were to be supplemented by a series of 'memorandums of understanding' between Europol and the central services of each of the non-EU states with which data were to be exchanged.

The above-mentioned statement shows that the story of efforts towards integration goes far back into the history of Europol. More recently, and beyond Europol, numerous integrating ties are in the making on the national level as well as between systems such as the ones we have discussed.

The crucially important Hague programme, adopted on 5 November 2004, introduced the so-called 'principle of availability', meaning that in principle *all data/intelligence held by a law enforcement agency in one state should be available to every other agency in the EU.* The principle of availability is defined as follows in the Hague programme:

> With effect from 1 January 2008 the exchange of … information should be governed by conditions set out below with regard to the principle of availability, which means that, throughout the Union, a law enforcement officer in one Member State who needs information in order to perform his duties can obtain this from another Member State

> and that the law enforcement agency in the other Member State which holds this information will make it available for the stated purpose, taking into account the requirement of ongoing investigations in that State.
>
> The methods of exchange should make full use of new technology and must be adapted to each type of information, where appropriate, through reciprocal access to or interoperability of national databases, or direct (on-line) access, including for Europol, to existing central EU databases, such as the SIS. (quoted in Bunyan, 2006a, p. 1)

Tony Bunyan comments: 'For a long time bilateral and multilateral agreements have been in place for law enforcement agencies in one EU member state to make requests to those in another EU state on specific cases .... The "problem" for the law enforcement agencies is that this procedure takes time, involves a formal request and sometimes judicial authorisation' (Bunyan, 2006a, p. 1). The Hague programme simplifies matters and alleviates this.[27] The European Commission followed up with a 'Communication from the Commission to the Council and the European Parliament on improved effectiveness, enhanced interoperability and synergies among European databases in the area of Justice and Home Affairs', COM (2005) 597 final (24 November 2005).[28] The Commission Communication is committed to the so-called 'interoperability', 'connectivity', 'synergy' and 'principle of availability' of IT systems, focusing specifically on the second generation of SIS (SIS II, see description earlier), the VIS, which is to share a technical platform with SIS II, and the European Asylum System (EURODAC). Only a few months before this (17 March 2005), the presidency stated in a Note to the Police Cooperation Working Party entitled 'Approach for enhancing the effective and efficient information exchange among EU law enforcement agencies' (7416/05 ENFOPOL 29),[29] i.e. (IV 11) in principle

> the JHA [Justice and Home Affairs Council] IT-systems should be widely accessible to the law enforcement authorities in order to combat terrorism and organised crime; ... [L]aw enforcement authorities [should] have access to the national law enforcement data of all Member States, in particular to identification, DNA and fingerprint data, on a hit/no hit basis. [L]aw enforcement authorities [should] have a direct access to national administrative systems of all Member States (e.g. registers on persons, including legal persons, vehicles, firearms, identity documents and drivers licences, as well as aviation and maritime registers).

The European Commission is also currently working on a proposal to interlink national DNA databases, and in the longer term proposes a 'European Criminal Automated Fingerprints Identification System ... combining all fingerprint data currently only available in national criminal

AFIS systems'. The interlinking of national DNA-data systems and other national databases containing fingerprints and information on car registers is particularly interesting. Direct access to such databases in other EU states will soon be coming up. The so-called Prüm treaty (named after a city not far from the town of Schengen, but on the German side of the border, where an agreement between seven EU states originally was signed on 27 May 2005[30]), provides the states concerned with such access, thus stepping up cross-border cooperation. The main purpose is to improve the exchange of information between the Contracting States, particularly by giving reciprocal access to national databases containing DNA profiles, fingerprints and vehicle registration data. The Council of the EU has agreed to incorporate this into the mainstream EU *acquis*. This, for example, would allow unregulated searching on a 'hit/no-hit' basis, followed by the automatic handing over of the file if there is a 'hit'. As Tony Bunyan has commented: '"Hit/no-hit" access would allow "fishing expeditions" to be carried out without any checks at all' (Bunyan, 2006a, note 13).

'European register(s) for travel documents and identity cards' – de facto EU population registers – are also planned along with the 'creation of an entry-exit system ... to ensure that people arriving and departing are examined and to gather information on their immigration and residence status' (information from Ben Hayes, Statewatch). For a critique of the Commission Communication, see European Data Protection Supervisor (EDPS), 'Comments of the Communication of the Commission on interoperability of European databases', Brussels 10 March 2006.

The importance of the Hague Programme and the Prüm Treaty can hardly be underestimated. In the end, they mean a more or less free flow of information across the borders of Europe, with cross border access to national databases as well as crosswise access to the major European bases discussed in this chapter. An entirely new situation is in the making. Its importance for example for protection of privacy will be great, to say the least. Generally, the planners of the various information systems are careful to point out that the information controlled by the systems is only to be used in relation to the particular goals of the systems. As indicated earlier, the goals in question are often vague and diffuse in the first place. As Simen Wiig (2007) has pointed out, now comes the added important fact that information in systems designed for one set of goals, will be open for scrutiny and use by other systems with other goals. Issues of privacy abound. And the integration of systems is taking leaps forward.

Weakening of State Ties

The second common feature of information systems is the weakening of ties between ordinary nation-state agencies and the increasingly integrated surveillance systems themselves. There is a development towards a generalized political sphere, above the reach of ordinary nation-state agencies, responsible for the surveillance systems.

National parliament members do not have the time or energy to dig deeply into the heap of documents pertaining to decisions about Schengen, Europol or what have you. Parliamentary debates become superficial and short, accepting the premises of ministries and even police agencies. In Norway, decisions of Parliament clearly follow the proposals of the government, and the government, in turn, seems to follow the signals and proposals which are central to international police culture and thinking. The mass media, supposedly controllers of it all, are not interested in going into the detail necessary for efficient control. The media, to a large extent a part of the entertainment industry, backs off: the dreary details and complexities of the surveillance systems are not in tune with the news criteria of the entertainment industry. A concrete example of the weakening of states ties in the context of the Hague Programme is given by Tony Bunyan:

> From the first draft of the Hague Programme (11 October 2004) the 'principle of availability' was set in stone. At the European Council (Summit) on 5 November 2004 the Hague Programme was simply nodded through without debate – the Prime Ministers had other more important matters to discuss. *Statewatch* had put the first draft online on 18 October but there was little or no time for parliaments or civil society to comment or intervene.
>
> It was a programme drawn up by officials and agencies, endorsed by Ministers and then Prime Ministers in secret meetings without any real democratic input. Thus was the justice and home affairs programme of the Council Presidencies and the European Commission for the next five years adopted. (Bunyan, 2006a, p. 3)

The decision-making persons and groups travelling back and forth between European cities (and especially to Brussels) gain additional discretionary power in relation to key issues. They are hard to compete with if you are closer to the grass roots and outside the higher echelons of power. Presumably, they are knowledgeable experts.

GLOBAL CONTROL WITHOUT A STATE?

Are we, then, developing an over-arching, far-reaching integrated global control system existing in and of itself, and also for itself, essentially without a state?

In 1997, Gunther Teubner edited *Global Law Without a State* (Teubner, 1997). Among the interesting contributions to the volume is Teubner's own introductory piece 'Global Bukowina: Legal Pluralism in the World Society' (Teubner, 1997, pp. 3–28). Teubner's main concern is the development of *lex mercatoria,* the transnational law of economic transactions, mostly transnational contract law, which he views as 'the most successful example of global law without a state' (*Ibid.*, p. 3). Global law, according to Teubner, has some characteristics which are 'significantly different from our experience of the law of the nation-state' (*Ibid.*, p. 7).

- The boundaries of global law are not formed by maintaining a core territory and possibly expanding from this, but rather by invisible social networks, invisible professional communities and invisible markets which transcend territorial boundaries.
- General legislative bodies are less important – global law is produced in self-organized processes of what Teubner calls 'structural coupling' of law with ongoing globalized processes which are very specialized and technical.
- Global law exists in a diffuse but close dependence not on the institutional arrangements of nation-states (such as parliaments), but on their respective specialized social fields – in the case of *lex mercatoria,* the whole development of the expanding and global economy.
- For nation building in the past, unity of law was a main political asset. A worldwide unity of law would become a threat to legal culture. It would be important to make sure that a sufficient variety of legal sources exists in a globally unified law.

In my own words, ideal-typically about *lex mercatoria*: transnational economic law is developed not by committees and councils established by ministries in nation-states and subsequently given sanction by parliaments, but through the work of the large and expanding professional lawyers' firms, the jet-set lawyers operating on the transnational level, tying vast capital interests together in complex agreements furthering capital interests. As *lex mercatoria* develops, it is not given subsequent primary sanction by national parliaments but is self-referential and self-validating, finding suitable 'landing points' in quasi-legislative institutions (Teubner, 1997, p. 17) such as international chambers of commerce, international law associations and all sorts of international business associations. It develops as a system of customary law in a diffuse zone around the valid formal law of nation-states, not inside valid formal law and not too far outside it. Eventually, it becomes regarded as (equivalent to) valid formal law or at least valid legal interpretation. It develops continuously, one step building on the other, in the end validating a law or a set of legal interpretations far from the law of the nation-states.

The independence of law and legal development is the crux of the matter. There is a great debate going on concerning the independence of global *lex mercatoria* – Teubner calls it a 30 years' war. I will not enter that war here, but simply ask the question: Do we, in recent developments in the late 20th and early 21st century, see signs of a developing independent global control system, a kind of frightening *lex vigilatoria* of political and social control? Global control without a state?

The question is complex. There are certainly ties between nation-states and say Schengen, the SIRENE exchange, EURODAC, communication control through retention and eventual tapping of telecommunications traffic data, the spy system Echelon, and so on. For one thing, some of them are established or proposed on the national level first. The above-mentioned story about how the EU directive on mandatory retention of telecommunications traffic data came about is an example. Secondly, some of the systems are established through various joint national efforts – some of them complex (meetings and memos over 10 years concerning communications control; the lengthy negotiations over Schengen), some simpler (framework decisions involving agreements of ministers from the nation-states) and some very simple (quick common positions cleared by governments). Thirdly, agreements such as partnerships in Schengen, Europol and EURODAC have to be sanctioned by national parliaments.

At the same time, there are signs suggesting that systems, such as the ones I have discussed, are not only increasingly becoming integrated or 'interlocked', but also increasingly becoming untied or 'de-coupled' from the nation-states. For one thing, as I have alluded to already, the parliamentary nation-state sanctioning of arrangements such as Schengen, Europol and EURODAC largely takes place without in-depth debates in public space, and, significantly, without parties and members of parliaments really knowing, to any degree of detail, the systems they are sanctioning. Parties and members must necessarily trust the work being done by various sub-committees and so on deep inside the EU structure, over and above agencies of the nation- states. There is neither time nor motive for anything else. An example is the scrutiny of the various *aquis*, enormous heaps of documents drastically reducing transparency for an ordinary parliament member (or even a researcher).

Furthermore, once the various interlocking systems are up and running, they interlock further through informal agreements and arrangements, rapidly expanding their practices – a kind of customary law, again in the diffuse zone of valid formal law. And the systems expand by internal sociological forces and logic, far from the control of nation-state institutions. In other words, the systems are increasingly integrated 'horizontally'. There seems to be an important relationship between the 'horizontal' integration or interlocking aspects of the various systems, and the 'vertical' weakening of ties or de-coupling aspects to nation-state agencies. The more integrated or interlocked the systems become ('horizontal' integration), the more independent of or de-coupled from national state institutions they will be ('vertical' weakening of ties), when the agendas for future developments and operations are set. Integration, interlocking, links the systems together in functional terms. Given moves are therefore simply regarded as 'necessary' or imperative, irrespective of the thinking which might be valid on the nation-state level. Interlocking at the system level also makes particular developments seem imperative from the point of view of the nation-state level. For example, the 'package' consisting of the SIS, Europol and EURODAC, in which all three systems are increasingly intertwined in terms of cooperation and goals, has made it increasingly 'obvious' and 'necessary' for Norway to participate in all three of them – if not without debate, at least with a minimum of debate. The question of Norwegian participation in the first of these, the SIS, created some critical debate. Norwegian participation in Europol and EURODAC hardly reached the newspapers or television at all.

Eventually, the interlockings and de-couplings are taken as *fait accomplis*, simply to be reckoned with. System functionaries – and there are thousands of them – take pride and find legitimacy in such developments. They become part and parcel of their system, defining their system as something they should foster, feeling great satisfaction when they manage to make the system function still better. These are entirely commonplace processes; this is how we all become more or less enveloped by the systems we are working in (Mathiesen, 2004b). A small example: in a discussion with Norwegian Schengen personnel some years ago, I ventured to guess that their workings were not all that rational after all – they probably took great pride and satisfaction in the computerized technical and complex activities they were involved in and were continuously developing. The response was instant – fumbling with papers, some blushing and some openly agreeing. To be sure, the various interlocking systems have their 'landing points', but, much like *lex mercatoria,* not in responsible parliamentary settings, but in quasi-legislative institutions – in this case especially, branches of the law enforcement agencies with their strongly vested interests.

CONCLUSION

A cautious conclusion for the time being: the various interlocking systems do not develop quite of their own accord, but with, what I would call, increasingly and strongly *diluted* ties to the institutions of the nation-states. While not global law fully without a state, a dilution of connections with the formal institutions of the nation-state is taking place. Most significantly, the institution of parliamentary sanction has become, at least in several European states, a perfunctory exercise with a silent public as a context.

But perhaps a 'state' is re-entering the scene on a different level? At least as far as the European control systems are concerned, the importance of the institutions of the EU is enhanced as the nation-state institutions fade. Any state, including a European one, requires certain institutions. One of them is policing (but not necessarily the kind we are witnessing today).

However, the European control systems, though largely emanating from the EU, also have tentacles far beyond the EU, interlocking horizontally with various systems of control in the US and other parts of the Western world. The EU-FBI attempts, pointed out so clearly by Statewatch, to develop transnational communication control over the past 10 years is a case in point.

Are we, then, facing once again a developing, unfinished, expanding global control, if not without a state so at least with increasingly diluted ties to state institutions? A *lex vigilatoria,* if not developing entirely of its own accord, at least with strong internal sociological forces leading the development, and control measures increasingly out of state control?

If so, we need to understand these sociological forces better if we are to oppose and contain them. A penetrating and critical research project exactly on this would be in order. Such a project could develop into counter-force. From a critical point of view, it is vital to stem this tide before it is too late.

NOTES

1. The civil liberties organization Statewatch does a great job in this respect, and constitutes a unique source of information on the information systems. See the Statewatch website http://www.statewatch.org, which continually updates the steady and great stream of documents and memoranda on the various systems, also providing important analyses of their intricate developments. See also *Statewatch Bulletin*, which is issued quarterly. The journal *Fortress Europe*, published by the Swiss/Swedish lawyer Nicholas Busch until his untimely death in 2005, has also been vitally important.

2. Convention of 19 June 1990, with later amendments.

3. See Council of the European Union SCH-EVAL 90 SIRIS 79 COMIX 427, 27 April 2007. The various data on Schengen provided in this paper will most likely increase substantially as Schengen membership of the Eastern European states takes effect.

4. From statewatch.org./news/2006/dec/eu-german-pres-policing-agenda.pdf

5. The UK police terrorism arrest statistics (excluding Northern Ireland) from 11 September 2001–31 March 2007 shows a total of 1,228 arrests (1,165 under the Terrorism Act, 2000). The statistics are a mixed bag and based on a very broad definition of 'terrorism'. Of the 1,228 arrests, 669–54% – led to release without charge. Close to 200 of the 1,228 arrests involved charges under other legislation than terrorism legislation, only 132 of the 1,228 involved charges under terrorism legislation only (*source:* Home Office in http://statewatch.org/news/2007/jul/05uk-terr-arrests.htm). There exists a 'handbook for police and security authorities concerning cooperation at major events with an international dimension', covering both public order and counter-terrorism aspects (10589/1/07 REV 1 ENFOPOL 119).

6. Art 2.2 is not infrequently used. It illustrates how Schengen arrangements have political functions: Art 2.2 is considered important to prevent demonstrators to cross national borders when summits and other major political events take place. For example, it was used during the World Bank meeting in Oslo, Norway in 2002. The arrangements cover border checks on the basis of Art 2.2 of the Schengen Convention, along with plans to put protestors under surveillance and deny entry to suspected troublemakers. Within Schengen, the security rationale surpasses the freedom rationale.

7. For interesting technical and legal explanations of the delay, including difficulties with the French site and delays in many of the national SIS II projects, see the European Commission's progress report 'On the Development of the Second Generation Schengen Information System (SIS II)' of 22 March 2007 (SEC 2007) 408. Although increased efficiency and integration of systems certainly takes place, we should not forget the obstacles involved in getting so many nations to cooperate.

Note that in the Schengen system there is a differentiation between asylum and visa matters, which were moved to the supranational (first pillar) area after the Amsterdam summit in 1997, and judicial- and police cooperation, which belong to the interstate (third pillar) area. The difference has important legal implications, for example pertaining to matters concerning protection of privacy and data security, and will apparently be maintained in the SIS II (Wiig, 2007, pp. 19–21).

On 31 May 2005, the EU Commission finally issued the proposed legislation to the SIS II. It consists of three proposals – two so-called Regulations and a Decision.

1. A proposal for a Council Decision on the establishment, operation and use of the second generation Schengen Information System (SIS II) COM (2005) 230, 2005/0103 (CNS). The proposed Decision is based on Title VI EU Treaty (police and judicial cooperation in criminal matters) which will govern the use of the SIS for third pillar purposes.
2. A proposal for a Regulation of the European Parliament and of the Council on the establishment, operation and use of the second generation Schengen Information System (SIS II) COM (2005) 236, 2005/0106 (COD). The proposed Regulation is based on Title IV EC Treaty (visas, asylum immigration and other policies related to the free movement of persons) which will govern the first pillar (immigration) aspects of the SIS II.
3. A proposal for a Regulation of the European Parliament and of the Council on the establishment, operation and use of the second generation Schengen Information System (SIS II) COM (2005) 237, 2005/–104 (COD). The proposed Regulation will be based on Title V (Transport) regarding specifically to the SIS data by authorities in charge of vehicle registration.

8. Eurojust's stated primary goal is to stimulate and make more efficient coordination of police investigations and prosecution, especially when two or more EU states are involved, see Wiig, 2007, p. 14; Council Decision 28 February 2002 (2002/187/JHA).

9. At the time of writing, picture and finger prints are exchanged as auxiliary information through the SIRENES, see later description. In the SIS II they are to be included in the SIS proper, which provides for more widespread and efficient exchange and search; see Wiig, 2007, p. 20.

10. For a number of further details on the SIS II, the reader is referred to the excellent Statewatch analysis (2007).

11. The SIRENE exchange was not mentioned in the original Schengen convention from 1990, and remained unmentioned until a change was introduced in the convention (Article 92.4) in 2005. The Article clearly shows the widespread and diffuse character of the information which may be exchanged: 'Member States shall, in accordance with national legislation, exchange through the authorities designated for that purpose (Sirene) *all supplementary information necessary* in connection with the entry of alerts and for allowing the appropriate action to be taken in cases where persons in respect of whom, and objects in respect of which, data have been entered in the Schengen Information System, are found as a result of searches in this System. Such information shall be used only for the purpose for which it is transmitted' (*italics* supplied).

12. Convention of 25 July 1995.

13. Proposal for rules applicable for work files, 4 January 1996 4038/93 Europol 2.

14. Council Act of 3 November 1998 adopting rules applicable to Europol Analysis Files [1999] OJ C26/1, Art 5.2.

15. The Dublin Convention of 15 June 1990, to which all EU member states are party, provides a mechanism for determining the state responsible for examining applications for asylum lodged in one of the member states in the EU.

16. Those apprehended for crossing borders illegally or residing without a permit in the EU are also compared, but supposedly not retained; see Statewatch analysis, 2007, p. 3.

17. For further details, see the Statewatch website http://www.statewatch.org.; readers of the Scandinavian languages may also see Mathiesen, 2000b, pp. 88–96, for details of this exciting story.

18. The Belgian expert on terrorism, Joseph Henrotin, is among the critics. Henrotin, the editor of the French journal *Défense et Sécurité Internationale* and frequently used expert commentator in Belgian and French media, has pointed out that increased cooperation and surveillance during the last few years has led Al Qaidas to change tactics. As large-scale terrorist attacks are becoming more vulnerable to detection in advance, tactics are changing to smaller scale attacks, approaching the size of the European terrorist attacks back in the 1970s. 'These smaller scale attacks are almost impossible to reveal in advance', says Henrotin. He refers to vigilance among ordinary people as the best way to control terrorism. Correct information and preparedness among people that terrorism may take place, will boost the morality among people and weaken the surprise effect of terror, he claims. 'You cannot attain security with surveillance cameras. They are not intelligent. The only intelligent controllers are people' (information and quotes from the Norwegian daily *Aftenposten* 3 July 2007).

19. Council doc. 3677/05, 3.2.06. Ref. http://www.statewatch.org/news/2006/feb/st03677-05.pdf

20. See the European Parliament Civil Liberties, Justice and Home Affairs Committee (LIBE) Visit within the framework of the EU-US Transatlantic Dialogue 1–18 April 2007.

21. *Source:* Statewatch analysis, 2007b, p. 1.

22. Same source.

23. http://www.statewatch.org/news/2007/may/uk-data-ret.pdf

24. See document 7421/97 JAI 14.

25. This comprises Justice and Home Affairs Ministers of the EU member states.

26. 6645/99 Europol 7.

27. The European Data Protection Authorities have produced a declaration, a common position and a checklist on the 'principle of availability', covering the collection, use and access to personal data for the purpose of law enforcement. The question is how helpful this will be in the context of massive data exchange between states.

28. The Communication has kindly been provided by Ben Hayes, Statewatch.

29. The Note has kindly been provided by Simen Wiig.

30. Five of the seven parties to Prüm were the five parties to the 1985 Schengen Agreement and the 1990 Schengen Convention. The Prüm initiative is of a unique nature within European cooperation. Seven states have agreed on a treaty, which is to be made relevant to all member states. The European Data Protection Supervisor (EDPS), who is partly critical, comments in the press release (EDPS/07/3 11 April 2007): '15 Member States propose to extend the application of the Treaty of Prüm, concluded between seven of them, to the whole EU without allowing for any major

revision'. Therefore, the EDPS goes on to say, 'the EDPS' suggestions mainly serve to improve the text without modifying the system of information exchange itself'. The example is important in showing how attempts, even by established and responsible agencies, to modify basic structures of information and surveillance systems may be curtailed by swift and general political decisions finalizing matters. For further critique, see also House of Lords, European Union Committee (2007b, published 9 May), which is critical on the same basis, and which in its foreword states that 'The Prüm Treaty is mainly concerned with the exchange of data. Inevitably this raises data protection issues. As so often, these tend to be overlooked'.

ACKNOWLEDGEMENT

Updated and greatly augmented article first published in Sarah Armstrong and Lesley McAra, eds, *Perspectives on Punishment: The Contours of Control,* Oxford University Press, 2006. Many thanks are due to Tony Bunyan and Ben Hayes of *Statewatch* for helpful information and advice, and to Simen Wiig, Master student at the University of Oslo, for useful information and comments.

REFERENCES

Bunyan, T. (2005). *Crossing the Rubicon: The emerging counter-terrorism regime*. Statewatch Report. London, Statewatch.

Bunyan, T. (2006a). The 'principle of availability'. Statewatch Analysis. London, Statewatch.

Bunyan, T. (Ed.). (2006b). *The war on freedom and democracy*. London: Spokesman Books.

Frigaard, I. (1996). Police cooperation: Current problems and suggestions for solutions in interstate police co-operation in Europe. Paper Delivered at the Fourth Schengen Colloquium of The European Institute of Public Administration. Schengen and the Third Pillar of Maastricht. Maastricht, 1–2 February.

Hayes, B. (2004). *From the Schengen Information System to SIS II and the Visa Information (VIS): The proposals explained.* Statewatch Report. London, Statewatch.

House of Lords, European Union Committee. (2007a). *Schengen Information System II (SIS II)*. Report with Evidence. London, The Stationery Office Limited.

House of Lords, European Union Committee. (2007b). *Prüm: An effective weapon against terrorism and crime?* London: The Stationery Office Limited.

Karanja, S. K. (2005). *SIS II legislative proposals 2005: Gains and losses* (pp. 81–103). Oslo: Norwegian Research Centre for Computers and Law, Yulex.

Karanja, S. K. (2006). *Schengen Information System and border control cooperation: A transparency and proportionality evaluation*. Unpublished Doctoral Dissertation. Faculty of Law, University of Oslo.

Mathiesen, T. (1999). *On globalisation of control: Towards an integrated surveillance system in Europe*. London: Statewatch.

Mathiesen, T. (2000a). On the globalization of control: Towards an integrated surveillance system in Europe. In: P. Green & A. Rutherford (Eds), *Criminal policy in transition* (pp. 167–192). Portland: Hart Publishing.
Mathiesen, T. (2000b). *Siste ord er ikke sagt: Schengen og globaliseringen av kontroll* [Last Word not Said: On Schengen and Globalization of Control]. Oslo: Pax Publishers.
Mathiesen, T. (2001). Die globalisierung der überwachung. In: C. Schulzki-Haddouti (Ed.), *Vom Ende der Anonymität* (pp. 11–24). Hannover: Verlag Heinz Heisse 2000.
Mathiesen, T. (2002). Expanding the concept of terrorism? In: P. Scraton (Ed.), *Beyond September 11. An anthology of dissent* (pp. 84–93). London: Pluto Press.
Mathiesen, T. (2004a). The rise of the surveillant state in times of globalization. In: C. Sumner (Ed.), *The Blackwell companion to criminology* (pp. 437–451). Oxford: Blackwell Publishing.
Mathiesen, T. (2004b). *Silently silenced. Essays on the creation of acquiescence in modern society*. Winchester: Waterside Press.
Statewatch analysis. (2007a; with H. Busch). *The dream of total control – status quo and future plans for EU information systems*. London: Statewatch.
Statewatch analysis. (2007b). *Mandatory retention of telecommunications traffic to be 'nodded through' in UK*. London: Statewatch.
Teubner, G. (1997). *Global law without a state*. Aldershot: Dartmouth Publishing.
Webber, F. (1995). Crimes of arrival: Immigrants and asylum-seekers in the New Europe. Paper at the 23rd Conference of the European Group for the study of deviance and social control. Crossmaglen, Northern Ireland, 1–4 September 1995. Published by Statewatch.
Wiig, S. (2007). Flyt og tilgjengelighet. En studie av det europeiske informasjonssamarbeidet innen politi, sikkerhets- og grensekontroll og dets konsekvenser for individers rettssikkerhet og rettigheter. [*Flow and accessibility. A study of the European cooperation regarding information within the police, security- and border control and its consequences for the legal protection and rights of individuals*.] Unpublished Masters thesis, Department of Criminology and Sociology of Law, University of Oslo.
Wright, S. (1998). *An appraisal of technologies of political control*. Luxemburg: European Parliament, Directorate General for Research, PE 166.499. Available online at: http://www.cryptome.org/stoa-atpc.htm
Zedner, L. (2007). Fixing the future? The precautionary principle as security technology. Keynote address given at the conference *Technologies of (In)security*, University of Oslo, 19 April.

PART II:
TECHNOLOGIES AND STRATEGIES

TECHNOLOGIES OF THE BODY, TECHNOLOGIES OF THE SELF: HOUSE ARREST AS NEO-LIBERAL GOVERNANCE

William G. Staples and Stephanie K. Decker

ABSTRACT

In this chapter, we argue that the practice of electronically monitored "house arrest" is consistent with Foucault's insights into both the workings of "disciplinary power" and "governmentality" and with the self-governing notions of a conservative, neo-liberal ideology, and mentality. Our interpretive analysis of a set of offender narratives identifies a theme we call "transforming the self" that illustrates the ways in which house arrest is experienced by some clients as a set of discourses and practices that encourages them to govern themselves by regulating their own bodies and conduct. These self-governing capabilities include "enterprise," "autonomy," and an ethical stance towards their lives.

On the one hand, we govern others and ourselves according to what we take to be true about who we are, what aspects of our existence should be worked upon, how, with what means, and to what ends. On the other hand, the ways in which we govern and conduct ourselves gives rise to different ways of producing truth.

– Mitchell Dean (1999, p. 18)

Surveillance and Governance: Crime Control and Beyond
Sociology of Crime, Law and Deviance, Volume 10, 131–149

ISSN: 1521-6136/doi:10.1016/S1521-6136(07)00206-0

> It's about remembering. Remembering things ... it conditions us not to forget. It immediately puts consequences on us if we do, if we do forget ... You are accountable for everything you do at every minute of everyday or they have to know where you're at. And it's your responsibility to make sure they know. It's not their responsibility to track you down ... You learn to be accountable for yourself.
>
> – "Julie," on living under house arrest, 2001

"Julie" is a 32-year-old European-American and a divorced parent of three young children, who works as a manager at a fast-food outlet. She is one of the two-dozen people we interviewed, living under house arrest in and around a Midwestern metropolitan area.[1] Julie had been convicted of a drug offense, and this was her second experience with house arrest, having been sentenced to "doing time at home" (Ball, Huff, & Lilly, 1988) in another state. All together she has spent more than two months being electronically monitored, this time following a three-month stint in a residential treatment center. Once on her own, Julie, along with her children, now lives with her mother. Being on house arrest means that Julie needs to respond to random phone calls from a Department of Corrections computer to verify she is home when she is scheduled to be. To do this, officials installed a small computerized unit in the house attached to her phone line. The calls demand that Julie complete a drill of answering the phone within the first three rings, saying her name and the time, and blowing into an alcohol tester that is built into the machine. While she is doing this, the device takes her picture and compares it to a reference photograph stored on a central computer.

The practice of electronically monitored "house arrest" has come to play an important role in the criminal justice system of the United States (Harrison & Karberg, 2003; Layton MacKenzie, 2006).[2] Currently, house arrest programs tether more than 70,000 individuals to central monitoring systems in the United States (Harrison & Karberg, 2003), with similar programs in Canada, United Kingdom, Sweden, The Netherlands, Australia, and New Zealand (Newman, 1999). The practice emerged in the context of the broader shift to community-based, "intermediate sanctions" associated with the "new penology" (Feeley & Simon, 1992) of the 1980s. This ideological and policy shift emphasized a new rationality centered on the sorting and classifying of offenders into finer categories of risk and dangerousness (Ericson & Haggerty, 1997). Intermediate sanctions – including boot camps, intensive supervision, community service, work-release, restitution, day-reporting centers, and day fines – were thought of as incremental punishments along the continuum between the extreme ends of probation and prison and therefore better suited to those classified as low risk and marginally dangerous.

In this chapter, we contend that the program of electronic monitoring we studied is consistent with Foucault's insights into both the workings of "disciplinary power" (Foucault, 1979/1995) and "governmentality" (Foucault, 1979, 1991), and with the self-governing notions of a contemporary, conservative, neo-liberal ideology and mentality (Rose, 1996, 1999; Barry, Osborne, & Rose, 1996; Dean, 1999). Our interpretive analysis of this set of offender narratives identifies a theme we call "transforming the self" that illustrates the ways in which house arrest is experienced by some "clients," as they are referred to, as a set of discourses and practices that encourages them to govern themselves by regulating their own bodies and conduct. These self-governing capacities may be characterized as "enterprise," "autonomy," and an ethical stance towards their lives (Foucault, 1991; Rose, 1996).

The bulk of the academic literature dealing with house arrest has focused on whether or not the treatment is effective in reducing recidivism and/or assessing its relative cost when compared to other sanctions (see Maxfield & Baumer, 1990; Glaser & Watts, 1992; Jolin & Stipak, 1992; Lilly, Ball, Curry, & McMullen, 1993; Sandhu, Dodder, & Mathur, 1993; Jones & Ross, 1997; Courtright, Berg, & Mutchnick, 1997a, 1997b, 2000; Ulmer, 2001).[3] Our goal, however, is to investigate and explore how it actually operates and is experienced by those subjected to it. Only a few studies (Holman & Quinn, 1992; Payne & Gainey, 1998, 2000; Maidment, 2002) have marginally explored what offenders make of the experience of house arrest, and only two (Ansay, 1999; Staples, 2005) have studied the experience in-depth. Through an examination of these offender narratives, we can see how the penalty of house arrest is made meaningful within the experiences of these individuals. This approach is consistent with the "law in everyday life" perspective (Sarat, 1998; Garth & Sarat, 1998; Byrant & Sarat, 1998; Ewick & Silbey, 1988) that considers law a social process and seeks to investigate how it operates in routine, everyday ways.

THEORIZING HOUSE ARREST

There is considerable evidence in the United States of the emergence of a new regime of social control; a regime that retains many of the modern themes and practices of the past, while, at the same time, has developed new methods of control and accountability that are both a product and reflection of postmodern culture (Staples, 1994, 2000) and a movement towards "conservative, neo-liberal" governance in criminal justice (O'Malley, 2002). House arrest with electronic monitoring is consistent with this new regime.

It incorporates a number of elements of modern "total institutions" (Goffman, 1961; Staples & Decker, 2007), while this new technique is, at the same time, quintessentially postmodern in design and implementation. But more, while the disciplinary regime of house arrest sets the boundaries or, in a number of the clients' words, the "structure" in their lives, it is largely left up to them to monitor themselves – to keep to their schedules, to work or go to school as required, meet with officials when scheduled, offer themselves up for drug tests, administer their own breathalyzers, and to literally turn themselves in if they deviate from the conditions of their contract. Thus, although house arrest operates "panoptically" – with clients "internalizing the gaze" of authorities – in a more complex way, their "participatory monitoring" (Corbett & Marx, 1991) actually helps to ensure that the entire system is working properly. They are, to paraphrase Poster (1996, p. 184), individuals plugged into the circuits of their own panoptic control.

We see the house arrest program then as a tactic of what Foucault (1991), in his later writings, called "governmentality," or the *practices* of governing, the techniques and technologies of how governing works, the rationalities and strategies invested to shape, guide, and direct the conduct of others, and that make human life a domain of power and knowledge (Foucault, 1991). As both a disciplinary strategy and a "technology of government," the practice of house arrest is designed to treat the client's body as an object to be monitored and assessed, broken down, analyzed, and improved upon. It is a program "... imbued with aspirations for the shaping of conduct in the hope of producing certain desired effects and averting certain undesired ones" (Rose, 1999, p. 52). It attempts, in other words, to normalize the participants and to produce docility. The individual, then, becomes both an object and subject of knowledge, not simply repressed, but shaped and formed within this discursive field and social practice. From this operation of power/knowledge come subjectivity and the formation of the self. Foucault (1997a, 1997b) also explored the idea of "technologies of the self" or patterns of practices that "permit individuals to effect by their own means, or with the help of others, a certain number of operations on their bodies and souls, thoughts, conduct, and way of being, so as to transform themselves ..." (p. 225). An individual, equipped with such technologies often created by various "experts" with skilled knowledge, may develop the self-governing capability that brings their conduct into alignment with broader moral, social, and political objectives (Rose, 1996, p. 155; Barry et al., 1996).

House arrest, then, as a technology of governing, tends to reflect the overarching political climate in the United States of the last several decades which, in turn, is echoed in the "new penology" and "actuarial justice" models (Feeley & Simon, 1992, 1994) that have emerged during the same time frame. As O'Malley (2002) argues, the US criminal justice system may be differentiated as a hybrid of neo-liberal principles – characterized by "... the rational choice subject, the superiority of markets to deliver efficiencies and goods, freedom of choice, a 'revised autonomy' of the enterprising self, the centrality of innovation and of enterprising individualism, the small and enabling state" – and backed up by a conservative, "three strikes" incapacitation mentality (p. 216). According to Rose (1996), this type of neo-liberal regime spawns and utilizes certain strategies, tactics, and regulations that encourage the self-governing capabilities of subjects. These capacities include (1) *Enterprise*, or the array of rules for the conduct of one's everyday existence that include energy, initiative, ambition, calculation, and personal responsibility (Rose, 1996, p. 154); (2) *Autonomy*, or the taking control of activities, defining a set of goals, and planning a course of action to satisfy the needs of existence through one's own powers (Rose, 1996, p. 155); and (3) *Ethics*, understood as the domain of practical advice as to how we conduct ourselves in the various aspects of our everyday existence and the ways by which we come to construe, decipher, and act upon ourselves in relation to the true and false, the permitted and forbidden, the desirable and undesirable (Foucault, 1991; Rose, 1996, pp. 30, 153).

The house arrest program discussed here embodies this hybrid form of neo-liberalism. It seeks to govern clients *through* the limited freedom it grants them but always with the constant threat that this freedom may be taken away. In other words, it governs its subjects through structuring and controlling the possible field of action where they are "free" to make the "right" choices in which they learn to govern themselves. Thus the clients are "not merely 'free to choose' but *obliged to be free*, to understand and enact their lives in terms of choice" (Rose, 1999, p. 87). They are treated as rational actors who respond to positive incentives and negative consequences. The program is touted by officials as being highly efficient and cost-effective as it deploys private sector technologies and makes program participants pay for the use of the equipment as well as their drug tests. Clients must work, most at least 40 hours a week, not at public sector "make work" jobs, but in private sector jobs that, if they do not already have one, are arranged for them. Thus, they are rendered docile not through their isolation and segregation from society but rather through their integration into everyday, labor/wage, and commodity exchange relations. The discourse of house arrest, articulated by both staff and clients, is one where the values of work, self-help, self-control, responsibility, and accountability are celebrated.

STUDYING HOUSE ARREST

In the fall of 2001, Staples conducted face-to-face, open-ended interviews with 23 clients in a house arrest program in a metropolitan area in the Midwest. The interviews took approximately 30–40 minutes to conduct. He used an interview guide of about a dozen general questions centered on their experiences with house arrest. The interviews were taped and later transcribed. The final sample included 23 individuals – 15 European-Americans (12 males and 3 females), and 8 African-Americans (1 female and 7 males). The mean age was 33 with a range of 18–74.

The house arrest program we studied is located in the same building as the Department of Corrections' 172-bed residential program we will call "The Center." House arrest, as practiced by this community corrections department, is a highly structured program involving mandatory employment, educational and treatment programs, frequent drug and alcohol testing, and offender fee payment that is touted by officials as an alternative to incarceration. Once assigned to house arrest, offenders sign a three-page 17-point contract that outlines all the conditions and behavioral expectations of the program. Clients must develop a daily schedule of approved activities (e.g., work, school, AA meetings, doctor's visits, etc.). Weekly, face-to-face meetings with their "House Arrest Officer" are required to review and update schedules and to verify their compliance through testing for drug or alcohol use, the collection of paycheck stubs, or signed verification of attendance at required treatment programs. All the clients we spoke with were court-ordered to maintain full-time employment. House arrest officers also make unannounced visits to places of employment and residences. In addition, a computerized Mitsubishi electronic monitoring system (MEMS) device is installed in the offender's home. The device functions to verify compliance with the house arrest contract by recording the offender's voice, taking his or her picture, and collecting breath samples for analysis. The house arrest officer determines the number of calls appropriate within a specific block of time and the system randomly calls the offender within those blocks.

When a call is made, a computer-simulated human voice tells the person answering the phone that the call is from the house arrest department, and a short pause gives the offender time to get to the phone in case someone else has answered. The system then instructs the offender to take their own photograph and to submit a breath sample through a straw inserted into the home monitoring unit. The photograph is displayed on a computer screen in the house arrest office next to a reference photo previously entered into the system so that staff can verify the identity of the person answering the call. The results of the alcohol breath test are recorded in the computer and displayed on screen with the photo of the offender, which includes name, date, and time. If a violation occurs, the computer immediately places a second call. If the second call results in a violation, an alarm alerts staff at the House Arrest office.

If a client does not respond to phone calls, there is a two-hour window for them to either call the house arrest office or show up in person. If they do not contact the office within this time frame, they are "violated" by their house arrest officer, meaning that they have broken their house arrest contract. Likewise, having a "hot UA" (a positive urine analysis) will result in an infraction. Program staff members frequently call upon local law enforcement to conduct residence checks and to look for those suspected of being AWOL, to provide arrest and detention services, to investigate allegations of abuse in the home, or verify an offender driving on a suspended license. The program requires offenders to pay a daily monitoring fee and all fees associated with the program, including drug testing.

TRANSFORMING THE SELF

For almost all the clients we spoke with, a sentence to house arrest produced a number of seemingly contradictory feelings and emotions. On the one hand, many complained, sometimes bitterly, about the intrusive and "disciplinary" aspects of the program such as the random phone calls and the rigid schedule as well as the cost of participating and the stress on them, their family, and other intimates (Staples, 2005). On the other hand, nearly all the clients began their interviews by declaring that, for a variety of reasons, house arrest was "better" than the alternative of sitting in the county jail. Some of those reasons included being able to sleep in their own bed, cook their own food, work and earn money, and to remain with family. Yet, another purported benefit of the house arrest program was what some described and what we call the tutelage aspects of the program: how it imposed a necessary "structure" onto their lives, how it taught them various life skills, and how it helped them become "responsible" and "accountable."

Therefore, we hear in the narratives of clients the adoption of various modes of evaluating and acting upon themselves as they seemingly develop the self-governing capabilities of "enterprise," "autonomy," and "ethics" (Rose, 1996).

For example, "Mark" is an African-American male in his early twenties who attended a local community college. He had been convicted of driving while intoxicated and manslaughter, the result of a car wreck. In his interview, Mark revealed a number of self-governing capacities he claims to have taken from the program. Asked about his daily schedule in the program, Mark said:

> Mark: I'm probably working more than I've ever worked in my life but I don't mind it since that's what [the program] gave me, was a work ethic, because I didn't have any since I was just a little spoiled kid who didn't do anything. And, that is just the whole reason they put you on it, to add that structure to your life, that you're saying, "I go to work, I come home, and that's the place I need to go, that's what I put that on my schedule and that's it." You're supposed to have that schedule for your life, and life is supposed to be full of appointments and schedules, and, but I mean life is supposed to have that structure and I guess that's what it's suppose to do.

When asked what the experience of house arrest meant to him, Mark said, "The whole meaning and purpose towards house arrest is to have a certain structure in your life and you can't have that structure without any type of guidelines." And finally, echoing an ethical stance on his own behavior, Mark told us, "Man, I don't feel any animosity towards being on house arrest 'cause I have to live up to my responsibilities which I guess this place really teaches that, you have to live up to certain things if you've made those mistakes, live up towards them. And so, I mean, I'm actually happy everyday."

"Marge," a European – American in her thirties and the mother of five children, who works in a fast-food restaurant, a job arranged by the program, told us that she hates to be monitored 24 hours a day, but "You know, it's better than sitting in jail." Marge went on to speak of the opportunity for self-sufficiency, lessons and skills learned, and personal accountability that the program permitted and encouraged: "And at least under house arrest you can provide for your family and if you're not out there doing what you're not supposed to be doing then it's not a problem anyway." She continued:

> Marge: And, it all, actually, sometimes people that have had problems in the past with maybe drugs or alcohol or situations like that, if they know what's more important then, house arrest can actually be a beneficial thing to a person. You know, for one, you're proving to the community that you can survive, be a part of the community without, you

know, I don't know, it's, it's not ... I'll be the first to tell you that I hate the phone calls, especially in the middle of the night, whatever, but it's a lot better than sitting in jail, at least you still have the opportunity to be with your family and provide for your family, and do the things that you need to do.

Asked to reflect on the relationship between house arrest and jail, Marge states:

> Marge: Even if you have been here at The Center, when you get out of here, I mean, you learn to appreciate a lot more, and, actually, this has all been a really good experience for me, because, I mean, it's taught me budgeting money, even with house arrest, I know have to have a payment in, you know, so I mean, this is part of my probation condition, so, you know, actually you have to sit down and budget your money too to make sure that you have the money for house arrest, you know, so, it's really been a learning experience, and, I've actually benefited from it.
>
> Staples: Is it a lot of pressure?
>
> Marge: Yeah. Basically, yeah. You know, because, you know you got to make sure that you are there when they call, you know and, if you've lived your life, you know, like I have, five children, I'm constantly busy, busy, busy. If you've lived your life that way, by a schedule, I mean it just totally turns your life upside down, you know. You have to make sure, and there's accountability again, responsibility, especially for your actions, you know, you wouldn't want to put yourself in that position so now you have to be accountable and responsible, and it is, it's hard at first, but like I said, once you get the hang of it and you're not doing anything wrong and you're doing what you're supposed to be doing, it doesn't become a problem anymore.
>
> Staples: Any final thoughts about the experience that you would like to share with me?
>
> Marge: I don't plan on ever, ever doing it again! (Laughs). I don't. I appreciate the fact that I'm here to talk to you and that I'm not in jail, but, the expense, is you know, just trying to keep up with appointments and scheduled times, and, it's definitely been a life learned experience, and I just do not want to have to face it again.

"Reggie" is an African-American male in his thirties, married with two children. He is a small business owner who had reluctantly agreed to plead guilty to a charge of sexual assault. He told us about how he ended up on house arrest and, like Marge, evokes a sense of personal responsibility that he linked to the community and the larger social order.

> Reggie: So I'm thinking, instead of him [the judge] giving me another thirty days and him taking me away from my family, I have a house note to pay, taking me from my house and my family is hurting my family also, so I think the judge was lenient and said, 'He doesn't need to be in jail, let's do, let's give him house arrest. And then by then I can get a better evaluation and he can be on his way.' So it's satisfying everyone, the judge, the community, you know everything.

And later in the interview, Reggie accepts his own liability for what has happened to him and sets his sights on moving forward and taking control of his own life:

> Reggie: I assume responsibility for being at that bar. So if I assume responsibility for being at the bar, I have no one to be mad at about this house arrest situation but me. That means I have to do what's necessary to get past it. Now, if I want to be mad about the situation and, "this isn't fair," then I'm not assuming responsibility and then therefore house arrest is going to be a pain in the ass for me. Let's get past it and go forward, this is what I need to do.
>
> Staples: So you are just constantly focusing on the alternative, which is just not very pleasant.
>
> Reggie: Right. And if you don't conform and meet the, uh, contract that you sign, can't say you didn't know, 'cause you signed it.

"Julie," the European-American in her thirties quoted at the beginning of this chapter, states when asked to describe her experience on house arrest that, "It was uh, [pause] structured. It was a lot better than being in jail, which was the alternative for me." Julie goes on to tell us about a specific skill she has developed and how this has led to personal accountability.

> Staples: Hmm, structure. Anything else about the experience that comes to mind?
>
> Julie: "It's [pause] remembering. Remembering things, um, 'cause a lot of times in the normal daily life it is difficult. So we forget things and it conditions us not to forget. It immediately puts consequences on us if we do, if we do forget. But when you're out on the street normally you forget 'oh well' little things. Here you pay the consequences if you forget anything. I forgot to find out one time here in The Center and lost my passes for three weeks. You forget something on house arrest, you could be called back here and sent straight back to The Center. They don't give you chances on house arrest." She continued, "It's immediate accountability. There is no second chances, there is no 'well, let me think about it.' No."
>
> Staples: Not a lot of slack, huh?
>
> Julie: Uh huh, no slack at all. You are accountable for everything you do at every minute of every day or they have to know where you're at. And it's your responsibility to make sure they know. It's not their responsibility to track you down. It's part of what it is going out, you learn to be accountable for yourself.

Later in the interview, Julie evokes the performative technology of the confession (Foucault, 1997a, 1997b) when she tells us how others know about her circumstances.

> Staples: So, a number of people know that you are on house arrest?

Julie: Yeah, everybody that I associate with, everybody that I talk to knows what's going on with me. I'm not ashamed to admit what I've done 'cause I'm paying for it. I took accountability for what I did and the responsibility for what I did and I'm trying to put it all behind me and move on and they are trying to help me. They are trying to help me put all behind me and making sure that I stay on the right track.

When Julie described the routine of house arrest, answering the phones, etc., she identified what she sees as the tutelage aspect of the program and the acceptance of docile subjectivity. She stated, "It's a conditioning program, that's what all this is. They're conditioning you to live like, better on life's terms. It does work. Doing the things you say you're going to do and uh, kind of uh, learning to respect authority." She goes on to compare the house arrest to a work release program she participated in another state: "That was plain, straight up, just a work release program. They didn't try to recondition you at all."

Finally, asked to summarize her experience Julie offered, "If people understand the concept of house arrest, it's a breeze. A lot of people try and put more into it than what's really there. They just basically need to read the rules and regulations. It helps. It says it all right there. It's guidelines is all." She went on to elaborate how she saw the need for the tutelage function to be embedded in the everyday lives of arrestees and how this offered them the opportunity to develop personal autonomy and control over their own lives.

Julie: It's just a wonderful program, an alternative to jail. I mean, this is a good program [The Center] also but the house arrest ... like I said, it keeps you controlled more in your environment you're used to. The environment that you need to be controlled in, the environment that you're going to be at out there. In here is not where you're going to be forever. That environment is where you need help controlling, not this one. I think the house arrest is a much better alternative to The Center, in my opinion. You know, this is much better than jail. But I really believe that a lot more people should go to house arrest because that's where they need to be controlled in their environment, in their own home. That's where they need the help, not in here ... You can't get a much better chance at organizing your life than house arrest. And I've done it; I know it can be done.

Similar to Julie's notions of transferring the lessons and skills learned from the program to affect her life when the structure is removed, "Duane," a 36-year-old African-American who works as a night grocer and has a wife and three children, offered the following when asked to describe his experience on house arrest:

Duane: Um [pause] it's very uncomfortable. This is the first time, hopefully the last, um, it's uh, it's an experience of that I feel that I can live by and learn from it, you know, apply it to my, you know, my life as I go on ... I think that uh, as far as me being, um, prompt. As far as being on time for something and being responsible, uh, as far as taking care of business. Uh, if I went and, you know, doing that before, um, I think it would kind of help me in the future to, uh, you know, to be more responsible. Uh, more business, as far as taking care business and that sort of thing.

Duane's "taking care of business" evokes the enterprising and autonomous skills necessary to navigate daily life. He goes on to take personal account of his past behavior and suggests an ethical narrative in the need to "get a life" (i.e., to govern oneself) or else be subjected to the disciplinary regime of authorities.

Duane: You know, uh, I guess, um, like I said, it's a valuable lesson. You know, no one can make my bed for me, I made it for myself. Whatever I brought on myself here is basically something that I have to deal with. Learn from it. You know, if you don't learn from your mistakes then, uh, better get a life then 'cause you'll always be in a predicament.

And, asked if he had any final thoughts, Duane turned to the energy, initiative, and ambition of "enterprise."

Duane: It's not bad, you know, just like with everything else, it can stand some improvement, on the way they, um, handle some situations. So far, so good with me. I want to be honest and just give it one hundred percent. I just hope that the program's honest with me and gives me a chance.

"Chris" is a 32-year-old European-American male and forklift driver, who was divorced with three kids but living with his parents. Like many of the others, Chris characterized the program "structure" imposed on him as both constraining and enabling.

Staples: So you feel it's effective in terms of watching or controlling you?

Chris: Yeah, sure. Yeah, sure it does. It, uh, like I said earlier, it just uh, it's effective in, uh, putting structure in your life. Putting, you know, living by a schedule, you know, instead of just. Basically, I mean it's not, it is kind of controlling. But, uh, that could be a good thing in a lot of areas.

Asked to reflect on the difference between jail and house arrest, Chris said:

Chris: Uh, basically what I get out of it. Instead of being incarcerated in jail, you're able to go to your house; you're able to work like a normal person. You're able to do the normal things as long as you have it on your schedule. But, it also keeps you from going out and, uh, doing the things that got you there in the first place. Which is for me, DUI.

Staples: So is it a blurring between the two?

Chris: Yeah, it's more of, uh, like you know really I think they ought to do it more often to people. Because instead having someone sit in a jailhouse where they can't make money. You know, some people need to be there because their attitude is, "I don't care-I'm going to go out and do the same thing I did before." Well, they need to be in the jail.

> But a lot of people, you know, realize that they made a mistake and they can get out and, you know, work and pay off fines or whatever. Child support, whatever they have. And uh, but at the same time still have structure in their life and be, you know, on a monitor system where, "Hey, we can tell if you've been drinking" or whatnot. I think it's a good system.

Here and below, Chris suggests that those who refuse to submit to authority and remain recalcitrant need to be incarcerated, whereas enterprising and self-governing individuals can employ the "structure" of the system to their own advantage and satisfy the needs of their existence through their own powers. He also connects the "structure" to his own ethical behavior and aspiring self-image.

> Chris: And so uh, it's a, but it keeps me, "What time are you going to be home?" "I'll be home at eleven fifteen," you know, and that makes me follow my schedule to where I have to go out and be back by eleven fifteen. And so, you're basically keeping your word.
>
> Staples: Any final thoughts?
>
> Chris: Well, let me think. Just basically I think it's a, uh, it's a good rehabilitation program. Basically, that's what I'd call it and uh, as a matter of fact where uh, you know, being uh, getting structure back in you life and uh living uh, living as you say you're going to live. Doing the things you say you're going to do and uh, kind of uh, learning to respect authority. Learning to, uh, humbling yourself. Basically, that's what I've learned from it, and, when I get off in seventeen days, it's going to take you know, transfer the structure that I had in this program into my everyday life when somebody's not monitoring me everyday. And uh, it helps you in doing that.

"Justin" is a European-American, 20 years of age, who works on a landscape crew and lives with his parents. He was originally assigned to a month house arrest following a drug conviction but a "hot" UA got him another month. Asked to describe his house arrest conditions, Justin focused, like many others, on the "structure" of the program but added that he perceives that the program is reserved for those who have demonstrated some enterprise and autonomy already. "Right, yeah, it's pretty relaxed. You got structure to it. So long as you talk to people you can get the schedule changes you need. It's not that bad," adding "Yeah, it's more of a deal for people who have, I guess, showed some kind of promise, you know. And you've held some employment for a while you look like you might possibly be straightening your life up. We just want to keep an eye on you for a little bit and make sure you're doing what you're supposed to be doing."

"Charles," an African-American, 32-year-old who lives with his fiancée, works at sales for a living. He did not reveal what he had done to be sentenced to house arrest, which he had served two weeks of a month sentence when we spoke.

> Staples: Maybe you could just describe for me what it's like living under house arrest.

> Charles: Well, it's, it beats the alternative, number one. Number two, it's really not bad 'cause you can pretty much, you know, be home to focus on things you need to focus on. A lot of times, you know when you don't have those restraints, definitely things going on and it's hard to keep up.

Asked to sum up his feeling about it, Charles focused on the themes consistent with enterprise, self-rule, and ethical behavior.

> Charles: The biggest thing about this whole program thing is its accountability. I mean, if a person's not used to being accountable for things, that's probably why they got in trouble. You know and I think that's something that maybe relies a lot on. I got to get back to accountability. I'm accountable for my actions, I'm accountable for my whereabouts, I'm accountable for everything I do. And that falls into responsibility so if you want to be responsible, you have to stay accountable.

"Dusty" is a European-American male in his twenties, who lives with a brother and roommate and, works at a car wash while trying to complete a technical degree at a local community college. He had served more than 60 days on house arrest and says that the routine of house arrest helps "Keeps me honest." Asked to elaborate on the idea of house arrest, Dusty suggested that the social control function actually creates a kind of space, an "opportunity" for some to reflect on and come to understand themselves.

> Dusty: Well, in my opinion, it's the type of situation where they're limiting your activities so you have an opportunity to think about what you're going to do where you have just a little bit of freedom, but also restrictions, keeping you from doing what you shouldn't be doing. Kind of see things from that perspective. You know, you'll maybe make some proper decisions. That's where I see it.

Later in the interview, Dusty declared that he has a course of action and a set of goals for himself now; a different way of being.

> Dusty: I have a better idea on what I want to do and what I'm going to change about myself. I had a drug problem myself and as far as my hang-up and places that I hung out, the people I hung out with, and a lot of times, I had a poor use of time. I just think it's, think maybe I'll make some better decisions as far as how I use my time and where I go ... Oh, you know I understand why they do it and it seems to make sense to me because uh, you know, I think initially if I wasn't on it, maybe, you know, I think being on it I have an opportunity to make some better decisions. Instead of being just left free from a restrictive environment I have opportunity, a window to think. I can't say I like it but I say it's probably the best thing for me.

Finally, we spoke with "Ernie," a 41-year-old African-American, married, who works as a general laborer in a factory. He had twice been convicted of drinking and driving and was half way through a 30-day sentence to house arrest.

> Staples: OK, maybe you could tell me what it's like living under house arrest. How do you experience the whole thing?
>
> Ernie: It's, it, it's better than being here [The Center] and it's not really that bad. You know, they call and you don't have to talk to anyone so you just talk to the machine which is great, you know, so you just blow in the machine. You know, no drugs, no alcohol present in the house, which ain't no problem. I've been sober for over nine years, you know, so I don't see that as a problem. I go to my AA meetings and I speak about it and I, you know, and my group is very supportive. Told me if I have any problems to let them know. My boss is very supportive. He told me he wanted me to make it through this.

Ernie, like Julie above, describes the transformative effects of the confessional technology of the 12-Step Program he participates in. Commenting on his experience at The Center, Ernie feels that it helped him gain self-control. "Yeah, in The Center, it's a wonderful program it's just you have to deal with so many different attitudes and personalities and you know, it taught me my patience. That's one thing this taught me." And later: "Umm, like I says, think before you act, think before you act on your impulse thing like I did."

In his final comments, Ernie offers his own words on how "modern individuals are not merely 'free to choose' but *obliged to be free*, to understand and enact their lives in terms of choice" (Rose, 1999, p. 87) and acquiring the skills to negotiate life's terms.

> Staples: Any final thoughts about the experience that you want to share?
>
> Ernie: You should weigh out the consequences 'cause sometimes you don't end up getting lucky like this, you just go to jail. You don't have the privilege of making money, you know, and another thing it taught me how to collect my money a little bit better. I was kind of doing that before, but now I think that I've got better at it 'cause you know, I've practiced it for three months. It's like embedded now.

CONCLUSIONS

The house arrest program we studied may well be seen as a "disciplinary technology," intent on the production of "docile bodies" through both "hierarchical observations," designed to instill the gaze of authorities in clients, and "normalizing judgments" that set the behavioral standards to be upheld (Foucault, 1979/1995; Staples, 2000). As practiced, and as participants have attested, it is an intrusive, regimented regime of control that is backed by the constant threat of incarceration. And yet, rather than simply coercing clients into behaving, this productive form of disciplinary power operates as a kind of training program intended to engender certain "self-steering mechanisms" (Foucault, 1997a, 1997b) by which participants may come to experience, understand, judge, and conduct themselves. In this "neo-liberal" model of governing, subjects are "... confronted with a field of incentives suggesting ways of utilizing individual skills and circumstances maximizing their own 'life chances' while minimizing their cost to the state" (Tuschling & Engemann, 2006, p. 452).

Did the house arrest clients we interviewed really develop the ostensibly self-governing qualities they professed to have adopted? Will they go on to be less troublesome citizens, newly reconstituted liberal democratic subjects? We cannot say; we can only report what they told us. Goffman (1961) suggests that those spending time in institutions tend to internalize the idea that they will be reformed or will learn something from the experience because they do not want to think that this time in their life has been simply wasted. These clients may have adopted a similar stance and want to believe that, despite how difficult they found the program, they met the challenges and it has changed them for the better. Furthermore, it may not be a coincidence that the clients who reflected on what they thought were the more productive qualities of the house arrest program were also the ones who had spent time in "The Center" and/or participated in various "12-Step" programs. Were they merely mimicking the therapeutic rhetoric of empowerment and transformation they had been conditioned to believe? Or were they just telling us what they thought we wanted to hear? Both are possible. Regardless, what seems important to us is that the discourse they have taken on, this narrative of truth they have come to employ to tell the story of their lives, is a crucial link in understanding the ways we are governed by others and the ways we attempt to govern ourselves.

NOTES

1. The names used here are fictitious. The research protocol was conducted within the ethical and procedural guidelines set out by the Human Subjects Committee of the University of Kansas, the American Correctional Association, and the American Sociological Association.

2. The original and commonly used form of electronically monitoring someone involves deploying a small radio transmitter that is typically attached to the ankle of the offender. A monitoring box is placed in the home and the arrestee cannot stray more than 150 feet or so from the box without triggering a violation that is recorded by a central computer. Sometimes called "tagging," the idea is credited to a New Mexico district court judge who was supposedly inspired by the use of a similar device in a 1979 Spiderman comic strip (Klein-Saffran, 1992).

3. The results have been mixed. A recent meta-analysis of this literature by Layton MacKenzie (2006, p. 322) concludes that, "A large body of research, including random assignment studies, consistently shows the failure of ... EM [Electronic Monitoring] programs to lower recidivism."

REFERENCES

Ansay, S. J. (1999). *When home is a prison cell: The social construction of compliance in house arrest*. Unpublished doctoral dissertation. University of Florida, Gainesville.

Ball, R. A., Huff, R. C., & Lilly, R. J. (1988). *House arrest and correctional policy: Doing time at home*. Newbury Park, CA: Sage Publications.

Barry, A., Osborne, T., & Rose, N. (1996). *Foucault and political reason: Liberalism, neo-liberalism and rationalities of government*. Chicago: University of Chicago Press.

Byrant, G. G., & Sarat, A. (Eds). (1998). *How does law matter*? Evanston: Northwestern University Press.

Corbett, R., & Marx, G. T. (1991). Critique: No soul in the new machine – Technofallacies in the electronic monitoring movement. *Justice Quarterly*, *8*(3), 399–414.

Courtright, K. E., Berg, B. L., & Mutchnick, R. J. (1997a). The cost effectiveness of using house arrest with electronic monitoring for drunk drivers. *Federal Probation*, *61*, 19–22.

Courtright, K. E., Berg, B. L., & Mutchnick, R. J. (1997b). Effects of house arrest with electronic monitoring on DUI offenders. *Journal of Offender Rehabilitation*, *24*, 35–51.

Courtright, K. E., Berg, B. L., & Mutchnick, R. J. (2000). Rehabilitation in the new machine? Exploring drug and alcohol use and variables related to success among DUI offenders under electronic monitoring – Some preliminary outcomes. *International Journal of Offender Therapy and Comparative Criminology*, *44*, 293–311.

Dean, M. (1999). *Governmentality: Power and rule in modern society*. Thousand Oaks, CA: Sage Publications.

Ericson, R. V., & Haggerty, K. D. (1997). *Policing the risk society*. Toronto: University of Toronto Press.

Ewick, P., & Silbey, S. (1988). *The Common place of law: Stories from everyday life*. Chicago: University of Chicago Press.

Feeley, M., & Simon, J. (1992). The new penology: Notes on the emerging strategy of corrections and its implications. *Criminology*, *30*, 449–474.

Feeley, M., & Simon, J. (1994). Actuarial justice: The emerging new criminal law. In: D. Nelken (Ed.), *The futures of criminology* (pp. 173–201). London: Sage Publications.

Foucault, M. (1979). Governmentality. *I&C no. 6* (R. Braidotti, Trans.), Autumn, pp. 5–21.

Foucault, M. (1979/1995). *Discipline and punish: The birth of the prison*. (A. Sheridan, Trans.). New York: Vintage Books.

Foucault, M. (1991). Governmentality. In: C. Gordon (Ed.), *The Foucault effect* (pp. 87–104). Chicago: University of Chicago.

Foucault, M. (1997a). The ethics of the concern for the self as a practice of freedom. In: P. Rabinow (Ed.), *Ethics: Subjectivity and truth* (P. Aranov & D. McGrawth, Trans.) (pp. 281–301). New York: New Press.

Foucault, M. (1997b). Technologies of the self. In: P. Rabinow (Ed.), *Ethics: Subjectivity and truth* (P. Aranov & D. McGrawth, Trans.) (pp. 223–251). New York: New Press.

Garth, B. G., & Sarat, A. (Eds). (1998). *How does law matter?* Evanston: Northwestern University Press.

Glaser, D., & Watts, R. (1992). Electronic monitoring of drug offenders on probation. *Judicature*, *76*, 112–117.

Goffman, E. (1961). *Asylums: Essays on the social situation of mental patients and other inmates.* Garden City, NY: Anchor.

Harrison, P. M., & Karberg, J. C. (2003). Prison and jail inmates at midyear 2002. *Bureau of Justice Statistics Bulletin*. U.S. Department of Justice Office of Justice Programs.

Holman, J. E., & Quinn, J. F. (1992). Dysphoria and electronically monitored home confinement. *Deviant Behavior*, *13*, 21–32.

Jolin, A., & Stipak, B. (1992). Drug treatment and electronically monitored home confinement: An evaluation of a community-based sentencing option. *Crime and Delinquency*, *39*, 158–170.

Jones, M., & Ross, D. (1997). Electronic house arrest and boot camp in North Carolina: Comparing recidivism. *Criminal Justice Policy Review*, *8*, 383–403.

Klein-Saffran, J. (1992). *Electronic monitoring versus halfway houses: A study of federal offenders*. Unpublished doctoral dissertation. University of Maryland, College Park.

Layton MacKenzie, D. (2006). *What works in corrections: Reducing the criminal activities of offenders and delinquents*. New York: Cambridge University Press.

Lilly, R. J., Ball, R. A., Curry, D. G., & McMullen, J. (1993). Electronic monitoring of the drunk driver: A seven-year study of the home confinement alternative. *Crime and Delinquency*, *39*, 462–484.

Maidment, M. R. (2002). Toward a 'woman-centered' approach to community-based corrections: A gendered analysis of electronic monitoring (EM) in Eastern Canada. *Women and Criminal Justice*, *13*, 47–68.

Maxfield, M. G., & Baumer, T. L. (1990). Home detention with electronic monitoring: Comparing pretrial and postconviction programs. *Crime and Delinquency*, *36*, 521–536.

Newman, G. (1999). *The global report on crime and justice*. New York: Oxford University Press.

O'Malley, P. (2002). Globalizing risk? Distinguishing styles of 'neo-liberal' criminal justice in Australia and the USA. *Criminal Justice*, *2*, 205–222.

Payne, B. K., & Gainey, R. R. (1998). A qualitative assessment of the pains experienced on electronic monitoring. *International Journal of Offender Therapy and Comparative Criminology*, *42*, 149–163.

Payne, B. K., & Gainey, R. R. (2000). Understanding the experience of house arrest with electronic monitoring: An analysis of quantitative and qualitative data. *International Journal of Offender Therapy and Comparative Criminology*, *44*, 84–96.

Poster, M. (1996). Databases as discourse, or electronic interpellations. In: D. Lyon & E. Zureik (Eds), *Computers, surveillance, and privacy*. Minneapolis: University of Minnesota Press.

Rose, N. (1996). *Inventing ourselves: Psychology, power, and personhood*. England: Cambridge University Press.

Rose, N. (1999). *Powers of freedom: Reframing political thought*. England: Cambridge University Press.

Sandhu, H. S., Dodder, R. A., & Mathur, M. (1993). House arrest: Success and failure rates in residential and nonresidential community-based programs. *Journal of Offender Rehabilitation, 19*, 131–144.

Sarat, A. (1998). *Crossing boundaries: Traditions and transformations in law and society research*. Evanston, IL: Northwestern University Press.

Staples, W. G. (1994). Small acts of cunning: Disciplinary practices in contemporary life. *Sociological Quarterly, 35*, 645–664.

Staples, W. G. (2000). *Everyday surveillance: Vigilance and visibility in postmodern life*. Lanham, MD: Rowman & Littlefield Publishers.

Staples, W. G. (2005). The everyday world of house arrest: Collateral consequences for families and others. In: C. Mele & T. Miller (Eds), *Civil penalties, social consequences* (pp. 139–159). New York: Routledge.

Staples, W. G., & Decker, S. K. (2007). *Between the 'home' and 'institutional' worlds: House arrest as postmodern social control*. Draft manuscript.

Tuschling, A., & Engemann, C. (2006). From education to learning: The emerging regime of learning in the European Union. *Educational Philosophy and Theory, 38*, 451–469.

Ulmer, J. T. (2001). Intermediate sanctions: A comparative analysis of the probability and severity of recidivism. *Sociological Inquiry, 71*, 164–193.

ACADEMIA, SURVEILLANCE, AND THE FBI: A SHORT HISTORY

Scott G. White

ABSTRACT

Whenever America has fought wars, civil liberties are compromised. Led by Hoover, whose career began in the Library of Congress, the FBI has historically conducted questionable surveillance, often spying illegally on American citizens. There is a history of FBI surveillance in the Academy, including surveillance in libraries. Researchers, students, and librarians have been the subjects of FBI surveillance efforts. Today, the Patriot Act has reignited concerns about FBI surveillance in academic institutions. Librarians have often led the fight against limits imposed on accessing information. This is a short history of the conflict between the Academia and FBI surveillance.

Soon after the terrorist attacks of September 11, 2001, a reference librarian from Florida's Delray Beach Public Library called the FBI because she recognized one of the hijack suspects as a former library user (Pressley & Blum, 2001; Rosenbaum, 2001). One week after, the FBI obtained a search warrant from a federal grand jury to seize two public use computers from the Broward County Library (Puzzanghera, 2002). Several of the hijackers used the library's computers to access web sites and e-mail, book airline tickets, and communicate with each other

Surveillance and Governance: Crime Control and Beyond
Sociology of Crime, Law and Deviance, Volume 10, 151–174

ISSN: 1521-6136/doi:10.1016/S1521-6136(07)00207-2

(Babson, Lebowitz, & Viglucci, 2001). Within hours of the attacks, investigators were themselves scouring the Internet to discover electronic traces or fingerprints left by the suspects.

During the ensuing weeks after September 11, public opinion polls indicated that Americans were willing to give up certain freedoms if it meant the country, and they, would be safer from the threats of terrorism. A Harris opinion poll published on October 3, 2001, demonstrated "overwhelming public support for increasing surveillance powers and, in spite of many concerns about potential abuses, confidence that these powers would be used properly" (Taylor, 2001, p. 1). Respondents were asked about ten specific proposals to increase surveillance of potential terrorists, or to establish new surveillance powers, and more than 90% of the public supported at least some of the measures. These included the use of facial-recognition technology, the establishment of a national I.D. system, increased camera surveillance in public areas, and law enforcement monitoring of Internet discussions (Taylor, 2001).

On October 11, 2001, the federal government responded to public concerns by passing the Uniting and Strengthening America by Providing Appropriate Tools Required to Intercept and Obstruct Terrorism Act (Patriot Act). The Patriot Act was conceived to help law enforcement authorities more effectively combat terrorism. In doing so, it helped expand the ability of government agencies to conduct surveillance, detain prisoners, trace money, and recover business records (Schulhofer, 2002; Solove, 2004). Many of the modifications made in the 2001 Patriot Act were proposed in the 1996 Anti-Terrorism and Effective Death Penalty Act. During the vote for the 1996 bill, the modifications were considered too restrictive of civil liberties, and were struck from the final version (Adas, 2002).

The passage of the Patriot Act was a bipartisan effort rarely witnessed in Washington over the last 30 years. The Patriot Act, House bill HR 2975, passed by a vote of 337 to 79 (Rodriguez, 2001). While many legislators, legal experts, and citizen's groups have since denounced the Patriot Act (Chang, 2001; Cole, 2003; Jaeger, McClure, Bertot, & Snead, 2004; Levin, 2007; Osher, 2002), there were few dissenters during the brief deliberation period leading up to the bill's passage. Russell Feingold (D-WI) was the only Senator to vote against the bill. The vote was held while many lawmakers were denied access to their offices because of the anthrax scare (Minow, 2002).

The Patriot Act was crafted to help monitor suspected terrorists and impede their ability to conduct business in a variety of areas. The Patriot

Act includes five main areas that address the amendment or modification of existing procedures.

1. Criminal Investigations: Tracking and Gathering Communications
2. Foreign Intelligence Investigations
3. Money Laundering
4. Alien Terrorists and Victims
5. Other Crimes, Penalties, Procedures (Doyle, 2002).

The strength of the Patriot Act, from a law enforcement perspective, lies in the easing of restrictions when conducting surveillance. It is not as difficult to conduct surveillance because it is not as difficult to get permission. Warrants are easier to obtain, they are more inclusive, allow investigators to follow an individual rather than listen in on a particular device, they do not need to be disclosed to the suspect and they do not expire (Chang, 2001; Doyle, 2002; Schulhofer, 2002; Solove, 2004). The law also makes it easier for investigators to track calls, e-mails, Internet communications, and cell phone use.

Given the expansive changes contained in the Patriot Act, there is a great deal written about how Orwellian it is (Foerstel, 2004), how it limits civil liberties (Chang, 2001), expands the ability to conduct "sneak and peak" warrants (ACLU, 2002), and generally expands the power of government surveillance (Jaeger, Bertot, & McClure, 2003). The Patriot Act met with immediate resistance from a variety of groups, but struck a particularly resistant chord with librarians and booksellers.

Librarians consider the Patriot Act the most current government attempt to spy on people's information retrieval habits. Much of the literature written by librarians focuses on the limits the law places on civil liberties (Kniffel, 2002; Strickland, 2002), freedom of access to information (Reynolds, 2002), academic freedom (Warwick, 2005), and unchecked government surveillance powers in libraries (Foerstel, 2004). Some also believe that the language of the law is written very broadly (Rackow, 2002), and that the executive branch of the government has taken advantage of that language (Baker, 2002).

Criticism of government behavior and the potentials for abuse of the law by government agencies like the FBI are written daily. However, if the history of the FBI is studied, there are many instances where questions concerning surveillance and lack of defined limits of the agency's investigative power exist. While planning the bureau in the early part of the 20th century, legislators had reservations about what the new agency's duties would be (Kessler, 2003). There was initial concern that agents would interfere with local law enforcement, and could be used as a tool by the federal government to police American citizens (Kessler, 2003).

> Despite the limitations on its power, questions arose very quickly about the extent of the bureau's authority and its methods. Yet whenever a new threat arose, those questions would be set aside and congress would entrust the bureau with new powers. (Kessler, 2003, p. 10)

Congress may have been initially wary of the new police force, but the role of the new agency continued to expand, especially as World War I loomed (Kessler, 2003; Powers, 2004). Over the last century, the FBI has become the most powerful law enforcement agency in America.

Throughout its history, especially during the trials of war, the United States government has taken measures that ultimately compromise civil rights and liberties. The 1798 Alien Sedition Act, the 1917 Espionage Act and Palmer raids, Japanese internment camps, and the domestic counter-intelligence programs (COINTELPRO) of the sixties are examples. These programs are remembered as embarrassing incidents in United States history. In some of these cases, reparations and apologies were necessary. All of these episodes also share FBI participation. The FBI was found to have conducted domestic surveillance for over fifty years (Davis, 1997; Foerstel, 2001, 2004; Kessler, 2003; Powers, 2004), some of it illegally.

In addition to the above episodes, the FBI has also conducted surveillance in academic institutions, and has spied on groups from several academic disciplines, including sociologists (Keen, 2004), anthropologists (Price, 2004), scientists (Badash, 2000, 2003; Jerome, 2003), and, of course, avowed communists (Lichtman, 2004). Prominent sociologists, including W.E.B. Dubois, C. Wright Mills, Talcott Parsons, Erving Goffman, and Edwin Sutherland had FBI files or were the subjects of FBI investigations (Keen, 2004). During the 1960s, when running an investigation called COINTELPRO, the FBI had informants in colleges throughout America. At one point, over 140,000 students attending over 30 colleges in the Philadelphia area were thought to be under some sort of government surveillance (Foerstel, 1991). During the 1970s, agents visited libraries to gather information about Eastern European users who were accessing what was considered sensitive technical information. The Library Awareness Program ran until the early 1990s.

Many of the surveillance activities conducted by the FBI under Hoover's watch were not discovered until after his death in 1972 (Davis, 1992; Powers, 1987). In 1971, a group calling itself the "Citizens Commission to Investigate the FBI" broke into an FBI office in Media, Pennsylvania and stole over 1,000 documents related to the COINTELPRO program. The group slowly released the documents to media outlets, which eventually led to congressional hearings (Davis, 1992, 1997; Kessler, 2003; Powers, 2004)

COINTELPRO was a domestic spying operation that included surveillance and infiltration of black radical groups, the Ku Klux Klan, and prominent national leaders like Dr. Martin Luther King. Much of the activity was clandestine and illegal (Davis, 1992), and included the surveillance of American citizens. Scholars have been able to determine the extent of FBI activity in the academy over the last two decades because of documents released by request under the Freedom of Information Act.

This chapter will highlight these cases. Also included will be a discussion as to why spying in libraries is considered an assault on the freedom to access information, and by extension, a limiting of civil liberties. Intellectual and academic freedom are ultimately similar concepts, with both supporting the pursuit of knowledge. Clandestine surveillance of library activity threatens intellectual freedom, and ultimately, academic freedom. This prospect does not sit well with librarians, and has garnered the attention of lawmakers, along with the general public.

In conclusion, a discussion of why such surveillance tactics can be easily abused will highlight the role the Internet can play in gathering information. There will be ramifications for the availability, the display and the retrieval of such information, especially if it is retrieved without a context that makes it easier to understand why the information was initially made available or accessed. FBI agents are paid to investigate wrongdoing. That is the context for the information they gather through surveillance they conduct. It is imperative that there is oversight of their methods to ensure that other contexts are understood, and protected, both for the subjects being investigated and the investigators themselves. Both academic and intellectual freedom depend on it.

THE CONTROVERSY: THE USA PATRIOT ACT

As details emerged about how the 9/11 attacks were planned and executed, it became apparent that the perpetrators lived relatively inconspicuous lives. The hijackers traveled freely about the country, attended American schools (the recognized mastermind of the attacks, Khalid Shaikh Mohammed, attended school in North Carolina (Mazzetti & Williams, 2007)), and were here legally in most cases. The hijackers spoke to each other through various electronic communication means. Conducting their own surveillance, they were able to identify airports where security measures were easily circumvented. They studied planes, learned to fly in American flight schools, and ultimately used this mass of knowledge and information to carry out their plans. Much of their communication, and potentially, a bulk of their planning may have been carried out in one of the most seemingly benign institutions in America, the public library.

The information culled from the seized library computers of hijacker activity were reported to have proved helpful to law enforcement (Fallis & Cha, 2001). Files left on computers used to communicate or download instructions can yield information about what Internet sites a user may have visited. Copies of e-mails, web sites visited, downloaded documents, etc., are stored in various places on a computer's hard drive, and are easily recoverable (Puzzanghera, 2002). At the time, it was believed that groups like Al-Qaeda used the Internet to help conduct their operations, as scholars debated concepts such as Netwar, and how it could impact future terrorist activity (Hoffman, 1998; Lesser, 1999; Zanini, 1999). It was hoped that examining the computers would help to identify co-conspirators.

Given these circumstances, special consideration of libraries, indeed most businesses, in the Patriot Act was not unreasonable. Of main interest were institutions that maintain personal information records to support business activities, including financial institutions, insurance companies, educational institutions, car, and locker rental businesses and hotels. Making sure that business transactions, even the most seemingly benign ones, were quickly and easily accessible was of paramount importance to investigators.

A provision of the Patriot Act includes the enhanced surveillance of business transactions, and has been interpreted by librarians as having special significance for the privacy of transactions in libraries. With the Patriot Act revisions, FBI agents can

- obtain a search warrant for "any tangible thing", including books, records, papers, floppy disks, data tapes, and computers with hard drives;
- force libraries to hand over circulation records, user information records, and Internet use logs, without a warrant;
- limit the demonstration of "probable cause" and require that the investigated library not reveal the FBI visit (Foerstel, 2004).

There is a history of FBI surveillance in libraries, and members of the Library community do not remember it fondly. The Library Awareness Program was developed so that librarians could be trained to recognize and alert FBI agents about suspicious activity of "Eastern European" users who were accessing technical data for a variety of subjects (Foerstel, 1991). While most librarians would be appalled to learn that terrorists, or any criminals, used their library to conduct illegal activity, providing users access to free information privately and democratically are cornerstones of library service.

The relaxed standards and broad powers the Patriot Act affords law enforcement agencies to conduct surveillance and gain access to library records threatens these beliefs, and is the issue that concerns librarians most (Foerstel, 2004; Klinefelter, 2004).

HOOVER, THE FBI, AND LIBRARIES

In 1942, well into J. Edgar Hoover's career as Director of the FBI, and in the midst of World War II, a controversy surrounding the papers of Boris Brasol, a former diplomat, author, private detective, and criminologist (Brasol, 1969), who was suspected to have ties to Fascist parties in Italy and Germany, drew Hoover into a showdown with the Librarian of Congress, Archibald MacLeish. Brasol was a controversial figure, credited with translating into English an infamous anti-Semitic work titled *Protocols of the Learned Elders of Zion* (Wallace, 2003), used by such luminaries as Henry Ford to perpetuate Jewish world domination conspiracy theories (Wallace, 2003). The work had been popular in Nazi Germany, helping to foment extreme anti-Semitism (Wallace, 2003). Brasol had deposited his papers with the Library of Congress with a stipulation that they not be released to the public until 1953. Hoover, along with FBI investigators, believed the papers had been donated with that stipulation because they contained evidence that could expose Mr. Brasol as a war criminal (Stielow, 1993).

Hoover understood how such information could be used in investigations, as he had worked for the Library of Congress from 1913 to 1917, where he developed ideas about the use of information, before moving to the FBI (*Directors, Then and Now*, 2007; Foerstel, 2004; Stielow, 1993). Macleish denied the request to see Brasol's papers. He adhered to the issues of privacy, fearing that if he acquiesced to Hoover's request, other potential donors would hesitate to donate their archives out of fear. Hoover dropped the matter, but continued to press for information gathering strategies that would root out potential foreign agents (Stielow, 1993) and communists (Kessler, 2003; Powers, 2004).

J. Edgar Hoover was and still is a legendary figure in the history of the FBI, having served as director for 48 years, from 1924 until his death in 1972. Hoover was instrumental in running the infamous Palmer raids during the Red Scare in the 1920s. Much of the information gathering he did for that operation was informed by work he did at the Library of Congress (Gentry, 1991; Powers, 1987; Stielow, 1993). He created an indexing system to collect names of perceived radicals, communist organizations, and communist publications. In two years, Hoover had amassed over 450,000 entries in his database (Gentry, 1991), giving him "a semi-monopoly over a sort of information so difficult to obtain, so extensive in coverage, and so commonly inaccessible as to make its independent verification almost impossible" (Powers, 1987, p. 69). The information Hoover gathered was used to arrest over 10,000 people, of which only a handful were prosecuted (Gentry, 1991; Powers, 1987). The Palmer raids were widely criticized at the time, but solidified Hoover's career in the FBI (Gentry, 1991; Powers, 1987).

As in the Brasol case, librarians are often at the forefront combating issues of censorship and limited access to information. Some librarians consider it their duty to uphold the freedom to access information and support intellectual freedom for any users of the library. These principles, believed by librarians to be established in the U.S. Constitution's 1st Amendment, are considered sacrosanct and immutable. Another principle concerns privacy when accessing information in libraries, without fear of reprisal from any entity.

While privacy between librarian and library user does not enjoy the legal protection similar to attorney-client privileges in criminal cases, there are some legal protections library users have when accessing resources in a library. For example, in most states, legal authorities need a court order to access the circulation records of individuals to examine what books they have checked out (Foerstel, 2004). Users sometimes exploit this freedom, but alternatives restricting access to information, or conducting widespread surveillance of library users' information retrieval habits, are not acceptable to librarians. An unfortunate result of these protections is that the hijackers of 9/11 conducted their business in libraries in Florida and Fairfax, Virginia in relative anonymity. This anonymity can partially be traced to state privacy laws developed in response to the FBI's Library Awareness Program (Foerstel, 1991, 2004).

LIBRARY AWARENESS PROGRAM

Amidst the controversy surrounding the Watergate break-in and cover-up, and after a series of disturbing revelations about the FBI and CIA, the United States Senate Select Committee to Study Governmental Operations with Respect to Intelligence Activities, or Church committee, led by Senator Frank Church (Blackstock, 1988; Johnson, 1988), was established in 1975. During these hearings, both the FBI and CIA were taken to task about methodologies and surveillance gathering techniques. The Church committee issued a series of reports suggesting FBI reforms for gathering intelligence, including the creation of Senate and House Intelligence committees, and legislation regulating surveillance practices (Johnson, 2004).

The FBI began its "Library Awareness Program" soon after the Church committee reports were issued. Conducted during the mid-1970s, and continued in the late 1980s and early 1990s (exact dates have never been confirmed), the Library Awareness Program attempted to recruit librarians to help identify Soviet spies in research libraries (Foerstel, 1991). The program tried to gauge who was accessing what was considered public, but sensitive material in specialized technical libraries (Foerstel, 1991).

The FBI had reason to believe that Soviet agents were targeting librarians to help gather publicly available information, usually technical in nature, and bundle it for transport overseas. From FBI memos, it can be gathered that librarians cooperated with the FBI and identified library users who were seeking such information, including those who were suspected Soviet intelligence agents (Foerstel, 1991).

If someone were seeking information at that time, it would be important to make contact with a librarian to complete his or her work. In the late 1970s and early 1980s, databases were developed that could more easily search for information, identifying books, and articles to support research topics. Initially, these searches resulted in charges based on connection times, so libraries usually had trained librarians conduct searches submitted by patrons to keep connection costs low. It would have been much easier for someone who was seeking difficult to obtain information to make contact with a librarian who conducted these searches. Of course, if we view the transaction from a librarian perspective, they believed they were doing their job.

The program was first exposed when a librarian from Columbia University was approached by two FBI agents. Paula Kaufman, director of Academic Information Services, wrote a letter describing the visit to the American Library Association (ALA), which in turn had contacted the *New York Times*, resulting in a front page article describing the program as a counterintelligence effort (Foerstel, 1991; McFadden, 1987). Milt Ahlerich, an Assistant Director at the FBI, quoted in the *New York Times*, explained the program.

> We alert those in certain fields of the possibility of hostile intelligence powers or their agents attempting to gain access to information that could be potentially harmful to our national security. In this regard, our New York office has contacted staff members of New York libraries to alert them to this potential danger and to request assistance. (McFadden, 1987, p. B2)

Ms. Kaufman, also quoted in the article, said,

> They (FBI agents) explained that they were doing a general "library awareness program" in the city (New York) and that they were asking librarians to be alert to the use of their libraries by persons from countries "hostile to the United States, such as the Soviet Union" and to provide the FBI with information about these activities, ... I explained that we were not prepared to cooperate with them in any way, described our philosophies and policies respecting privacy, confidentiality and academic freedom, and told them they were not welcome here. (McFadden, 1987, p. B2)

ALA continued to press for congressional hearings, urging libraries across America to write letters to government representatives expressing outrage at the program. On June 20, 1988, the first hearing took place before the Subcommittee on Civil and Constitutional Rights, led by California representative Don Edwards (D-CA) (Foerstel, 1991). It was a new experience for the FBI. The days when clandestine surveillance was tolerated without question were over. FBI agents could not understand the strong negative reaction exhibited by the librarians they contacted. One after another, including librarians at New York University, the Brooklyn Public Library, the University of Maryland and others refused FBI requests to collect or release information. In each case, the librarians in charge demanded to see court orders (Foerstel, 1991). When confronted with such requests, FBI agents usually did not pursue establishing further contact.

Suspicion concerning FBI surveillance activity was still fresh as a result of information uncovered during the Church committee hearings. When agents tried to develop human intelligence assets, as they were when they approached Ms. Kaufman, they could no longer guarantee cooperation, either by appealing to a sense of patriotism, or through fear and coercion, as was often the tactic during the Red Scare of the 1920s and 1940s/50s. This is a significant development in the course of FBI history. Most intelligence and counter-intelligence efforts, especially related to issues concerning communist or anarchist activity, relied on the recruitment and development of human intelligence assets, or informants. American law enforcement in general is a reactive entity, relying on citizens to report crimes, and often relying on citizens to help give them information for the investigation and solving of crimes. For many years, the FBI used informants to develop cases and maintain files about hundreds of thousands of American citizens. After meeting Ms. Kaufman, the reliance on such assets would be put in jeopardy.

The FBI could not have been aware of how difficult it would be for librarians to take part in the Library Awareness Program. The tasks were difficult to perform because it meant tracking and interpreting the importance of what people read and accessed in the library. The sheer number of information requests a day in a library would make it impossible to identify patrons. When the FBI began the Library Awareness Program in the late 1970s, surveillance targets switched from academics who displayed radical tendencies either by publishing or participating in radical action networks to library users who were accessing what was deemed sensitive material. This was a paradigm shift in that no longer was it necessary to conduct oneself in a deviant manner, according to government standards, but now one could be considered deviant based on information they were accessing in libraries. The identification of surveillance targets also became harder because students usually outnumber faculty on campuses 20 to 1.

The Library Awareness Program seemed to go away on its own. The fall of the Iron Curtain, signaling the symbolic end of the Cold War, in the early 1990s probably hastened its demise as the threat of Russian espionage decreased. In 1988, FBI Director William Sessions wrote a series of memos and provided written guidelines for the program for review by Congressman Edwards and others (Foerstel, 1991). While never mentioning that they would cease the program, Sessions indicated that the FBI would tread carefully when conducting similar surveillance exercises in libraries.

The FBI never technically broke a law, because, as far as is now known, no information culled from the program was used in investigations or as proof in criminal proceedings. Several Freedom of Information Act requests about the Library Awareness program have returned heavily excised documents that add little to what is already known (Foerstel, 1991, 2004). It is generally agreed that the Library Awareness Program was a public relations disaster for the FBI, and it is also generally agreed that librarians won that round of the information access battle. With the advent of the Patriot Act, the tide may be turning, as surveillance of libraries is now again an issue.

The tenets of academic freedom are ostensibly the same as intellectual freedom. Both principles aspire to protect individuals from unjust persecution because of the line of inquiry they choose to pursue, whether conducting research or teaching a class. Academic freedom is more formalized (Fuchs, 1963), and allows faculty and researchers to pursue inquiry without fear of dismissal based on intellectual reasons. Academic freedom has been the subject of Supreme Court cases, and has been upheld more often than not (Fuchs, 1963; Rabban, 1990). There are some restrictions on both academic and intellectual freedom. Seditious speech and speaking of violent overthrow of the government are not permitted (Woolston, 1942), and can garner the attention of authorities, usually through an informant alerting agencies.

During the Red Scare of the 1950s, the FBI used many informants to spy on people and groups they thought were guilty of sedition. Many academics underwent severe scrutiny during this period. While Hollywood had its blacklists, academic institutions also suffered under McCarthyism. Faculty members had their academic freedom curtailed, faced censorship, and lost their positions because of FBI coercion and intimidation. With the help of the Freedom of Information Act, researchers and scholars were able to gain access to government files that had remained secret for 50 years (Foerstel, 1999). Within these files scholars have discovered that members of the Academy were the subjects of domestic surveillance, much of it conducted by the FBI, often in support of investigations into activities considered un-American (Keen, 2004; Price, 2004). In most cases, individuals who were under suspicion for a variety of transgressions, including seditious speech, cavorting with communists in foreign nations, or being a member of a radical group qualified to come under FBI surveillance.

EDWIN SUTHERLAND, J. EDGAR HOOVER, AND THE UCRs

Edwin Sutherland, a famous criminologist, developed the theory of differential association, helping to move the study of criminology from the positivistic based biological and psychological theories to more sociological constructions. He believed that criminal behavior may be learned and not inherited (Keen, 2004), an idea that shook the criminal justice world at the time because it was normally thought that crime was centered in poor, urban environments and committed by people in the lower classes. Much of this work is outlined in Sutherland's books, *Criminology* and *White Collar Crime*.

On August 2, 1940, Sutherland wrote a letter to then FBI director J. Edgar Hoover, asking him to send copies of the 1938 FBI's Uniform Crime Reports (UCRs), which Sutherland had not received (Keen, 2004). The reports, still in wide use today, contain a compilation of crime data from police agencies across the nation. They are integral to crime studies, and are an important tool for criminologists when conducting crime research. Hoover had heard about an address Sutherland had made in defense of parole. In his address, Sutherland both argued for parole and called into question statements Hoover himself had made about parole statistics (Keen, 2004). Unbeknownst to Sutherland, Hoover had removed him from the UCR mailing list as retaliation for his comments.

The above episode is a microcosm of what have been historical problems with FBI surveillance, information gathering techniques, and subsequent action strategies. Hoover had received the information about Sutherland's address second hand from FBI agent Herold Reinecke, who had himself received a copy of Sutherland's address from *Indianapolis Times* reporter Albert G. McCord. According to a memo retrieved by Keen (2004) from Sutherland's FBI file, Hoover acknowledged receiving the report from Reinecke, replying in a letter, "For your information, Professor Sutherland is on the Editorial Board of the *Journal of Criminal Law and Criminology* and utterances such as he made are to be expected" (Keen, 2004, p. 192).

Not only was Sutherland's name struck from the UCR mailing list, but his remarks, having caught the attention of the Director of the FBI, resulted in the creation of a personal FBI file. Sutherland was punished by Hoover, as informally as it was, for ideas he discussed in public with which Hoover did not agree. There were no hearings or communications to Sutherland.

In today's technology-laden world, where many paper reports also exist online, it may seem insignificant to mention Hoover's retaliation. However, in an age where paper copies of reports were the only means of access to information, and access to many government reports was difficult because of copying costs, his actions were significant and intended to cause harm to Sutherland. Hoover ignored Sutherland's entreaties to be placed back on the list for over two years. Sutherland, not knowing what had happened, tried to handle the situation as someone normally would, but was stonewalled by the FBI (Keen, 2004).

The parallels between the Sutherland and Hoover controversy and FBI investigatory practices today, especially in terms of terror investigations, are striking. For many years, especially under Hoover's watch, the FBI conducted clandestine investigations (Davis, 1997; Powers, 2004). A major difference is that now secrecy has been embedded in the legislative code, so such surveillance gathering techniques can be employed without scrutiny. In some ways, the Patriot Act serves the function of allowing the FBI to revisit surveillance practices it employed successfully from the late 1920s to 1970.

THE FBI AND THE ACADEMIA

David Price (2004) takes exception to the way Americans have remembered McCarthyism, and does not believe that the public understands the role of the FBI. Hoover was an avowed anti-communist (Keen, 2004; Price, 2004), as demonstrated by his role in the Palmer raids. He considered Communist groups capable of infiltrating other American radical movements, such as the labor movement and civil rights organizations to affect societal shifts towards Marxism (Price, 2004).

> The red scares of the 1940s and 1950s have in some revealing ways been condensed in what remains of America's popular memory. First, almost two decades of disparate episodes of red-baiting have come to be incorrectly recalled as limited to the early 1950s. Second, most Americans conceptualize these attacks as being primarily on individuals with links to Communist organizations, rather than with groups advocating racial equality, arms reduction, or the progressive labor movement. Third, the wide range of public loyalty hearings is simply associated with what is imagined as the isolated ranting of Senator Joseph McCarthy of Wisconsin. (Price, 2004, p. 13)

As laws were passed to combat the threat of communism, including the 1939 Hatch Act, barring communists from working in the government, and the 1940 Smith Act, making it a crime to discuss violent overthrow of the government, or belong to an organization that did, the FBI became the government agency responsible for investigating violators of the law (Price, 2004).

> In June 1940, Attorney General Jackson created the Custodial Detention Program empowering the FBI to compile lists of American citizens and resident aliens to be affiliated with subversive organizations. With this, the FBI's budget jumped from $8.7 million to $14.7 million the following year ... as the organization was given broad powers to conduct break-ins, tap wires, and use confidential informants to gather hearsay statements pertaining to the beliefs and actions of a wide range of Americans. (Price, 2004, pp. 13–14)

The academic profession is often filled with individuals who study or are members of what can be considered subversive organizations. The control of ideas is at the crux of FBI activities in the Academy. Left leaning academics, interested in questioning the state of things, were prime targets for surveillance. This is why faculty members enjoy the privileges of academic freedom and tenure. While there are problems inherent in the tenure system, it does its job protecting faculty from dismissal for ideas that are not popular, but that may be true. As an example, Galileo comes to mind. A life changing revelation can take some time to digest, makes the establishment uncomfortable, however, accurate the revelation may turn out to be.

Well-known social scientists were the subject of FBI surveillance from the 1920s to the 1960s. W.E.B. Dubois and E. Franklin Frazier were investigated for both being affiliated with Communist organizations, and for being active in radical black movements. Delivering a speech to a perceived radical group, or writing a letter to a known member would result in an FBI investigation (Keen, 2004). Talcott Parsons was investigated after an FBI informant said that he was the leader of a communist group while a professor at Harvard. Parsons' case "reveals the power of a few anonymous informants, acting on whatever personal motives, to bring on the full force of an FBI investigation on the basis of ridiculous charges" (Keen, 2004, p. 137). Parsons was eventually interviewed by the FBI, and denied all of the allegations, which were never corroborated in an extensive investigation lasting two years. However, even after being cleared, Parsons was still the subject of FBI surveillance into the 1960s (Keen, 2004).

Other sociologists who were investigated by the FBI included Ernest Burgess, a founder of the Chicago school, William Felding Ogburn, a top quantitative social statistician who was also a member of the Chicago school, Robert and Helen Lynd, who co-wrote *Middletown*, and the noted anthropologist, Margaret Mead. In many of these cases, informants led the FBI to the subjects. Once tipped as to an individual's position or affiliation with a suspect or subversive organizations, an investigation would ensue. In the case of Dubois, FBI agents had contact with people in the State Department, mulling the idea of revoking Dubois' passport. Any travel that Dubois did was heavily monitored (Keen, 2004).

Eventually, after interviewing friends and colleagues about an individual, word would get back to the subject that they were under surveillance. What's odd is that many of the sociologists mentioned, such as Ogburn, Parsons, or Pitirim Sorokin, never publicly discussed their investigations (Keen, 2004). This may have meant that they were not aware of the surveillance, or did not want to mention it for fear of reprisal, especially professionally.

Most troubling to scholars studying this time period is the potential effect such surveillance activities had on scholarship. The leading professional organizations often chose to ignore the pressures that McCarthyism was placing on academic endeavors, choosing to pursue government controlled funding opportunities at the expense of academic freedom (Price, 2004).

> Functionally, the public humiliation rituals of the (McCarthy) hearings advertised the painful consequences of engaging in activism and highlighted the message that scholars were being watched, or better yet, announced that any anthropologists who wished to receive private or governmental grants or employment should avoid anything that could be easily interpreted as a subversive analysis. The costs of such deviance were easily contrasted with the virtue of compliance. (Price, 2004, p. 345)

Jerome Rauch, an anthropologist at Columbia University in the 1940s and 50s, felt the sting of McCarthyism, having been censured by members of his profession after publishing an essay in the *Journal of Negro*

Education (Rauch, 1955). His premise was that funding relationships between social science research projects and governmental agencies "transformed academic research into applied science, and reconfigured social research into ideology" (Price, 2004, p. 346). He reasoned that government agencies, such as the military and intelligence complexes, private funders, like Standard Oil, and private foundations, like the Viking Fund, could directly influence research agendas (Price, 2004). Since Rauch was censured, his line of inquiry was effectively stopped. The phenomenon of abandoned scholarship for fear of reprisal or lack of funding because of ideological reasons is not measurable. After being told he was unemployable by a colleague at Columbia, Rauch quit the anthropology department and enrolled in library school (Price, 2004).

At the height of McCarthyism, college and university librarians watched closely as leading academics were harassed and censured. In 1953, the American Library Association created the Freedom to Read statement. Not many groups were interested in contesting the oppressive censorship and political power of McCarthyism, but librarians felt it necessary to voice some support for those being persecuted (Foerstel, 1991). One of the principles in the statement, number 6, reads:

> It is the responsibility of publishers and librarians, as guardians of the people's freedom to read, to contest encroachments upon that freedom by individuals or groups seeking to impose their own standards or tastes upon the community at large; and by the government whenever it seeks to reduce or deny public access to public information. (ALA, 2007b)

The Freedom to Read statement speaks against any type of oppression concerning ideas, or the investigation of those who seek information about subjects that are not as widely accepted or unpopular in the culture. Whenever a book is banned from a library because of content, it is almost always one group's position against another regarding what is considered an acceptable idea within the community. Often these debates are considered in moral or religious contexts, sometimes in cultural contexts. If the debate considers a legal context, law enforcement agencies wield tremendous power. If the power structure favors one group over another, as is the case with the FBI in its government role, then there is no debate.

A major problem with FBI activity in the Academia was that often an investigator's interest was in ideas that could be interpreted as radical, rather than criminal. As an example, see Professor Rauch's experience. Operating in secrecy with no oversight for many years from the 1920s to the 1960s, FBI agents were able to access information and produce whatever stories they wanted about surveillance targets to justify their investigations. This idea is chilling to librarians. In the case of the Library Awareness program, librarians chose to challenge the power. More recently, a group of librarians chose to challenge the power structure again.

SECTION 215 AND THE GAG RULE

In 2005, a Connecticut library consortium, Library Connection, received a National Security Letter (NSL) from the local FBI office, ordering that the library turn over records about all Internet activity tied to a particular Internet Protocol (IP) address (Carson, 2006; Oder, 2006). IP addresses are embedded in web data packets that carry information throughout the Internet. IP addresses are usually unique to a specific location, and can tie illegal activity to specific computers. Investigators can track what other IP addresses connected to the suspect IP, or what websites the suspect IP address visits. Since so much of the information provided by libraries is available on the web, they have become de facto Internet providers.

National Security Letters are administrative tools used by the FBI to request information to support investigations. They are not reviewed or issued by courts, but can be issued directly by over five-dozen supervisors in the FBI, usually those in charge of field offices (Gellman, 2005; Strickland, 2003). The letters were also protected by the gag rule in section 215 of the Patriot Act.

Any encroachment on the rights of individuals to access information freely and without fear of reprisal is considered by library professionals as limiting civil liberties. The American Library Association (ALA), in its Core Values Statement concerning librarianship, lists confidentiality/privacy as its second out of eleven core values to which library professionals try to adhere. "Protecting user privacy and confidentiality is necessary for intellectual freedom and fundamental to the ethics and practice of librarianship" (ALA, 2007a). Much of the professional literature written by librarians concerning the Patriot Act highlights its potential for limiting civil liberties and blocking access to information (Cohen & Minow, 2006; Martorella, 2006; Phillips, 2005; Pike, 2007).

One of the more troubling aspects of the law for librarians was the secrecy written into the legislative code of the Patriot Act, in the form of a "gag rule" in section 215 (Schulhofer, 2002). Businesses or libraries contacted during terrorism related investigations were prohibited from disclosing any details about an FBI visit, as it could have compromised investigations. In the original Patriot Act, section 215 read as follows:

> (Sec. 215.501.(d)) No person shall disclose to any other person (other than those persons necessary to produce the tangible things under this section) that the Federal Bureau of Investigation has sought or obtained tangible things under this section. (USA PATRIOT Act, 2001)

In a previous part of Sec. 215, the law states:

> Sec. 215.501.(c)(1) Upon an application made pursuant to this section, the judge shall enter an ex parte order as requested, or as modified, approving the release of records if the judge finds that the application meets the requirements of this section.

Examining the language in these two sections, one could surmise that a judge is required to grant a surveillance order if it meets pre-determined criteria, and no one can talk about the order after the surveillance is executed.

The text of Section 215 made it unclear what would happen if someone were to disclose that the FBI had asked for information from a business or library when conducting a terrorism investigation (Albitz, 2005; Jaeger et al., 2003). Many librarians believed this rule concentrated broad power in the hands of the executive branch of the government, and did not adequately protect library users from unwarranted FBI surveillance (Foerstel, 2004; Klinefelter, 2004). While it is often necessary for investigations to remain secret, Section 215, as originally written, diminished the ability to conduct proper judicial and legislative oversight of FBI surveillance activities (Swire, 2005).

In its original form, Section 215 of the Patriot Act conflicted with state privacy laws requiring court orders to allow librarians to disclose information about personal library records (Foerstel, 2004). Confusion arose regarding how librarians should respond to a request for information from the FBI. For example, could managers discuss the order with their directors, or the directors with a legal entity representing the organization? The gag rule suggested that when someone is asked for information, they could not disclose any information concerning the request. The potential for abuse by investigators is large, but could remain undetected, given the way the law is written.

After the National Security Letter was issued to the Connecticut consortium, the librarians were not sure how to proceed. They were unclear about who they could talk to concerning the request. To complicate matters, their names were not properly redacted from FBI memos related to a lawsuit brought by the consortium regarding the constitutionality of the program. Through investigations, reporters from the *Washington Post* (Gellman, 2005) and the *New York Times* (O'Connor, 2006) were able to find out the identities of the librarians. Even after being identified, the librarians were still forced to adhere to the gag rule (Cowan, 2006). After much deliberation, the 'John Does' of the case decided to challenge the constitutionality of the National Security Letters, specifically the gag rule. They refused to release the requested information, and along with the American Civil Liberties Union, the group sued the government in *Doe v. Gonzales* (*Doe v. Gonzales*, 2005; Strickland, 2003).

In its ruling, the District Court in Connecticut ruled that the gag rule supporting National Security Letters was unconstitutional (Raab, 2006), and hindered the free speech of the four John Does. The government dropped its appeal (O'Connor, 2006), and the four were identified as George Christian, Barbara Bailey, Peter Chase, and Janet Nocek. The National Security Letter challenged by the librarians was one of only two to be challenged in four years. Over 120,000 had been served in that time (Raab, 2006). The FBI dropped its case, indicating that they had found the information they were requesting in the letters through other avenues.

As this case progressed, the Patriot Act was undergoing scrutiny by legislators who were struggling with its renewal. Librarians were still concerned about the power the law gave investigators to access private users records. In March 2005, the Patriot Act was amended. While not exactly what librarians were looking for, the Patriot Act amendments concerning NSLs included (Doyle, 2005):

- Lifting the gag rule and allowing recipients of National Security Letters to consult an attorney
- Giving recipients the ability to challenge the letters in court.
- Having the Director of the FBI sign the letter, ostensibly providing an extra layer of administrative review
- Give judges the ability to dismiss frivolous requests.

In some ways, *Doe v. Gonzales* represented another small victory for librarians over big government surveillance.

CONCLUSION

On December 11, 2002, librarians held a teleconference discussing the Patriot Act and its ramifications for patron privacy. Some librarians suggested defiance of the Act, although the moderators indicated forcefully that in no way should librarians impede execution of the law. One idea that kept recurring, however, was that the fewer records libraries kept, the less information the government could obtain as a result of a search (Clymer, 2002). This is counterintuitive to what the government had in mind when the legislation was created and passed. An Internet management system now clears the caches of computers immediately at the public library in Delray Beach mentioned in the introduction, and erases log-on information after 24 hours (Privacy and Security, 2003). At the University of California, Berkeley, library staff destroy Internet sign in sheets every day and erase information from computer servers daily. According to a Berkeley librarian, "We're not being actively anti-government. We're trying to protect people's privacy. We're trying to provide something we have always promised our patrons – that they could use the library in comfort, safety and security" (Privacy and Security, 2003).

Before the explosion of Internet communication, much of what the FBI tried to glean came from informants or colleagues of their targets. New surveillance technology has made it possible to bypass the use of human informants and allow agents to spy on targets directly. Using programs such as Carnivore, a data mining tool used by the FBI to monitor Internet use, especially effective in combating child pornography rings, activity of suspected criminals can now be monitored directly by law enforcement personnel.

One of the keys to information gathering is not just to evaluate the information in the context it has been acquired, but in the context that it has been requested. Looking at the raw data from search engines, or seeing what books or websites a library user has viewed tells an investigator little about the nature of the research unless there is awareness of the research assignment or project itself, especially in university environments. Technology that allows users to customize their home pages, save search histories and maintain a set of active links can provide a great deal of information about an individual and their information seeking habits. Without understanding the context of the research, one can draw a myriad of possibilities about the intent of information seekers in an Internet search environment. Investigators can easily construct nefarious outcomes from the tapestry of keywords searchers might use to find information. More effective communication with library professionals will produce more useable, reliable information to help in preventive investigations in the future.

Ultimately, the potential for abuse while conducting surveillance is where most concerns lie about the Patriot Act. As indicated in this chapter, there is a history of abuse by the FBI when conducting surveillance. This abuse cannot be tolerated again. People's ability to obtain and read materials, gather information, and make judgments about their lives with that information is a necessary part of a democratic society.

There have been many successful terrorist attacks carried out throughout the world since September 11, 2001. It may be that security measures taken by the government have thwarted other attempts to do harm to Americans in our country. However, even in the face of fear, and even in an environment where the toughest security measures are necessary, there are still sacrosanct areas where the government does not tread without resistance. The freedom to read and access ideas without fear of reprisal is one of them.

REFERENCES

ALA. (2007a). *ALA core values statement*. From http://www.ala.org/ala/oif/statementspols/corevaluesstatement/corevalues.htm. Retrieved August 4, 2007.

ALA. (2007b). *The freedom to read statement*. Retrieved August 27, 2007, from http://www.ala.org/ala/oif/statementspols/ftrstatement/freedomreadstatement.htm

ACLU – American Civil Liberties Union. (2002). How the USA Patriot Act expands law enforcement "sneak and peek" warrants. *Privacy and Technology: General Bulletin*, March 7. From http://www.aclu.org/Privacy/Privacy.cfm?ID = 9769&C = 39. Retrieved December 6, 2002.

Adas, J. (2002). New York Congressman Nadler calls USA Patriot Act extreme danger to civil rights. *Washington Report on Middle East Affairs*, *21*(6), 57.

Albitz, B. (2005). Dude, we are my civil rights? *Journal of Academic Librarianship*, *31*, 284–286.

Babson, J., Lebowitz, L. & Viglucci, A. (2001). Terrorist tickets booked in library. *Courier Mail*, News, September 19, p. 7. Retrieved March 11, 2003, from Lexis/Nexis database.

Badash, L. (2000). Science and McCarthyism. *Minerva*, *38*, 53–80.

Badash, L. (2003). From security blanket to security risk: Scientists in the decade after Hiroshima. *History and Technology*, *19*(3), 241–256.

Baker, N. V. (2002). The law: The impact of antiterrorism policies on separation of powers – Assessing John Ashcroft's role. *Presidential Studies Quarterly*, *32*(4), 765–778.

Blackstock, N. (1988). *COINTELPRO: The FBI's secret war on political freedom*. New York: Pathfinder.

Brasol, B. (1969). *The elements of crime: Psycho-social interpretation*. Montclair, NJ: Patterson Smith.

Carson, B. M. (2006). Legally speaking – John Doe and the Patriot Act. *Against the Grain*, *18*(4), 70–72.

Chang, N. (2001). *The USA Patriot Act: What's so patriotic about trampling on the Bill of Rights?* From http://www.ccr-ny.org/v2/reports/docs/USA_PATRIOT_ACT.pdf. Retrieved August 18, 2007.

Clymer, A. (2002). Threats and responses: privacy; librarians get advice on handling government requests for information on readers. *New York Times,* December 12, A30. Retrieved March 11, 2003 from Lexis/Nexis Academic Universe.

Cohen, H., & Minow, M. (2006). Intellectual freedom in libraries: Then and now. *Advances in Librarianship, 30*, 73–101.

Cole, D. (2003). *Enemy aliens, double standards and constitutional freedoms in the war on terror.* New York: The New Press.

Cowan, A. L. (2006). Librarian is still John Doe, despite Patriot Act revision. *The New York Times*, March 21, p. B3.

Davis, J. K. (1992). *Spying on America: The FBI's counterintelligence program.* Westport, CT: Praeger.

Davis, J. K. (1997). *Assault on the Left: The FBI and the sixties antiwar movement.* Westport, CT: Praeger.

Directors, Then and Now. (2007). Federal Bureau of Investigations website. From http://www.fbi.gov/libref/directors/hoover.htm. Retrieved July 26, 2007.

Doe v. Gonzales, 386 F. Supp. 2d 66 (D. Conn. 2005).

Doyle, C. (2002). *The USA Patriot Act: A Sketch.* CRS Report for Congress, Order code RS21203, April 18. Congressional Research Service, Library of Congress: Washington, DC.

Doyle, C. (2005). *USA Patriot Act: Background and comparison of House- and Senate-approved reauthorization and related legislative action.* CRS Report for Congress, Order code RL33027, August 9, 2005. Congressional Research Service, Library of Congress: Washington, D.C.

Fallis, D. S., & Cha, A. E. (2001). Agents following suspects' lengthy electronic trail; Web of connections used to plan attack. *Washington Post*, Final Edition, October 4, p. A24.

Foerstel, H. N. (1991). *Surveillance in the stacks: The FBI's library awareness program.* New York: Greenwood Press.

Foerstel, H. N. (1999). *Freedom of information and the right to know: The origins and applications of the Freedom of Information Act.* Westport, Connecticut: Greenwood Press.

Foerstel, H. N. (2004). *Refuge of a scoundrel: The Patriot Act in libraries.* Westport, CT: Libraries Unlimited.

Fuchs, R. F. (1963). Academic freedom: Its basic philosophy, function, and history. *Law and Contemporary Problems, 28*(3), 431–446.

Gellman, B. (2005). The FBI's secret scrutiny: In hunt for terrorists, bureau examines records of ordinary Americans. *The Washington Post*, November 6, p. A01.

Gentry, C. (1991). *J. Edgar Hoover: The man and the secrets.* New York: Norton.

Hoffman, B. (1998). *Inside terrorism.* New York: Columbia University Press.

Jaeger, P. T., Bertot, J. C., & McClure, C. R. (2003). The impact of the USA Patriot Act on collection and analysis of personal information under the foreign intelligence surveillance act. *Government Information Quarterly, 20*, 295–314.

Jaeger, P. T., McClure, C. R., Bertot, J. C., & Snead, J. T. (2004). The USA Patriot Act, the foreign intelligence surveillance act, and information policy research in libraries: Issues, impacts and questions for libraries and researchers. *The Library Quarterly, 74*, 99–121.

Jerome, F. (2003). *The Einstein file: J. Edgar Hoover's secret war against the world's most famous scientist.* New York: St. Martin's Press.

Johnson, L. K. (1988). *A season of inquiry, congress and intelligence.* Chicago: Dorsey Press.

Johnson, L. K. (2004). Congressional supervision of America's secret agencies: The experience and legacy of the Church committee. *Public Administration Review*, *64*(1), 3–14.

Keen, M. F. (2004). *Stalking sociologists: J. Edgar Hoover's FBI surveillance of American sociology*. New Brunswick: Transaction Publishers.

Kessler, R. (2003). *The bureau: The secret history of the FBI*. New York: St. Martin's Press.

Klinefelter, A. (2004). The role of librarians in challenges to the USA Patriot Act. *North Carolina Journal of Law and Technology*, *5*(2), 219–226.

Kniffel, L. (2002). USA Patriot Act raises red flag over liberties. *American Libraries*, *33*(1), 20.

Lesser, I. O. (1999). *Countering the new terrorism*. Santa Monica, CA: Rand Corporation.

Levin, B. (2007). Trials for terrorists. *Journal of Contemporary Criminal Justice*, *23*, 195–218.

Lichtman, R. M. (2004). Louis Budenz, the FBI, and the "list of 400 concealed communists": An extended tale of McCarthy-era informing. *American Communist History*, *3*, 25–54.

Martorella, G. (2006). Libraries in the aftermath of 9/11. *The Reference Librarian*, *94*, 109–137.

Mazzetti, M., & Williams, M. (2007). In tribunal statement, confessed plotter of September 11 burnishes image as a soldier. *New York Times*, March 16, p. A15.

McFadden, R. (1987). F.B.I. in New York asks librarians' aid in reporting on spies. *New York Times*, September 18, p. A1.

Minow, M. (2002). The USA Patriot Act. *Library Journal*, *127*(16), 52.

O'Connor, A. (2006). Librarians win as U.S. Relents on secrecy law. *The New York Times*, April 13, p. B1.

Oder, N. (2006). Gag order on "John Doe" lifted. *Library Journal*, *131*(8), 22.

Osher, S. A. (2002). Privacy, computers and the Patriot Act: The fourth amendment isn't dead, but no one will insure it. *Florida Law Review*, *54*, 521–542.

Phillips, H. (2005). Libraries and national security law: An examination of the USA Patriot Act. *Progressive Libraries*, *25*, 28–42.

Pike, G. H. (2007). The Patriot Act illuminated. *Information Today*, *24*(5), 17–18.

Powers, R. G. (1987). *Secrecy and power: The life of J. Edgar Hoover*. New York: Free Press.

Powers, R. G. (2004). *Broken: The troubled past and uncertain future of the FBI*. New York: The Free Press.

Pressley, S., & Blum, J. (2001). Hijackers may have accessed computers at public libraries; Authorities investigating possible Internet communications. *The Washington Post*, September 17, p. A4.

Price, D. H. (2004). *Threatening anthropology: McCarthyism and the FBI's surveillance of activist anthropologists*. Durham: Duke University Press.

Privacy and Security. (2003, February 3). Editorial. *San Francisco Chronicle*, p. D4. Retrieved March 13, from Lexis/Nexis Academic Universe.

Puzzanghera, J. (2002). Terrorist's Internet use may betray them; FBI is following up on tracks they left behind. *Milwaukee Journal Sentinel*, News, September 21, p. 14A. Retrieved March 10, 2003, from Lexis/Nexis Academic Universe.

Raab, C. (2006). Fighting terrorism in an electronic age: does the Patriot Act unduly compromise our civil liberties. *Duke Law and Technology Review*, *3*. From http://www.law.duke.edu/journals/dltr/articles/2006dltr0003.html. Retrieved August 9, 2007.

Rabban, D. M. (1990). A functional analysis of "Individual" and "Institutional" academic freedom under the first amendment. *Law and Contemporary Problems*, *53*(3), 227–301.

Rackow, S. H. (2002). How the USA Patriot Act will permit governmental infringement upon the privacy of Americans in the name of "intelligence" investigations. *University of Pennsylvania Law Review*, *150*, 1651–1696.

Rauch, J. (1955). Area institute programs an African studies. *The Journal of Negro Education, 24*(Fall), 409–425.

Reynolds, T. (2002). Collateral damage: Information, intellectual freedom, and the war on terrorism. *Alki, 18*, 28–30.

Rodriguez, C. D. (2001). *Preserving the balance between security and liberty,* October 17. FDCH Press Release.

Rosenbaum, D. (2001). A nation challenged: questions of confidentiality; Competing principles leave some professionals debating responsibility to government. *The New York Times,* November 23, p. 7B. Retrieved March 22, 2007 from Lexis/Nexis Academic Universe.

Schulhofer, S. J. (2002). *The enemy within: Intelligence gathering, law enforcement, and civil liberties in the wake of September 11.* New York: The Century Foundation Press.

Solove, D. J. (2004). Reconstructing electronic surveillance law. *George Washington Law Review, 72*, 1701–1747.

Stielow, F. (1993). The FBI and library spying: A World War II precedent. *American Libraries, 24*(8), 709–711.

Strickland, L. S. (2002). Information and the war against terror, III: New Information-related laws and the impact on civil liberties. *Bulletin of the American Society for Information Science and Technology, 28*(3), 23–27.

Strickland, L. S. (2003). Civil liberties vs. intelligence collection: Foreign intelligence surveillance act speaks in court. *Government Information Quarterly, 20*, 1–12.

Swire, P. P. (2005). Reply: Section 214. Pen register and trap and trace authority under FISA; section 215. Access to business recods under FISA. In: S. A. Baker & J. Kavanagh (Eds), *Patriot debates: Experts debate the USA Patriot Act.* Chicago: ABA Publishing.

Taylor, H. (2001). Overwhelming public support for increasing surveillance powers. *The Harris Poll, #49*, October 3, 2001. From www.harrisinteractive.com. Retrieved November 15, 2002.

Wallace, M. (2003). *The American axis: Henry Ford, Charles Lindbergh, and the rise of the Third Reich.* New York: St. Martin's Press.

Warwick, S. (2005). Will the academy survive 9/11? Scholarship, security and United States government policy. *Government Information Quarterly, 22*, 573–593.

Woolston, H. (1942). Free speech in wartime. *American Sociological Review, 7*(2), 185–193.

Zanini, M. (1999). Middle Eastern terrorism and netwar. *Studies in Conflict & Terrorism, 2*(3), 247–256.

“WHAT IF SHE’S FROM THE FBI?” THE EFFECTS OF COVERT FORMS OF SOCIAL CONTROL ON SOCIAL MOVEMENTS☆

David Cunningham and John Noakes

ABSTRACT

This chapter examines the effects of covert forms of social control on social movement participants. Current social science literature addresses the effect of surveillance on social movement organizations, but stops short of exploring the experience of surveillance for political activists. We begin by reviewing how state social control has been incorporated into paradigmatic social movement models. Drawing on examples from the FBI's counterintelligence programs and the growing literature emphasizing the emotional components of social movement mobilization processes, we then demonstrate the range of direct and indirect costs exerted by social control agents on both organizational and individual targets.

☆Authors are listed alphabetically; both shared equally in the writing of this chapter.

Surveillance and Governance: Crime Control and Beyond
Sociology of Crime, Law and Deviance, Volume 10, 175–197

ISSN: 1521-6136/doi:10.1016/S1521-6136(07)00208-4

Reflecting on her time on the Hollywood Blacklist, Mary Chesney Lardner, actress and wife of Ring Lardner, Jr., one of Hollywood Ten, described how the persistent threat of infiltration and surveillance affected her ability to make friends and build trusting relationships. "The terrible thing about the whole business," she told historians Bud and Ruth Schultz (2001, pp. 115–116), "is what it does to people." In 1962, for example, Lardner, having received a lead role in a Broadway production, began to become friendly with her understudy.

> We were in a dressing room in Boston, and I suddenly felt, "What if she's from the FBI?" I had nothing to conceal whatsoever ... yet I had this sense that maybe she was put there .... You're worried or you're fearful because there are things that have to be protected. Children have to be protected, your livelihood has to be protected. I went from being an open, free, healthy outgoing person to suddenly having these walls around me. I think it does that to you. (Schultz & Schultz, 2001, pp. 115–116)

While the *fact* of FBI surveillance and infiltration is now well established, Lardner's story directs attention to the *experience* of being targeted.

Among other things, Lardner's testimony suggests that surveillance *is* something to worry about, even, as the old bromide goes, if you have done nothing wrong. In his now classic 1974 *American Journal of Sociology* article, "Thoughts on a Neglected Category of Social Movement Participant: The Agent Provocateur and the Informant," Gary Marx (1974, p. 408), argued that "undercover agents can seriously distort the life of a social movement; they can serve as a mechanism of containment, prolongation, alteration, or repression." Since then scholars of social movements have recognized the role of repression and social control in the dynamics of social movements. Yet, with some notable exceptions (see Marx, 1974, 1979, 1988; Cunningham, 2003, 2004; Davenport, 2005a; Noakes, 1998, 2000, 2003), political surveillance, infiltration, counter-intelligence, and the work of *agents provocateurs*, remain a neglected category of sociological research.

When Marx first broached the topic he suggested several reasons for sociologists' inattentiveness, including: the difficulty of observing secret activities; skepticism about the reports of either former undercover agents or their alleged targets; a hesitancy on the part of researchers to attribute any of the dynamism of protest activity to social control agents; and the inability of sociologists to see the complex world of social movements from the comfortable confines of their offices. In the years since 1974 many of these barriers have become less problematic. Most notably, surveillance and counterintelligence activities have become less secret – at least as historical phenomena – because of the work of journalists, biographers, historians, and other scholars who have collected a voluminous evidentiary record of the extensive and intensive intelligence and counterintelligence operations targeting social movement organizations and political activists in the United States (see Churchill & VanderWall, 1990; O'Reilly, 1989; Gelbspan, 1991; Schultz & Schultz, 1989, 2001; Friedly & Gallen, 1993; Carson & Gallen, 1991; Wiener, 1999). As a result, social movement scholars can access considerable information on undercover operations without leaving their office and, while a healthy skepticism is still a necessary scholarly tool, there is little doubt that surveillance and infiltration of social movement organizations in the United States have had a considerable impact on the movements and their participants.

In this chapter we examine the effects of covert forms of social control on social movement participants. The current sociological literature addresses the effect of surveillance on social movement organizations, but stops short of exploring the experience of surveillance for political activists. One earlier effort to develop a typology of the functions of covert forms of state social control (Cunningham, 2003), for example, details a set of political opportunity costs imposed on social movement organizations targeted for surveillance and counterintelligence activities, including the damage to its public image, internal dynamics, access to organizational resources, relationship with other groups, and capacity to protest. While this catalog also recognizes the costs incurred by individuals, it focuses only on the straightforward ways in which previous participants become deterred from joining in protest activities. Here, we draw on the growing literature emphasizing the emotional components of social movement mobilization processes, to consider the emotional costs borne by the targets of covert forms of social control and their impact on social movement mobilization and persistence.

COVERT SOCIAL CONTROL AND SOCIAL MOVEMENT THEORY

In retrospect, Marx' article remains notable for two reasons: it shed light on the dynamics of repression, while also more broadly helping to define a significant shift in social movement theory. Throughout the 1950s and 1960s, social movements were frequently seen as the product of grievances generated by a structurally induced sense of rising expectations, status inconsistency, or relative deprivation (Lenski, 1954; McAdam, 1982; Smelser, 1962). Importantly, any resulting contentious activity was viewed as a means to a fundamentally psychological end; protestors were not seen as acting to right a perceived injustice, but rather to therapeutically alleviate the "strain" created by structural changes (McAdam, 1982). In this sense, social movements were a manifestation of "irrational" collective behavior, in sharp contrast with the "rational" weighing of costs and benefits that analysts suggested defined institutional political venues.

Marx' treatment of social movement actors was decidedly different. In his characterization of informants and provocateurs as operating strategically within social movement organizations engaged in the pursuit of instrumental goals, he anticipated much of the thinking that was to define the political process approach that emerged as paradigmatic in the 1980s and 1990s. Political process models emphasize the broad environments within which social movements operate, focusing on how structural shifts at times create opportunities for aggrieved actors to make change as well as the material, social, and cultural resources that these actors mobilize strategically to produce contentious political action (McAdam, 1982; Tarrow, 1998; Tilly, 1978). Within this framework, movement actors were increasingly viewed as operating instrumentally, engaging in contention to pursue interests ignored or marginalized within institutional political venues.

An important work by Bert Klandermans and Dirk Oegema (1987) clearly extended this instrumental logic of social movement activity to organizations' attempts to recruit participants. Looking at a particular campaign by Dutch peace activists to mobilize attendees for a public demonstration, Klandermans and Oegema found that, while 74 percent of the local population agreed with the goals of the demonstration, only 4 percent ultimately attended the event. One of the key factors explaining why the vast majority of sympathizers did not attend was their perception that the event engendered costs and/or risks. These disincentives were sometimes specific to the individuals surveyed (such as constraints imposed by competing commitments, illness, or procuring child care), but more frequently were tied to individuals' perceptions of the event itself. Those who felt that there was any likelihood of violence, discomfort, or even passive policing activity would be much less likely to attend, regardless of their views of the cause. In other words, perceived costs (including those tied to overt or covert policing, as well as the likelihood of a hostile oppositional presence at the event) had a demonstrable chilling effect on participation in a mass peace demonstration.

This study provided convincing evidence that members – as well as prospective members and supporters – of activist groups made decisions in the same way we assume participants in the mainstream political process did: by rationally weighing associated costs and benefits. Forms of state repression, therefore, can be understood as having an impact on social movements through their effect on the cost-benefit calculus undergone by activists or potential participants.

Subsequent research on state-dissident interactions has provided insight into certain aspects of such effects. Drawing on studies focused on how the interdependence of actors influences engagement in collective action (e.g. Granovetter, 1978), Gerald Marwell and Pamela Oliver have demonstrated how the structure of activist networks impacts movements' ability to mobilize in the face of costs such as those borne by surveillance and state repression (Oliver, Marwell, & Teixeira, 1985; Marwell & Oliver, 1993). One implication is that, when repression is successful at dissuading even a small number of central individuals from expressing or lawfully acting on their political beliefs, it can create very large effects on organizations' ability to mobilize sympathizers. The process plays out as one individual's choice not to attend an event or provide material support makes it considerably more likely that those to whom he or she is tied also will not contribute, which in turn reduces the likelihood that their contacts stay home, and so on, in what Oberschall (1994) refers to as a "multiplier effect." Therefore, the immediate impact of state action on individuals' decision to participate generates an indirect and more global chilling effect as a potentially large pool of contributors quickly erodes to a smaller, more marginalized grouping of participants.

In contrast, particular mobilization sites lend themselves to an alternate process, in which the costs incurred by repression are perceived as a solidarity-inducing benefit by a tight-knit activist core. McAdam (1988) and Loveman (1998) have both demonstrated how such "embedded social networks" within movements can provide settings for activists' willingness to engage in "high-risk" activism, enduring jailings and physical harm for the sake of the cause. Zwerman and Steinhoff (2005) have further highlighted the identity processes through which "micro-cohorts" of committed activists escalate their participation in the face of repression, even as the movement's mass following erodes.

Others have focused on the flip side of the interaction, attempting to explain how authorities allocate repression. Marx (1979), Carley (1997), and Cunningham (2003) have created typologies of repressive actions engaged in by state agencies. Cunningham (2004) and Davenport (2005a) have also dealt with the patterning of such allocations, finding that state action against perceived dissidents is related to the organizational structure of policing agencies as well as the characteristics of neighborhoods where potential targets reside. Earl (2003) has reviewed a number of studies that focus on the ways in which broader structural elements matter as well, though she finds that the literature has tended to elide various dimensions of contention, including repressing agents' relationship to political elites, the proactive versus reactive character of repressive action, and the repertoire of repressive strategies and tactics employed by state agencies. Such conceptual gaps indicate that we have far to go to understand fully how the "costs" imposed by state repression operate, or how they ultimately impact activist targets.

In post-September 11th America these questions once again loom large. In 2002, for example, Attorney General John Ashcroft amended the rules governing FBI domestic spying, allowing data-mining and the searching of public databases and the Internet, practices that had been forbidden since 1976 (US Department of Justice, 2002). Moreover, reports from various demonstrations since the Seattle WTO protests indicate that surveillance and infiltration are once again common tactics employed by local and state police against social movement organizations (SMOs) (Boghosian, 2004). Such accounts hark back to the Cold War Era, when the FBI engaged in decades-long surveillance operations against perceived threats to national security, and established a set of formal counterintelligence programs designed to actively disrupt and neutralize these threats.

These counterintelligence programs provide an unusual window into the range of impacts of state repression. Because they entail a specific intent to harm activists or movement organizations, their initiation requires a strategic deployment of state resources toward some defined end. Below, we review a typology of such "ends" sought by the FBI in their counterintelligence programs, and argue that each of these impacts are also produced – either directly or indirectly – when authorities engage in ostensibly "passive" surveillance activities. As such, we provide a catalog of costs borne by organizational targets of repression, which in turn provide a basis to interrogate how the presence of surveillance impacts the mobilization process and affects social movement dynamics.

In a general sense, surveillance falls under the rubric of *intelligence*, i.e. activities designed to gather information about a target or suspect. In some cases, intelligence actions are postliminary, or oriented to events that have already occurred, with the intention of procuring confessions or other evidence of wrongdoing. But intelligence can also be anticipatory, intended to identify ongoing or likely future activity. *Counterintelligence*, in contrast, is intended to actively hinder or prevent a target's ability to carry out planned actions, or to encourage acts of wrongdoing. Its agents sometimes pursue the former goal in a narrow sense, focusing on a particular unwanted action, but also can seek to exploit any available opportunity to neutralize a target's ability to act generally. Both of these classes of action can fall under the broader umbrella of *state social control*, by which we mean any activity initiated by state authorities that seeks to raise the cost of action for predetermined targets.

Counterintelligence is intended to have an active effect on its targets, even as it frequently employs covert means. Intelligence gathering, on the other hand, is generally portrayed by the FBI and other policing agencies as a passive act. Marx (1974), however, makes the point that many infiltration techniques, especially the placement of informants, by definition have some impact on their settings:

> His [or her] presence can make a movement seem stronger than what it is. If nothing else, he may provoke the kind of information he is looking for. He may pass on to authorities false, exaggerated, or misinterpreted information. This may move through several police agencies and bureaucratic levels, and can lead to police actions with self-fulfilling effects. Even an informer who does not concoct information may 'provoke' violence if his identity becomes known to the infiltrated group. They may attack him, and this may lead to counterattacks from authorities. (p. 405)

Such effects are only heightened in contexts where targeted groups are aware of the possibility of infiltration or other forms of surveillance. As the executive director of the National Lawyers Guild, a longtime target of FBI intelligence, puts it: "we work with the assumption that everything we do is being monitored by the government. Unfortunately, we've become used to surveillance" (Moynihan, 2007).

Taking this point seriously has a significant analytic implication: in the absence of total passivity (both actual and perceived – a condition that may in fact exist only as an ideal), it makes little sense to draw a sharp distinction between intelligence and counterintelligence activities when assessing the impacts of state social control. Certainly, state agents themselves have recognized the fact that intelligence tactics frequently impose costs on their targets. As former FBI special agent Jack Ryan has reported:

> (W)hen the FBI investigated somebody, it could be very intimidating. This used to be our number one technique. We worked a lot of cases that we never intended to prosecute because we just weren't able to. But we would investigate the devil out of them. And it just hurt the groups. It dried up their sources of funding; people backed off." (quoted in Schultz & Schultz, 2001, p. 377)

Similarly, an oft-cited 1970 FBI memo argued that interviews by agents and other intelligence-gathering techniques had the deeper function of "enhanc[ing] the paranoia endemic in such circles and ... to get the point across there is an FBI agent behind every mailbox."

COINTELPRO AND THE IMPACT OF STATE REPRESSION

Given this de facto overlap, we begin by examining one well-known set of state counterintelligence programs, the FBI's COINTELPROs. The FBI operated these programs with the expressed goal to "expose, disrupt, misdirect, discredit, or otherwise neutralize" individuals or groups targeted for their supposed subversive potential (Cunningham, 2004, p. 6). Toward those ends, agents in all Bureau field offices were ordered to propose counterintelligence actions strategically designed to meet the program's goal by producing an outcome that would impose particular costs on their targets. As such, internal FBI memos document the proposals that lay behind each of the literally thousands of counterintelligence actions carried out during the 15-year life of the program. Important for our purposes, these memos include clear statements of the intended function of each action, and subsequent progress reports detail the "tangible results" realized after actions have fully played out.

The first COINTEL program was initiated against the Communist Party-USA (CP-USA) in 1956. Throughout much of the 20th century, membership in the CP-USA had been an offense prosecutable under the Smith Act, which forbade individuals from advocating the overthrow of the government by force or organizing or belonging to a group that had such a goal. During the mid-1950s, the US Supreme Court ruled the Smith Act unconstitutional, and as an alternative the FBI established the framework for COINTELPRO, which would use covert political repression to do the work that had previously been handled through the courts.

But consistent with our earlier point about the blurred distinction between such actions and more conventional surveillance activities, this program against the CP-USA operated in concert with the FBI's existing intelligence work. Bureau agents used information they obtained through the covert surveillance of CP-USA members and their associates to spread misinformation and falsehoods about individual members' motives and the group's methods and goals. Agents also developed a large number of informants, frequently utilizing them as *agents provocateurs*, or operatives working to systematically break down trust and cohesion within their target groups (Churchill & VanderWall, 1990; Cunningham, 2004: Marx, 1974). These activities were confined to the Bureau itself, but initiated with the consent of the Eisenhower administration.

Through the early 1960s, the COINTELPRO against the CP expanded in scope; in 1961 a second program was initiated against the Socialist Workers Party (SWP), followed by COINTELPROs against "White Hate Groups" in 1964, "Black Nationalist/Hate Groups" in 1967, and the New Left in 1968. The White Hate Groups program, directed mostly against a range of KKK-affiliated organizations, broadened the FBI's reach in significant ways; it set a precedent through which the Bureau began to deal with purely domestic threats to national security, rather than only those which were posited to be somehow linked to hostile foreign powers and thus, by definition, involved in subversive activities.

In Table 1, we present six central functions of FBI counterintelligence operations, as identified by Cunningham (2003), along with illustrative examples taken from the FBI's counterintelligence actions against the SWP.[1] The SWP operation, exposed as part of the proceedings of a lawsuit first filed against the FBI in 1973, provides unusually deep insight into the dynamics of covert social control. Pretrial motions led to the release of hundreds of thousands of pages of state documents to the plaintiffs, which were presented alongside detailed accounts by SWP targets of the impact of those actions on their political activities and lives more generally.

Each of the examples in Table 1 demonstrates how the FBI's counterintelligence activities were intentionally designed to produce particular types of costs for their targets. The SWP's lawsuit against the FBI provided a rare window into both sides of that process: the motives of FBI agents, codified in their assessments of the "tangible results" realized by each counterintelligence act, and the actual and perceived impact of those acts on SWP members themselves. Courtroom claims made by SWP adherents and their representatives provided first-hand evidence of the latter, with presiding judge Thomas Griesa agreeing that "there can hardly be a more compelling case" for demonstration of harm. The Bureau, argued Griesa, "embarked on a series of actions with the express purpose of harming the SWP by causing internal mistrust and strife, by weakening its alliances with other groups, by hampering its scheduled activities and by other means." Moreover, he concluded, "[t]here was no legal authority or justification for such operations" (quoted in Jayko, 1988, p. 112).

Table 1. Intended Functions of FBI COINTELPRO Actions, with Illustrative Examples from COINTELPRO-SWP[a].

1. *Create a negative public image*
   In 1963, agents in the FBI's Chicago field office exploited the public visibility of Clifton DeBerry, a SWP member and candidate for public office in New York. After learning that DeBerry had lapsed in his child support payments to his ex-wife, agents obtained his speaking schedule from an informant placed in the SWP. They then worked with the Cook County (IL) Department of Welfare to secure an arrest warrant and ensure that DeBerry was detained immediately before he was to give a speech to the Militant Labor Forum. The short-term impact of that event was limited when DeBerry was immediately able to post bond and return for his scheduled speech, but agents then proceeded to exploit DeBerry's later conviction for non-support, providing information about his personal problems to cooperative media contacts to generate negative publicity about DeBerry and the SWP generally.
2. *Break down internal organization*
   To exacerbate simmering racial tensions within the SWP in 1969 stemming from black SWP leader Paul Boutelle's claims that the organization had a "patronizing" attitude toward African-Americans, FBI agents in the New York field office sent an inflammatory anonymous letter to Boutelle. Purporting to be from "Your nasty friends [in the SWP]" the letter suggested that Boutelle and his "fellow party monkeys" leave the SWP, so that the (white) membership could work on "the job Trotsky had in mind for us." Due to this letter and a follow-up anonymous missive suggesting that his exit from the SWP would not result in any significant "brain-drain," Boutelle angrily discussed the attacks at subsequent meetings, intensifying conflicts within the membership.
3. *Create dissension between groups*
   In 1962, the SWP became a central supporter of a group of African-Americans arrested after an incident of racial violence in Monroe, NC. To disrupt the group's participation in the coalition organization the "Committee to Aid the Monroe Defendants" (CAMD), FBI agents sent anonymous letters and phone calls to an SWP member, to promote the (false) accusation that an unnamed African-American partner organization was misusing resources intended for CAMD.
4. *Restrict access to organizational resources*
   The FBI again targeted CAMD as a vehicle to stigmatize the SWP and reduce support for the Monroe defendants. Agents from the New York

Table 1. (*Continued*)

field office sent an anonymous letter to the NAACP criticizing that organization's support of CAMD. The criticism was due to the SWP's role, which the FBI painted as "domination by ... the Trotskyist branch of the communist movement." Likely as a result, the NAACP ceased providing CAMD with financial support.

5. *Restrict organizational capacity to protest*

To disrupt the planning of a 1969 anti-war parade jointly sponsored by the SWP, the Young Socialists Alliance, and the Student Mobilization Committee to End the War in Vietnam, FBI agents in the New York office constructed and distributed a leaflet purportedly composed by radical leftists at Columbia University. The leaflet criticized the organizing coalition for their general lack of militancy, including their unwillingness to attack the police and incur the necessary "battle wounds." As the organizers had been stressing the importance of a peaceful event, such accusations generated "difficulties in managing the march."

6. *Hinder the ability of targeted individuals to participate in group activities*

To harm the efforts of SWP member John Franklin, a candidate for Manhattan Borough President in 1961, agents in the FBI's New York field office furnished information about Franklin's past criminal record to the *New York Daily News*. Due to the resulting negative news story, which the FBI proceeded to circulate anonymously, Franklin discontinued his involvement with the SWP.

[a]Detailed descriptions of each of the FBI actions discussed below can be found in the COINTELPRO-SWP files, available on microfilm from Scholarly Resources International. Summaries of each action are also included within the text of Judge Griesa's decision in the *Socialist Workers Party v. Attorney General* lawsuit discussed above, which is reprinted in full in Jayko (1988, pp. 19–134).

Griesa's finding that at least 17 COINTELPRO acts resulted in clearly demonstrable harm led to an award of $42,500 in remunerative damages. However, an equally telling aspect of the trial was the ruling that the FBI's intelligence actions during the same period – electronic surveillance, informant presence, "surreptitious entries," and the groundless placement of SWP members on the Bureau's "Security Index" – also had deleterious effects on the SWP. A total of $221,500 in damages was awarded to the SWP for harms associated with these intelligence actions.

This judgment, and the associated testimony of SWP members clearly documenting sweeping and durable harms endured as the target of surveillance and other intelligence activities, is instructive. Consistent with many analyses of social movements, the typology presented in Table 1 focuses primarily on the direct, measurable costs imposed at the structural and organizational level. However, it seems clear that any full analysis of the effects of covert social control must also account for the psychological and relational costs of repression. In the next section, we extend the existing typology by assessing as well the indirect costs imposed by covert forms of social control on the personal emotions of social movement participants and the collective emotions of social movement organizations.

EMOTIONS, REPRESSION, AND SOCIAL MOVEMENTS

The dominance of structural models of social movement dynamics during the last quarter or so of the 20th Century has been well documented and well critiqued (see Goodwin & Jasper, 1999). Among the fruits of this critique is a renewed attention to the role of emotions in contentious politics. In their rush to distance themselves from the emphasis on the irrationality of political protesters in dominant social movement models of the 1950s and 1960s, social movement theorists "somehow ... managed to ignore the swirl of passions all around them in political life" (Goodwin & Jasper, 1999). Research on a range of related issues, including the role that various emotions play in recruitment (Gould, 2001; Nepstad & Smith, 2001) and the redefinition of collective identity (Britt & Heise, 2000; Stein, 2001; Taylor, 1996, 2000), the emotional costs of movement participation (Adams, 2003; Kemper, 2001; Taylor, 2006), and how the successful reframing of political discourse can prime broader populations for collective action (Kane, 2001; Polletta, 2002; Young, 2001), has revealed a previously "seen but not considered" aspect of social movement dynamics (Goodwin & Jasper, 1999).

Within this literature, however, little is written about the relationship between emotions and repression or social control. The role of emotions in the decline of movements tends to be discussed in terms of reciprocal emotions among movement members, such as how the love that grows during the intense periods of movement mobilization can lead pairs of activists to withdraw from the public sphere in search of private bliss (Goodwin, 1997) or how jealousy between aspiring leaders can pull organizations apart. Gould (2002) also considers the ways in which broader morals and norms provide a means by which demonized activists develop an "emotional common sense" that shapes their subsequent protest capacity. There is little discussion, however, of how social control agents attempt to manipulate emotions within groups as a means of hindering organizational actions. In one of the few pieces of research that deals, if only obliquely, with the emotional dynamics of repression, Goodwin and Pfaff (2001) argue that for high-risk movements to mobilize, the fear of social control and other threats must be mitigated and managed. Despite this work, sociologists have still largely "seen but not considered" the complex indirect and longer-range emotional impacts of explicit social control efforts.

To examine the relationship between emotion and covert forms of social control, we draw on the testimonies of American political activists collected in two volumes by historians Bud and Ruth Schultz. Given space constraints, we focus on three emotions: fear, trust, and paranoia. We contest the benign view of the emotional impact of covert forms of social control, arguing instead that the targets of surveillance, infiltration, and counterintelligence activities often suffer a significant emotional toll that shapes social movement dynamics.

Our interest in the emotional impact of surveillance on social movement members is piqued in part by the research on the effects of covert observation in other contexts, in which the threat or realization of surveillance "exerts a powerful influence over behavior, beliefs, and feelings" (White & Zimbardo, 1980, p. 59). During social psychological experiments, for example, participants who feared that their beliefs would be revealed to third parties (whether police officials or simply people who disagreed with them) were less likely to express contentious opinions (White & Zimbardo, 1980; Nail, Van Leeuwen, & Powell, 1996). There is also considerable evidence in studies of workplace monitoring to support the argument that the threat or presence of surveillance influences people's behavior and feelings. Social psychological studies of workplace surveillance suggest that when a worker is under surveillance or believes they are under surveillance they follow rules more carefully and are more subservient (Thibaut, Friedland, & Walker, 1974; Hashimoto, 1979; Aiello & Kolb, 1995; Nabeker & Tatum, 1993; Tang & Baumeister, 1984; Robins & Webster, 1989), experience greater stress (Garson, 1988; Carayon, 1993), feel a loss of personal control (Stanton & Barnes-Farrell, 1996; Brigham & Corbett, 1997), are more likely to be aggressive toward a supervisor (Greenberg & Barling, 1999), have a decreased sense of procedural justice (Alge, 2001), and feel degraded as if their personal privacy has been invaded

(Alge, 2001; Gilliom, 2005). These emotions are exacerbated when the subject feels inappropriate information has been collected, that personal information will be shared with others without their permission, and when revelation of the information can lead to negative consequences (Tolchinsky, McCuddy, Adams, Ganster, & Fromkin, 1981).

Studies of the effects of closed circuit television surveillance support these findings. Residents of both low-income housing projects and high-income gated communities reported feelings of fear, anxiety, and discomfort as a result of being constantly monitored (Monahan, 2006). This finding is supported by Gilliom's (2005) study of welfare recipients, who resisted the "fearsome presence" of persistent "record keeping, observation, and verification." Similarly, high-income residents of a gated community reported that they felt "somewhat protected, somewhat violated" by the camera's constant surveillance and bemoaned the artificial boundaries between communities and the isolation borne of the residents, fears of being constructed as non-conformists. Citizens interviewed by Yesil (2006) indicated that the uncertainty of surveillance in public spaces caused stress as potential subjects worried about how the system would respond to things they have no control over, such as being unintentionally proximate to someone suspicious.

Based on these findings of the emotional effects of surveillance in relatively benign situations, it is reasonable to assume that covert forms of social control affects the feelings and emotions of social movement participants who are targeted individually or who are members of targeted SMOs. There is evidence of such effects in our above discussion of how covert forms of social control can create dissension between groups. The FBI would "plant provocateurs to create conflict," Cleveland Sellers remembers; "[t]hey know how to stir up difficulties in personal relationships, how to play people off against each other" (quoted in Schultz & Schultz, 1989, pp. 256–257). Sellers' comments, however, point us beyond the structural issue of how internal dissension can be a barrier to movement persistence. Dissension created by "stir[ring] up personal relationships" can spark emotions such as dejection, fear, paranoia, depression, and anger, which can have their own independent effects on movement maintenance.

Consistent with Goodwin and Pfaff's (2001) emphasis on the strategic importance of overcoming fear within activist circles, the testimonies in Schultz and Schultz (1989, 2001) make clear the centrality of fear in response to covert forms of social control. Persistent surveillance created fear and anxiety among activists and potential activists, either for their own well-being or that of those near to them. As African-American postal worker Arthur Drayton, who lost his job in the 1940s because of alleged loyalty issues, put it, federal agents "did frighten the hell of me ... [t]hey were hounding me so" (quoted in Schultz & Schultz, 1989, p. 173). Drayton's fear was compounded by the fact those around him did not believe his complaints of government harassment, so its persistence "drove [him] to despair" (quoted in Schultz & Schultz, 1989, p. 173). Lesbian activists in the women's movement in Kentucky during the 1970s had to confront the fear of being outed to their parents or employers by FBI investigators doing background checks on activists (Schultz & Schultz, 1989). Even those who claimed to have adjusted to the presence of undercover agents themselves, such as Paul Robeson, Jr., son of long-time FBI target Paul Robeson, expressed fear for the well-being of his children (Schultz & Schultz, 2001).

The fear inspired by persistent covert activities also extended to potential recruits. Before going to Berkeley to begin college, Jackie Goldberg, a future member of the Berkeley Free Speech Movement, remembers she "was terrified of getting duped by Communists" and of "signing my name to anything or being associated with anyone" (quoted in Schultz & Schultz, 2001, p. 295). Though Goldberg overcame her fears, not all did. Participants in Vietnam Veterans Against the War (VVAW) began dropping out, Scott Camil, the VVAW's field coordinator in Florida, remembers, when government agents began interviewing their employers and those of their family members:

> the government has a lot of power and a lot of money. They can be very disruptive to your life. They got people thrown out of school. They went to employers of ex-GIs in the VVAW and got them fired for being "communists." I started getting phone calls: "Hey, man, I can't come to meetings anymore. The FBI went to where my mom works. She almost lost her job." (quoted in Schultz & Schultz, 1989, p. 323)

John Lewis remembers that federal government agents' investigations of organizing activities in a largely unsuccessful civil rights movement campaign in Albany, GA in the early 1960s "incurred great fear and misgiving on the part of those we were trying to organize" (quoted in Schultz & Schultz, 2001, p. 169).

James Jasper (1998, p. 402) has argued that "[t]rust and respect are examples of affects with enormous impact on political action." Generalized trust in the government, for example, lowers the incentive for mobilization. Trust among members and between members and potential recruits helps build collective identity, is helpful in the recruitment of members, and fosters the solidarity helpful in the persistence of movements.

Much like the activists in Goodwin and Pfaff (2001) who argued that it was necessary to manage fear, many activists reported that one had to assume there were infiltrators in their organization and move on or risk being paralyzed by paranoia. But the following statements by activists who discovered that there were infiltrators in their organizations suggest that the issue of trust is not so simply solved:

> Well, it shook me. I vowed never to get in another organization, not from the fear of getting killed or arrested or anything like that, but because I just didn't trust people. I always believed that there were – and I still do to this day believe there were – so many more informants than we'll ever know about. It makes you really kind of leery of trusting people. It's like you have been robbed, beaten, and raped. (Akua Njeri, formerly Deborah Johnson, of the Chicago area Black Panther Party; quoted in Schultz & Schultz, 2001, p. 236)

> She was the spy that hurt me the most because I was closest to her. Sheli Lulkin was a hard worker, a very capable person ... And she made personal friends with Sylvia Kushner, our executive secretary. Sylvia was not in a good position financially; we couldn't pay her very much. And Sheli was easy with money. She treated Sylvia to concerts and operas ... ." (Jack Spiegel of the Alliance to End Repression, quoted in Schultz & Schultz, 2001, p. 421)

> Emerson Poe was one of my best friends. My girlfriend and I used to babysit for his wife and him. When she had a miscarriage, we took care of his kid while they were at the hospital. We were that close. Poe had been right with me as assistant regional coordinator of VVAW. He worked with us when we met with the lawyers, talked strategy, and he even helped us select a jury. ... It blew me away. I couldn't believe he could be an informer. ... I don't understand why he went along with them. I don't understand. (Scott Camil of Vietnam Veterans Against the War, quoted in Schultz & Schultz, 1989, p. 328)

As these testimonies indicate, undercover informants infiltrated more than just political organizations, they also infiltrated activists' personal lives. It was this invasion of his privacy that most upset the VVAW's Camil. Not only did he not understand how one of his best friends been an informant, but "it really hurt [his] head a lot" when he found out that a woman "had been assigned by the government to sleep with" him (quoted in Schultz & Schultz, 1989, p. 331).

Finally, there is evidence that covert forms of social control can cause paranoia, particularly when they are persistent over a long period of time. Paul Robeson, Jr. remembers the toll that surveillance took on his father:

> Dad's isolation in the US was like a quarantine: You cut off access to the media, you cut off access to artistic outlets, cut off access to everything – hoping he'll die on the vine. Then you work on him psychologically ... . The system ... [is a] paranoia producer, quite consciously. (quoted in Schultz & Schultz, 2001, p. 149)

Paranoia is also produced within organizations, as Akua Njeri confirms, creating the potential for nearly anyone to be suspicious of nearly anybody else:

> I never thought [Chicago BPP member] O'Neal was an FBI informant. But there had been some people in the party who suspected him because he would disappear for long periods of time. Most people just chalked that up as paranoia. You've got to understand that at that time the police were always watching us, our phones were tapped. And, of course, we knew they had infiltrated us. It reached a point where it could be said about anybody, "Oh, he might be a pig ..." Of course, after the fact, everything falls into place ... (quoted in Schultz & Schultz, 2001, p. 236)

Such suspicions can make many movement operations difficult.

EXPLAINING THE IMPACT OF COVERT SOCIAL CONTROL: IMPLICATIONS FOR FUTURE RESEARCH

One lesson from these accounts is that the costs of social control activity should not be understood as confined simply to the organizational-level effects emphasized in the typology summarized in Table 1. Though agencies such as the FBI may allocate repressive actions with the intention of creating those sorts of discrete structural barriers to protest, analysts need also to recognize the sorts of longer-term, often indirect emotional impacts described in the preceding section. From purely a data standpoint, this realization means that we cannot rely solely on the accounts or records of social control agencies to capture the full scope of the costs of repression. While individual social control agents have acknowledged awareness of the long-range emotional toll posed by their actions, such motives are rarely documented in the internal files of policing or other social control agencies. Thus, we need to be sensitive as well to the perceptions of the targets of repressive activity. Only by focusing on both sides of the social control dynamic, as well as how the actions of agents and their targets iteratively interact, can we understand the full range of effects – both short- and long-term, intended and unintended – engendered by covert repression.

Equally important, taking this premise seriously can move us closer to a richer assessment of *how* social control efforts raise the costs of subsequent mobilization attempts. The impact of state repression on mobilization has long been of interest to social scientists focused the so-called "repression-mobilization nexus." Since the 1970s, work on this problem has become increasingly paradoxical, as analysts have found support for a variety of hypotheses: in some cases, repression appears to straightforwardly reduce subsequent protest activity, while in others it seems to spur threat-based mobilizations (see, e.g. Khawaja, 1993; Lichbach, 1987; White, 1989). Initially, researchers responded to such contradictory baseline findings by positing various more complex curvilinear relationships (Moore, 1998; Opp & Roehl, 1990; Rasler, 1996), but with similarly inconsistent results.

Such difficulties have sparked a number of efforts to fine-tune analyses. Davenport (2005b) notes that recent research has placed renewed emphasis on alternative data sources, more robust indicators of repression, longer-term temporal dynamics, and the role played by a range of contextual factors. A parallel effort has refined the "nexus" question itself, to examine how social control impacts the "quality" rather than only the "quantity" of protest (Chang, 2007). This approach is promising, as it recognizes that the imposition of social control can produce a range of reactions from activist targets, impacting their ability and propensity to form coalitions, redefine strategic and tactical priorities, generate innovative frames, and so on.

This latter effort aligns squarely with recent calls for analyses centered on the dynamic processes and mechanisms through which more traditional variable-centered causal relationships operate (Hedstrom & Swedberg, 1998; McAdam, Tarrow, & Tilly, 2001). In this case, such an approach would shift our primary concern from the aggregate effect that a covert social control act might produce, and toward an assessment of *how* such an act might generate a range of context-dependent outcomes. Significantly, this approach subverts the traditional aim of "discovering uniformity" across cases, and instead "aims at explaining change and variation," as constituent mechanisms and processes combine in varying ways to yield distinct outcomes (Tilly, 2005, p. 212).

The cases we have described above, centered on the range of direct and indirect outcomes produced by the FBI's intelligence and counterintelligence efforts, illustrate the potential payoff of a mechanism-centered approach. By pairing an examination of social control agencies' intentions with the perceptions of protest targets, we can produce a more complete picture of the intended and unintended effects of covert social control. Put another way, such an approach can allow us to more expansively consider the concept of "costs." By viewing costs in a way that accounts for direct and indirect, and short- and long-term, processes, we can construct a richer conceptualization of the term.

But equally important, by shifting our emphasis away from efforts to identify causal regularities associated with the imposition of costs, we can move toward a process-oriented emphasis on the myriad ways in which cost-related factors combine within particular contextual settings to produce a range of outcomes. While such an approach will not advance an agenda centered on ambitious but overgeneralized assertions of how "covert social control impacts mobilization," it does provide a basis for demonstrating the range of mechanisms through which social control activity imposes particular costs on its intended targets. By then assessing the manner in which particular contexts influence the combinations of salient mechanisms, we can produce a textured account that systematically explains the diversity and variation intrinsic to the real-world settings within which political contention occurs.

NOTE

1. These functions were identified through an analysis of 5,527 memos that have been publicly released through Freedom of Information Act requests related to two COINTEL programs (White Hate Groups and the New Left). The FBI's actions in these programs served three additional functions not included here. One of those was purely intelligence related, i.e. to *gather information* for future prosecutions and/or counterintelligence actions, and is thus incorporated into our discussion below. The other two – to *displace conflict* and to *control target group actions* – were each initiated in fewer than one percent of the 961 total actions carried out in those two programs (see Cunningham, 2003), and due to their marginality have been excluded from our discussion here.

REFERENCES

Adams, J. (2003). The bitter end: Emotions at a movement's conclusion. *Sociological Inquiry*, *73*, 84–113.

Aiello, J. R., & Kolb, K. J. (1995). Electronic performance monitoring and social context: Impact on productivity and stress. *Journal of Applied Psychology*, *80*(3), 339–353.

Alge, B. (2001). Effects of computer surveillance on perceptions of privacy and procedural justice. *Journal of Applied Psychology*, *86*(4), 797–804.

Boghosian, H. (2004). *The assault on free speech, public assembly, and dissent: A national lawyers guild report on government violations of first amendment rights in the United States*. Great Barrington, MA: The North River Press.

Brigham, M., & Corbett, J. M. (1997). Email, power and the constitution of organizational reality. *New Technology, Work and Employment*, *12*(1), 25–35.

Britt, L., & Heise, D. (2000). From shame to pride in identity politics. In: S. Stryker, T. J. Owens & R. W. White (Eds), *Self, identity, and social movements* (pp. 252–268). Minneapolis, MN: University of Minnesota Press.

Carayon, P. (1993). Effect of electronic performance monitoring on job design and worker stress: Review of the literature and conceptual model. *Human Factors*, *35*(3), 385–395.

Carley, M. (1997). Defining forms of successful state repression of social movement organizations: A case study of the FBI's COINTELPRO and the American Indian movement. *Research in Social Movements, Conflict, and Change*, *20*, 151–176.

Carson, C., & Gallen, D. (1991). *Malcolm X: The FBI file*. New York: Carroll & Graf Publishers.

Chang, P. Y. (2007). Unintended consequences of repression: Alliance formation in South Korea's Democracy Movement, 1970–1979. Paper presented at the ASA collective behavior and social movement section workshop on movement cultures, strategies, and outcomes, Hempstead, New York.

Churchill, W., & VanderWall, J. (1990). *The COINTELPRO papers: Documents from the FBI's secret war against dissent in the United States*. Boston: South End Press.

Cunningham, D. (2003). State vs. social movement: The FBI's COINTELPRO against the new left. In: J. Goldstone (Ed.), *States, parties, & social movements: Protest and the dynamics of institutional change* (pp. 45–77). New York: Cambridge University Press.

Cunningham, D. (2004). *There's something happening here: The new left, the klan, and FBI counterintelligence*. Berkeley: University of California Press.

Davenport, C. (2005a). Understanding covert repressive action: The case of the U.S. government against the Republic of New Africa. *Journal of Conflict Resolution*, *49*(1), 120–140.

Davenport, C. (2005b). Repression and mobilization: Insights from political science and sociology. In: C. Davenport, H. Johnston & C. Mueller (Eds), *Repression and mobilization* (pp. vii–xli). Minneapolis, MN: University of Minnesota Press.

Earl, J. (2003). Tanks, tear gas, and taxes: Toward a theory of movement repression. *Sociological Theory*, *21*(1), 44–68.

Friedly, M., & Gallen, D. (1993). *Martin Luther king, Jr.: The FBI file*. New York: Carroll & Graf Publishers.

Garson, B. (1988). *The electronic sweatshop*. New York: Simon & Schuster.

Gelbspan, R. (1991). *Break-ins, death threats and the FBI: The covert war against the central America movement*. Boston, MA: South End Press.

Gilliom, J. (2005). Resisting surveillance. *Social Text*, *83, 23, 2*, 71–83.

Goodwin, J. (1997). The libidinal constitution of a high-risk social movement: Affectual ties and solidarity in the Huk Rebellion, 1946 to 1954. *American Sociological Review*, *62*, 53–69.

Goodwin, J., & Jasper, J. M. (1999). Caught in a winding, snarling vine: The structural bias of political process theory. *Sociological Forum*, *14*(1), 27–54.

Goodwin, J., & Pfaff, S. (2001). Emotion work in high-risk social movements: Managing fear in the U.S. and East German civil rights movements. In: J. Goodwin, J. M. Jasper & F. Polletta (Eds), *Passionate politics* (pp. 282–302). Chicago: University of Chicago Press.

Gould, D. B. (2001). 'Rock the boat, don't rock the boat, baby': Ambivalence and the emergence of militant AIDS activism. In: J. Goodwin, J. M. Jasper & F. Polletta (Eds), *Passionate politics: Emotions and social movements* (pp. 135–157). Chicago: University of Chicago Press.

Gould, D. B. (2002). Life during wartime: Emotions and the development of ACT UP. *Mobilization*, *7*(2), 177–200.

Granovetter, M. (1978). Threshold models of collective behavior. *American Journal of Sociology*, *83*(6), 1420–1443.

Greenberg, L., & Barling, J. (1999). Predicting employee aggression against co-workers, subordinates and supervisors: The role of person behaviors and perceived workplace factors. *Journal of Organizational Behavior*, *20*(6), 897–913.

Hashimoto, A. (1979). Measuring the effect of police surveillance on the prevention of traffic accidents. *Accident Analysis and Prevention*, *11*(4), 261–270.

Hedstrom, P., & Swedberg, R. (Eds). (1998). *Social mechanisms: An analytical approach to social theory*. New York: Cambridge University Press.

Jasper, J. M. (1998). The Emotions of protest: Affective and reactive emotions in and around social movements. *Sociological Forum*, *13*(3), 397–424.

Jayko, M. (Ed.) (1988). *FBI on trial: The victory in the socialist workers party suit against government spying*. New York: Pathfinder.

Kane, A. (2001). Finding emotion in social movement processes: Irish land movement metaphors and narratives. In: J. Goodwin, J. M. Jasper & F. Polletta (Eds), *Passionate politics* (pp. 251–266). Chicago: University of Chicago Press.

Kemper, T. (2001). A structural approach to social movement emotions. In: J. Goodwin, J. M. Jasper & F. Polletta (Eds), *Passionate politics* (pp. 58–73). Chicago: University of Chicago Press.

Khawaja, M. (1993). Repression and collective action: Evidence from the west bank. *Sociological Forum*, *8*(1), 47–71.

Klandermans, B., & Oegema, D. (1987). Potentials, networks, motivations, and barriers: Steps towards participation in social movements. *American Sociological Review*, *52*(4), 519–531.

Lenski, G. E. (1954). Status crystallization: A non-vertical dimension of status. *American Sociological Review*, *19*(4), 405–413.

Lichbach, M. (1987). Deterrence or escalation? The puzzle of aggregate studies of repression and dissent. *Journal of Conflict Resolution*, *31*, 266–297.

Loveman, M. (1998). High risk collective action: Defending human rights in Chile, Uruguay, and Argentina. *American Journal of Sociology*, *104*(2), 477–525.

Marwell, G., & Oliver, P. E. (1993). *The critical mass in collective action*. New York: Cambridge University Press.

Marx, G. T. (1974). Thoughts on a neglected category of social movement participant: The *agent provocateur* and the informant. *American Journal of Sociology*, *80*(2), 402–442.

Marx, G. T. (1979). External efforts to damage or facilitate social movements: Some patterns, explanations, outcomes, and complications. In: J. D. McCarthy & M. N. Zald (Eds), *Dynamics of social movements*. Cambridge, MA: Winthrop.

Marx, G. T. (1988). *Undercover*. Berkeley, CA: University of California Press.

McAdam, D. (1982). *Political process and the development of black insurgency, 1930–1970*. Chicago: University of Chicago Press.

McAdam, D. (1988). *Freedom summer*. New York: Oxford University Press.

McAdam, D., Tarrow, S., & Tilly, C. (2001). *Dynamics of contention*. New York: Cambridge University Press.

Monahan, T. (2006). Electronic fortification in Phoenix: Surveillance technologies and social regulation in residential communities. *Urban Affairs Review*, *42*(2), 169–192.

Moore, W. H. (1998). Repression and dissent: Substitution, context, and timing. *American Journal of Political Science*, *42*(3), 851–873.

Moynihan, C. (2007). FBI files on lawyers guild show scope of surveillance. *New York Times*, June 25, p. A19.
Nabeker, D. M., & Tatum, B. C. (1993). The effects of computer monitoring, standards, and rewards on work performance, job satisfaction, and stress. *Journal of Applied Social Psychology*, *23*, 508–536.
Nail, P. R., Van Leeuwen, M. D., & Powell, A. B. (1996). The effectance versus the self-presentational view of reactance: Are importance ratings influenced by anticipated surveillance? *Journal of Social Behavior and Personality*, *11*(3), 573–584.
Nepstad, S. E., & Smith, C. (2001). Rethinking recruitment to high-risk/cost activism: The case of nicaragua exchange. *Mobilization*, *4*, 40–51.
Noakes, J. (1998). Bankers and common men in bedford falls: How the FBI determined that *it's a wonderful life* was a subversive movie. *Film History*, *10*, 311–319.
Noakes, J. (2000). Official frames in social movement theory: The FBI, HUAC, and the communist threat in Hollywood. *The Sociological Quarterly*, *41*(4), 639–656.
Noakes, J. (2003). Racializing subversion: The FBI and the depiction of race in early cold war movies. *Ethnic and Racial Studies*, *26*(4), 728–749.
Oberschall, A. (1994). Rational choice in collective protests. *Rationality and Society*, *6*(1), 79–100.
Oliver, P. E., Marwell, G., & Teixeira, R. (1985). A theory of the critical mass, I. interdependence, group heterogeneity, and the production of collective goods. *American Journal of Sociology*, *91*(3), 522–556.
Opp, K-D., & Roehl, W. (1990). Repression, micromobilization, and political protest. *Social Forces*, *69*(2), 521–547.
O'Reilly, K. (1989). *Black Americans: The FBI file*. New York: Carroll & Graf Publishers.
Polletta, F. (2002). *Freedom is an endless meeting: Democracy in American social movements*. Chicago: University Chicago Press.
Rasler, K. (1996). Concessions, repression, and political protest in the Iranian revolution. *American Sociological Review*, *61*(1), 132–152.
Robins, K., & Webster, F. (1989). *The technical fix: Education, computers, and industry*. New York: St. Martin's Press.
Schultz, B., & Schultz, R. (1989). *It did happen here: Recollections of political repression in America*. Berkeley, CA: University of California Press.
Schultz, B., & Schultz, R. (2001). *The price of dissent: Testimonies to political repression in America*. Berkeley, CA: University of California Press.
Smelser, N. J. (1962). *Theory of collective behavior*. New York: Free Press.
Stanton, J. M., & Barnes-Farrell, J. L. (1996). Effects of electronic performance monitoring on personal control, task satisfaction, and task performance. *Journal of Applied Psychology*, *81*(6), 738–745.
Stein, A. (2001). Revenge of the shamed: The Christian right's emotional culture war. In: J. Goodwin, J. M. Jasper & F. Polletta (Eds), *Passionate politics* (pp. 115–131). Chicago: University of Chicago Press.
Tang, T. L., & Baumeister, R. F. (1984). Effects of personal values, perceived surveillance, and task labels on task preference: The ideology of turning play into work. *Journal of Applied Psychology*, *69*(1), 99–105.
Tarrow, S. (1998). *Power in movement: Social movements and contentious politics*. New York: Cambridge University Press.
Taylor, J. (2006). The emotional contradictions of identity politics: A case of a failed human relations commission. *Sociological Perspectives*, *49*(2), 217–237.

Taylor, V. (1996). *Rock-a-bye baby: Feminism, self-help and post-partum depression.* New York, NY: Routledge.

Taylor, V. (2000). Emotions and identity in women's self-help movements. In: S. Stryker, T. J. Owens & R. W. White (Eds), *Self, identity, and social movements* (pp. 271–299). Minneapolis, MN: University of Minnesota Press.

Thibaut, J., Friedland, N., & Walker, L. (1974). Compliance with rules: Some social determinants. *Journal of Personality and Social Psychology, 30*(6), 792–801.

Tilly, C. (1978). *From mobilization to revolution.* New York: Longman.

Tilly, C. (2005). Repression, mobilization, and explanation. In: C. Davenport, H. Johnston & C. Mueller (Eds), *Repression and mobilization* (pp. 211–226). Minneapolis, MN: University of Minnesota Press.

Tolchinsky, P. D., McCuddy, M. K., Adams, J., Ganster, D. C., & Fromkin, H. L. (1981). Employee perceptions of invasion of privacy: A field simulation experiment. *Journal of Applied Psychology, 66*(3), 308–313.

US Department of Justice. *The attorney general's guidelines on general crimes, racketeering enterprise and terrorism enterprise investigations*, 30 May 2002.

White, G. L., & Zimbardo, P. G. (1980). The effects of threat of surveillance and actual surveillance on expressed opinions toward Marijuana. *Journal of Social Psychology, 111*(1), 49–61.

White, R. W. (1989). From peaceful protest to Guerrilla war: Micromobilization of the provisional Irish Republican Army. *American Journal of Sociology, 94*(6), 1277–1302.

Wiener, J. (1999). *Gimme some truth: The John Lennon FBI files.* Berkeley, CA: University of California Press.

Yesil, B. (2006). Watching ourselves: Video surveillance, urban space and self-responsibilization. *Cultural Studies, 20*(4–5), 400–416.

Young, M. P. (2001). A revolution of the soul: Transformative experiences and immediate abolition. In: J. Goodwin, J. M. Jasper & F. Polletta (Eds), *Passionate politics* (pp. 99–114). Chicago: University of Chicago Press.

Zwerman, G., & Steinhoff, P. (2005). When activists ask for trouble: State–dissident interactions and the new left cycle of resistance in the United States and Japan. In: C. Davenport, H. Johnston & C. Mueller (Eds), *Repression and mobilization* (pp. 85–107). Minneapolis, MN: University of Minnesota Press.

PLURAL POLICING AND CCTV SURVEILLANCE

Michael McCahill

ABSTRACT

This chapter aims to make a contribution to recent debates on the 'governance of security' (Johnston & Shearing, 2003) by drawing upon empirical research conducted by the author and other writers on 'plural policing' and the construction of closed circuit television (CCTV) surveillance networks. The chapter attempts to avoid the tendency in some of the 'governmentality' literature to 'airbrush out the state' (Hughes, 2007, p. 184), whilst at the same time showing that the aims and intentions of dominant state forces and elites are not always realised in practice. The chapter also tries to avoid any simplistic notion of a shift in policing strategies from 'crime fighting' to 'risk management'. The aim instead is to show how the construction of surveillance networks is blurring the boundaries of the 'public–private' divide along the 'sectoral', 'geographical', 'spatial', 'legal' and 'functional' dimensions (Jones & Newburn, 1998), giving rise to a plural policing continuum.

Surveillance and Governance: Crime Control and Beyond
Sociology of Crime, Law and Deviance, Volume 10, 199–219

ISSN: 1521-6136/doi:10.1016/S1521-6136(07)00209-6

INTRODUCTION

As a number of writers have shown, before the early nineteenth century, 'policing' referred not to the 'law enforcement' or 'crime fighting' activities of the modern professional police force, but to a very general schema of regulation, inspection and guardianship carried out by a wide range of 'public' and 'private' actors operating in civil society rather than within the confines of the central state (Crawford, 2003; Garland, 1996; Ericson & Haggerty, 1997; Johnston, 1992; McMullan, 1998; Pasquino, 1991; Zedner, 2006). It was only during the nineteenth and twentieth centuries that a 'state-centred' view of governance came to dominate criminological thinking and that the word 'police' was eventually used to refer to the order maintenance activities of uniformed officers. Today it is argued that the balance between 'public' and 'private' provision has shifted once more to the extent that a 'pluralized, fragmented and differentiated patchwork has replaced the idea of the police as the monopolistic guardians of public order' (Crawford, 2003, p. 136).[1] For Johnston and Shearing (2003), this most recent challenge to the state monopoly of 'policing' is part of a paradigm shift in policing and social control signified by the move away from a 'punishment' mentality towards a 'risk' mentality. In the former paradigm, the central state exercises its security responsibilities through the employment of specialised professionals, such as police officers, whose main concern is the apprehension and punishment of suspected wrongdoers (Johnston & Shearing, 2003, p. 14). Under the 'risk' paradigm, on the other hand, security is exercised under plural auspices beyond the central state and is characterised by a 'corporate' mentality 'that emphasises proactive prevention rather than reactive punishment, and actuarial calculation rather than conventional moral prescription' (*ibid.*, 2003, p. 14). Moreover, this future-oriented approach is no longer the exclusive preserve of corporate security, because the state police have also become more and more preoccupied with risk-based policies and practices (Ericson & Haggerty, 1997).

In some respects, it could be argued that the emergence of a pluralised and differentiated patchwork of policing has been reinforced by the introduction of closed circuit television (CCTV) surveillance systems which operate across both state and extra-state institutions and are monitored by a diverse range of actors, including police officers, police civilian staff, special constables, private security and local authority personnel (Norris & Armstrong, 1999, p. 57). According to 'plural' policing theorists, CCTV is also 'an exemplar of "actuarial" technology' which encourages the targeting of 'aggregate populations' as well as individual suspects (Johnston & Shearing, 2003, p. 69). The use of such systems also reflects the shift towards 'pre-emptive' policing that involves interventions being made by the police before any deviant act has taken place (Graham, 1998). The central aim of this chapter is simply to introduce some empirical complexity to these debates by drawing upon research conducted by the author and other writers on the construction and operation of CCTV surveillance networks. It will be argued that the adoption of 'proactive' or 'risk-based' policing strategies is not an inevitable consequence of utilising CCTV. Much depends, for example, on the technological sophistication of the system and the organisational environment in which CCTV is placed. Plural policing theorists also suggest that the 'risk' mentality emphasises 'actuarial calculation' rather than 'moral prescription' and targets whole groups and populations. However, this chapter argues that there is no easy distinction to be made between 'public' and 'private' actors in moral or ideological terms and that the gaze of both 'private' and 'state'-based surveillance systems is firmly fixed on the 'usual suspects'. The chapter concludes by suggesting that some of the abstract theoretical debates on the relationship between 'policing' and 'state power' need to be accompanied by empirically grounded accounts of 'plural' policing in practice.

SURVEILLANCE AND PROACTIVE POLICING

While 'proactive' policing is seen by some writers as a central feature of the new 'risk paradigm', it can hardly be described as a recent development in policing practice. As early as the eighteenth and nineteenth centuries, policing experts such as Fielding, Colquhoun and Chadwick 'regarded prevention as more important than retrospective functions of arrest, detention and prosecution that later came to characterise modern policing' (Zedner, 2006, p. 85). As McMullan (1998) has pointed out, 'personality training, character formation, surveillance and risk-management were all part of Chadwick's general and indivisible police' (p. 119), and Fielding (1758) believed that 'it is better to prevent even one man from being a rogue than apprehending and bringing forty to justice' (p. 102). More recently, Hillyard (1987) argued that during the 1970s policing in Northern Ireland shifted from a 'reactive' to a 'pro-active' or 'pre-emptive' approach where large sections of the community were 'regularly and systematically monitored and surveilled' (p. 290). Also, Williams (2003, pp. 13–14) points out that while most writers have dated the arrival of CCTV in the UK to the 1980s, surveillance cameras with remote pan, tilt and zoom facilities were already being used by the Metropolitan Police during the 1960s to monitor crowds at the state visit of the Thai royal family.

The first explicit connection to be made between CCTV and 'proactive' policing, however, came in 1993 when the Audit Commission, worried that society was in danger of losing the 'battle against crime', called for a massive expansion of proactive, intelligence-led policing and singled out CCTV as having a major role to play in crime prevention. One year later, the Conservative government announced the first of four CCTV Challenge Competitions to support the expansion of open-street city centre CCTV. As Norris and McCahill (2006) have pointed out, bids had to be in the form of a partnership with matched funding from other sources particularly the business sector, the local authority, the police or other Government Agencies. The competition would only fund up to 50% of the total capital costs and no contribution would be made for the annual running costs which could be as much as £250,000 (Webster, 1998). In this respect, the introduction of CCTV devolved the responsibility for crime control on to local partnerships and dovetailed neatly with the governments' ideological demands for privatisation of the public sector. These developments have continued under 'New Labour' where the use of partnerships in the delivery of CCTV has become entrenched in the Crime and Disorder Act (Norris & McCahill, 2006; Webster, 2004).

However, while the central state clearly played a key 'steering' role in the construction of public space CCTV networks through the use of Home Office funding mechanisms, it is also clear that 'state' police officers were not always proactive in drawing upon the knowledge and intelligence capacities of those doing the 'rowing' in the CCTV control rooms. A number of research reports, for example, have shown that initially CCTV surveillance systems were not readily incorporated into police information gathering and intelligence systems (Goold, 2004; Gill & Spriggs, 2005; Fussey, 2007). In his study of six open-street CCTV surveillance systems, Benjamin Goold (2004) found that while many senior police officers publicly welcomed the efforts of local authorities to bring CCTV to their towns, many adopted a 'wait and see' approach. This meant that they had little say over the design of the systems, camera placement or operational procedures. Thus, while some police officers, in his study, stated that they would have preferred the use of overt cameras to 'pre-empt' crime by acting as a deterrent, local authority officials favoured the use of (covert) cameras to apprehend and prosecute offenders. The first major national evaluation of public-space CCTV also found that the police were often reluctant to become involved in the construction of CCTV systems.

Police reluctance was due to a number of factors, including 'a concern that the new CCTV system would increase workload or place a strain on already stretched financial resources' (Gill & Spriggs, 2005, p. 68). These findings were supported in a study carried out by the Metropolitan Police in 2001 which found that many police officers regarded CCTV as 'a demand on their time or as a tool to be used reactively during investigations to crimes' (Anon., 2006a, p. 25). Meanwhile, another study found that the 'state' police were entirely unaware of the existence of a public CCTV scheme in an inner-city area of Metropolitan City (Fussey, 2007).

The use of surveillance systems for 'proactive' policing is also highly dependent on the organisational environment in which CCTV is placed. Many open-street CCTV systems, for example, are monitored by local authority employees or private security officers who, as Norris and Armstrong (1999) pointed out, are not always in a position to demand intervention by the police. Thus, in their study of 3 public-space open-street CCTV systems, they witnessed just 45 deployments out of a total of 600 hours of observation, with an arrest being made in only 12 incidents (Norris & Armstrong, 1999, p. 168). Meanwhile, in his study of six open-street CCTV systems in the Southern Region Police area, Goold (2004) found that police officers were less than enthusiastic about their working relationships with some of the CCTV operators, particularly those in local authority-led schemes who lacked the police officers' 'sixth sense' and often passed on jobs perceived by the police as 'rubbish'. As one police constable explained, 'at the end of the day, however hard civilians try ... they still don't have the gut feeling and never will have the gut feeling that a police officer is so often trained to have' (Goold, 2004, p. 134). Thus, in three local authority-controlled CCTV control rooms, the author found that there was one deployment of a police officer in 243 targeted surveillances (*ibid.*, p. 164). Similarly, in the 13 CCTV systems studied by Gill and Spriggs (2005), only three control rooms used two-way police radio links which were regarded by the authors as essential for mobilising a rapid response to monitored non-compliance (Gill, 2006).

Despite some of the issues raised in the studies reviewed above, more recent evidence suggests that the state police have begun to make a number of organisational changes that may facilitate the use of CCTV for a 'pre-emptive' approach to policing. For instance, in London there are plans to link each local authority CCTV control room to one of the three new purpose-built police communications centres at Hendon, Bow and Lambeth. This will facilitate greater co-operation between CCTV control rooms and the police Borough Operational Command Units (BOCUs), with

24-hour operations including 'radio communications and CCTV feeds' (Anon., 2006a, p. 26). The CCTV control rooms will also utilise the new 'Airwave' radio system (the secure police cellular radio system), which means that CCTV operators will 'be able to provide responding officers with advance information about what's going on at a particular incident – perhaps warning them of potential risks before they arrive' (*ibid.*, p. 26).

As Norris and McCahill (2006, p. 114) point out, 'proactive' interventions are also facilitated by the growth of computer simulation whereby cameras can be linked with computer software that can analyse the flow of digitalised images to plot possible future outcomes. This information can then be used to signal when intervention is necessary, before an event takes place (Graham, 1998; Norris, 2003). Although research has shown that the use of automated systems in the UK is not extensive,[2] the use of such systems by the police is set to expand rapidly over the next few years, particularly on the roads where the police are beginning to utilise Automatic Number Plate Recognition (ANPR) schemes as a mainstream policing tool (Kelly, 2006, p. 34). As Norris (2006, p. 3) has explained, 'by coupling the camera to a computer it is possible to read automatically the license plates of passing cars and check them against the records held by the DVLC and databases held on the Police National Computer (PNC)'. In 2003, the Home Office announced a national pilot of the ANPR scheme which involved 23 police forces setting up 50 ANPR-enabled intercept teams consisting of officers operating from cars and motorcycles. Norris goes on to point out that during the

> First nine months of operation over twenty million vehicle registrations marks were read and 900,000 of these were flagged on police databases as being of interest to them. As a result over 130,000 vehicles were stopped and over 10,000 people arrested. (Norris, 2006, p. 3)

Meanwhile, the Association of Chief Police Officers (ACPO) is now keen to encourage local authority CCTV operators to add ANPR capability to their existing town centre camera networks (Kelly, 2006, p. 36).

ACTUARIAL JUSTICE

As well as promoting 'proactive prevention' over 'reactive punishment', plural policing theorists argue that the 'risk' mentality targets whole groups and populations as well as individual offenders (Johnston & Shearing, 2003; Ericson & Haggerty, 1997). With around 4.2 million CCTV cameras operating in publicly accessible spaces in the UK (McCahill & Norris, 2003a), it could be argued that the construction of CCTV surveillance networks reflects this dynamic in that surveillance practices that were once targeted at 'serious' criminals have become extended to target the majority of the population. Similar developments have taken place with other surveillance technologies, such as the introduction in 1995 of a National DNA Database, set up by the Forensic Science Service (FSS) on behalf of the ACPO. Although the database was originally established as a forensic source to help identify those involved in serious crimes (i.e. murder and rape), an amendment to the Criminal Justice and Public Order Act (1994) allows samples to be taken without consent from any person convicted or suspected of a recordable offence (McCahill & Norris, 2003b).

However, while such technologies clearly expand the population that is subject to monitoring, the gaze of 'state-based' surveillance systems is still fixed firmly on the 'usual suspects'. For instance, 40% of black males have their profiles stored on the National DNA Database, compared with 9% of white males (*The Guardian*, 2006). Also, over half of those subject to the surveillance practices (e.g. electronic monitoring, drug testing, voice verification and face-to-face monitoring by probation officers) of the Intensive Supervision and Surveillance Programme (ISSP) are unemployed with poor literacy skills (Youth Justice Board, 2004). Meanwhile, the introduction of Anti-Social Behaviour Orders (ASBOs) in the UK has been used disproportionately against the most vulnerable. For instance, in his research Campbell (2002) found 'that more than 60 per cent of the ASBOs examined in the survey were associated with people manifesting very obvious social or mental problems, chiefly linked to drug or alcohol abuse' (in Hughes, 2007, p. 127). Similarly, the operation of open-street CCTV systems leads 'to the over-representation of men, particularly if they are young or black' (Norris & Armstrong, 1999, p. 196). These figures are supported by Goold (2004, p. 157) who found that black people were four times more likely to be targeted by CCTV operators than we would expect from their numbers in the total population.

It is not only state-based surveillance systems that focus narrowly on the 'usual suspects'. At one shopping mall in the North of England, for example, the CCTV operators targeted 'known criminals', 'suspected drug addicts' and those 'wanted' for the breach of bail conditions (McCahill, 2002). Similarly, in her study of the use of CCTV in three different leisure centres, Wakefield (2003) reported how private security officers often aided the police in imposing bail conditions on offenders and how this reflected 'a shared sense of those who were deemed "outsiders" or "troublemakers" ... so that the movements of many individuals were tracked by the police outside the centres and the security staff within' (Wakefield, 2003, pp. 218–219). Nor are the strategies pursued by commercial actors completely 'instrumental' and gutted of morally toned or punitive language (Ericson & Haggerty, 1997). The employment of 'tough-looking', ex-military personnel, trained in the use of force means that '*de facto*, many private security personnel now occupy important positions of trust and authority' (Morgan & Newburn, 1997, p. 139; Wakefield, 2003). Moreover, measures introduced under the Police Reform Act (2002), including the accreditation of non-state policing personnel from private security firms, means that 'accredited officers will be granted certain "public" powers over and above those exercised by ordinary citizens' (Crawford, 2006, p. 464).

Empirical research has also shown that the language used by private security officers in mass private property to describe 'out-groups' is often the same as the police officers' description of 'police property'. CCTV operators in two shopping malls in a Northern City, for example, targeted young working class males who were described as 'scrotes', 'scumbags', 'shit', 'druggies', 'G heads' (i.e. 'glue sniffers') and 'scag heads' (McCahill, 2002). Meanwhile, in Goold's (2004) study, CCTV operators in police-controlled systems adopted police terminology referring to suspects as 'scrotes', 'scumbags' and 'tealeaf's'. Most of the operators also appeared comfortable using the police 'IC' categories, a system used by the police to classify individuals according to their apparent ethnic background (Goold, 2004, p. 139). In practice, therefore, the gaze of CCTV cameras in both 'public' and 'private' spaces is mediated by the selective concerns of the operators who tend to disproportionately target young working class males (see McCahill & Norris, 2002; McCahill, 2002; Smith, 2004; Lomell, 2004; Urbaneye, 2004).

THE EXCLUSIONARY IMPULSE

The use of CCTV surveillance systems to enforce exclusionary strategies of control is now well documented in the context of 'mass private property' (McCahill, 2002; Wakefield, 2003). For instance, in her study of 3 leisure centres, Wakefield (2003) found that 578 people were excluded over a 5-week period. In his study of two shopping malls in a Northern City, McCahill (2002) found that 'when a guard was deployed to deal with a group of teenagers there was a fifty–fifty chance that someone would be ejected' (p. 135). Historically, these strategies of control have not been an option that is open to central state actors. As Johnston and Shearing (2003) point out, one of the main reasons that state policing in 'public' space came to focus primarily on 'bandit catching' rather than 'risk minimization', was the differential impact of law on the policing of public and private property. In privately owned spaces, such as shopping malls, for example, 'social relations are underpinned by contracts and the specific conditions defined therein' (*ibid.*, p. 83). This means that anyone entering 'private' space

> May have to waive rights they would normally possess as citizens in public places. Under such conditions, private police may have the legal authority to exercise powers that they would not be able to exercise publicly. In public places, by contrast, rights-bearing citizens make no such trade-offs and the police are constrained by legal rules protecting the rights and liberties of individual citizens. Historically, then, laws protecting citizens in public places made the exercise of risk-based policing by state police problematic. (*ibid.*, p. 83)

Over the last decade, however, a number of developments have changed this situation and have facilitated the use of 'exclusionary' and 'risk-based' strategies of social control in 'public' space. In the early 1990s, for example, the growth of Town Centre Management (TCM) schemes led some writers to argue that town centres were increasingly being replaced by pseudo-public spaces like those in shopping malls, where commercial imperatives dominate and control strategies are intended to 'attract a spending public in preference to those perceived as anti-social and "unaesthetic"' (Reeve, 1996, p. 18). The same author went on to argue that while there was no evidence in the mid-1990s that TCM schemes were actively discriminating against undesirables in order to massage the social space of the town centre into something more socially acceptable to consumers:

> The institutional apparatus – the formalised involvement of the private sector in TCM; the technology, such as CCTV; and hiving off of the responsibility for town centre security and policing to the private sector – are in place or beginning to be in place to allow the effective implementation of the values of the private interest in town centres above those of the more disinterested state. (Reeve, 1996, p. 33)

Today, most of the control strategies identified by Reeve have become a common feature of the UK's urban landscape where the use of exclusionary strategies of social control have narrowed the gap between the way people's movement and behaviour is controlled in 'public' space and 'mass private property' (Wakefield, 2003, p. 32). The growth of 'public–private' partnerships in town centres, for example, means that the private sector has been closely involved in both the construction and operation of public space CCTV systems (McCahill, 2002; Coleman, 2004). The use of Retail Radio

Networks, for example, which allow 'private' actors to initiate the targeting practices of 'public' CCTV systems is now a common feature in UK cities. In the first major national evaluation of public area CCTV, it was reported that several CCTV control rooms had Retail or Pub Radio schemes 'which accounted for the reporting and observation of significant numbers of incidents' (Gill & Spriggs, 2005, p. xiii). Also, five out of six of the CCTV control rooms in Goold's (2004) study of open-street CCTV had 'established radio links between all of the major retailers and shopping centres' (p. 145). In Lincoln, a Business Improvement Group 'provides two-way radios to nearly 200 pubs, clubs and shops' in the town centre, with 'incidents initiated by radio contact accounting for 70% of the control room staff's activities' (Anon., 2006b, p. 16). Similarly, the CCTV system in Liverpool is monitored by a private security company who liaise with the 'state' police and in-house store security via a retail radio link which 'makes possible the monitoring of persons through both open public space and private shop space' (Coleman, 2004, p. 159). Meanwhile, Norris and Armstrong (1999) found that one of the CCTV control rooms in their study became 'almost entirely co-opted by and subservient to the demands of the private security nexus of the town centre store detectives' (p. 160).

Recent years have also seen the introduction of new legislation which allows shopping mall managers and town centre managers to use the courts and criminal justice agencies for exclusions, through the use of ASBOs or exclusions attached to bail conditions. In their study of 'plural policing' in Liverpool, Crawford, Lister, Blackburn, and Burnett (2005) found that local business leaders worked closely with the local ASBO Unit and Youth Offending Service to encourage 'the use of exclusion from certain stores in the city centre as an aspect of such contracts' (p. 51). It is estimated that 'between April 1999 and June 2005, some 6,500 ASBOs were issued in England and Wales' (Crawford, 2006, p. 465), and there is increasing evidence that CCTV surveillance systems are being used to provide evidence for the enforcement of these measures. The CCTV operators in the control room operated by Newark and Sherwood District Council, for example, work closely with a council-employed ASBO officer who provides the control room staff with details of any ASBOs issued to local residents. As the control room manager stated, 'CCTV can provide us with irrefutable evidence that an ASBO has been broken' (Anon., 2005a, 2005b, p. 21). Far from being a purely 'beyond-the-state' technology, therefore, CCTV is being used to enforce ASBOs, the breach of which can result in 'the ultimate sovereign sanction – the prison' (Crawford, 2006, p. 455).

Like the shopping mall, where CCTV operators deploy 'private security guards' to deal with 'nuisance behaviour' and enforce the local norms of the mall, many town centre CCTV systems are now integrated with the deployment of 'private street patrollers' who enforce local bylaws in the new spaces of consumption. In York, for example, 'private street patrollers' are deployed by state police control room staff 'to attend low-level incidents, notably relating to groups of youths gathering in public places' (Crawford et al., 2005, p. 48). Similarly, research conducted in London as part of the URBANEYE project found that 'private' street patrollers in one borough worked closely with the Metropolitan Police and the Council's CCTV control room. The street patrollers in this borough were on the Retail Radio Network and were deployed by CCTV control room staff to enforce local by-laws and other acts including litter, abandoned vehicles and dog by-laws. During the first four weeks the patrollers dealt mainly with incidents involving fly tips (24 cases), untaxed vehicles (22), drunk and disorderly people (19), damaged paving slabs (15), vagrants (13), reports of drug dealing (12), abandoned vehicles (11), injured people (9) and shoplifters (7) (McCahill & Norris, 2003c, p. 10). Meanwhile, while the policing activities in town centres increasingly resemble those which take place in the shopping mall (Reeve, 1996), retail establishments are being penetrated by the criminal justice system, with state police officers patrolling shopping malls (McCahill, 2002; Crawford, 2006, p. 463) and plans announced by the Home Office to install short-term 'retail jails' on high streets and in shopping malls to detain offenders for up to 4 hours (Travis, 2007).

FROM 'LOSS PREVENTION' TO 'LAW ENFORCEMENT'

As Shearing and Stenning (1987) have argued, the policing strategies adopted by 'private security' focus not so much on the apprehension and punishment of known offenders, but on the goals of 'private justice' and 'loss prevention'. For instance, rather than 'draw on the slow and costly criminal justice process in pursuit of sanctions' (Wakefield, 2005, p. 532), private actors will choose the most cost-effective option by trying to prevent problems before they occur. Moreover, when deviations from the rules of the organisation do take place, these are likely to 'be dealt with administratively by warnings rather than by criminal prosecution' (Ericson & Haggerty, 1997, p. 269; Henry, 1987). As Jones and Newburn

(1998) point out, private policing on the school or college campus, for example, will be 'concerned with the "internal policing" problems associated with the students within the college, as much as protecting the college from the threat of external crime' (Jones & Newburn, 1998, p. 189). Similarly, the use of CCTV surveillance and private security patrols in the industrial workplace will be used to monitor compliance with health and safety regulations, time keeping, access restriction, unauthorised breaks and so on (McCahill, 2002, p. 161). Also, minor problems, including minor criminal offences, identified by local authority employees responsible for monitoring the CCTV systems in public housing schemes may be reported to the Neighbourhood Housing Manager rather than the local police (McCahill, 2002, p. 175).

Meanwhile, in shopping malls, the central concern with 'loss prevention' is clearly evident in the use of CCTV surveillance systems. At one shopping mall in a Northern City, for example, the author found that almost a quarter (23%) of the 'targeted surveillances' initiated by CCTV operators in the control room, were related to 'health and safety', 'lift breakdowns' and 'accidents' which could involve 'comeback' in the shape of 'insurance claims' (McCahill, 2002). As the Security Manager stated, following an accident caused by water spillage on the mall, 'if [the customers] go for compensation we've got to have something to throw back at them' (McCahill, 2002, p. 138). The management's views on the importance of such incidents were communicated to the front-line CCTV operators and subsequently received a rapid response from all security officers. During accidents, for example, all the patrol guards on this site were instructed to keep off the air (i.e. not use their radios) by the CCTV operators until the incident had been resolved. The rapid response to such incidents was in stark contrast to the lack of interest shown by CCTV operatives to messages received in the control room via the City Centre Radio Link concerning 'known offenders' (McCahill, 2002).

While the concern with 'private justice', 'commercial image' and 'loss prevention' is clearly evident in the policing of 'semi-public' and 'private' space, the CCTV systems and policing practices on these sites can be easily and routinely co-opted for traditional policing. In previous research, for example, the author has shown how many of these sites have external CCTV cameras which can prove useful for criminal investigations instigated by the state police (McCahill, 2002). The Security Manager of a newspaper publishing plant, for example, explained how the police brought a tape to the control room and asked him if he would use one of the sites external cameras to record people visiting a second-hand electrical shop suspected of receiving stolen goods. Meanwhile, a CCTV operator employed by a frozen foods factory explained how the police had used external cameras on the site to monitor a person living in a nearby residential area who was suspected of drug dealing. When asked if the police came to the control room to watch the monitors, the CCTV operator replied:

> No, they gave us a tape and we were taping it 24 hours a day on one channel. They were watching this house here (the security officer zooms in on a house across the road). The guy used to go out on Friday night and come back on Monday morning and apparently what it was he was drug dealing so we caught him on camera. We were recording what time of the day and night people came in, the registration numbers of cars, what people looked like, and as I say it was all on tape. (McCahill, 2002, p. 83)

'State' police officers also use 'private' surveillance systems themselves to monitor the suspect population. The Bull Ring Shopping Centre in Birmingham, for example, 'has a video link to the nearby police station which enables the police to take control of one of the centre's external cameras which cover some areas and roads' (Anon., 2006c, p. 14). On other sites, state police officers sit in 'private' control rooms and use CCTV cameras located in 'private' space to monitor suspects in 'public' space, as the security manager of a shopping mall in a Northern City explained:

> We used to get the CID down here a lot. What they'd do is they'd sit up here and watch the "druggies" on the phone'. Kevin zooms in on one of the public phone booths in the Shopping Centre and says, 'what they'd do is zoom in and read the telephone numbers that the "dealers" were dialling. (McCahill, 2002, p. 117)

The localised knowledge of 'private security' officers can be extremely useful for 'state' police officers. In one case study, for example, the author has reported how the CCTV control room situated in a shopping mall (Housing Estate Mall) in an extremely deprived area of a Northern City came to act as an intelligence-base for the state police. On this site, the majority of the security officers were local people with extensive knowledge of the area and its inhabitants. The main CCTV operator, for example, was a woman called Dawn who had lived on the estate for 10 years. Dawn had previously worked as a Special Constable in the operations room of the police station which was 50 yards from the shopping centre. The security manager had also lived on the estate for a number of years and went to the local school which was two minutes walk from the shopping centre. The security manager stated how his local knowledge was useful, because it meant that he knew the identities of most of the 'trouble causers' who visited the centre. During one incident, for example, he knew the name of the person under surveillance because he played in the same darts team as the suspects' dad.

Other members of the security team drank in the same pubs and played for the same Sunday League football teams as those who frequented the mall. Informal liaison on this site often took the form of 'joint operations' whereby state police and private security officers would use the surveillance system to monitor, apprehend and punish 'known offenders' for past crimes. This included CCTV operators ringing the beat officer on his mobile telephone when 'wanted' persons (e.g. people who had breached bail conditions) had entered the shopping centre, and security officers liaising with the local pharmacist responsible for dispensing methadone to pass the names and addresses of 'wanted' persons to the police so that they could be arrested (McCahill, 2002). As this example shows, while the mobilisation of local knowledge can play an important part in 'the construction of just and democratic forms of security governance' (Johnston & Shearing, 2003, p. 140), it can also be used in a highly discriminatory way by private security officers wanting to 'settle old scores', or by the state police for 'bandit catching'.

The introduction of CCTV surveillance systems in civic institutions 'below' the state can also be co-opted by central state actors as the control rooms that house new surveillance systems come to serve as an 'intelligence base' for traditional policing. For instance, many Local Authority housing departments in the UK have installed concierge-operated CCTV systems that monitor the communal areas of high-rise flats (McGrail, 1998; McCahill, 2002). In one study, informal liaison took place between the police and local authority employees responsible for monitoring the images displayed by a network of 222 CCTV cameras that had been installed in the city's 16 high-rise flats (McCahill, 2002). As the local authority employees responsible for monitoring the system explained:

> We can watch the "drug flats" and count how many visits there are to those flats. We've got a list there where we jot down the number of visits which we can then show to the police (Concierge Worker).
>
> Well we gather information for the police. They'll say, "where does so and so live?" or "we think this flats dealing keep an eye on it". I'll keep a record then of who visits the flat and pass it on to the intelligence field officer (Senior Concierge). (McCahill, 2002, p. 176)

In this study, liaison between the concierge staff and the police was soon formalised by the signing of an Information Sharing Agreement between members of the Housing Department and the Northern City Police Force. Under the agreement, senior housing staff and senior operational police staff started to meet regularly to discuss appropriate cases and to share information to be better able to take civic tenancy enforcement or criminal action, respectively. Some of the requests made by the police included that the Housing Department would give information (including photographic information) to individual police officers on request; would make available free of charge empty houses and flats for the use of police officers as vantage points to mount covert observations; and an agreement that Northern City Divisional Commanders would always be informed of the council's intention to use private detectives to gather evidence in specific cases, and that this evidence would be available to the police on request (McCahill, 2002). Similar developments have been reported in Portsmouth where the digitally recorded images displayed by the CCTV cameras operating in 45 council housing blocks, have been used in the enforcement of ASBOs (Cohen, 2005). These developments reinforce the 'formalization of secondary social controls' (Jones & Newburn, 1998; Jones & Newburn, 2002), whereby problems that would have once been dealt with informally by caretakers or housing officers, are now dealt with by 'security-oriented personnel' equipped with sophisticated surveillance systems and working alongside state police officers.

The incorporation of 'non-state' and 'commercial' security into state policing networks, however, is by no means straightforward. As we saw earlier in the case of Housing Estate Mall, many of the security officers on this site were local people with extensive knowledge of the estate and its inhabitants. The degree of familiarity between the security officers and the local suspect population was reflected in the security officers' use of local nicknames to describe the 'surveilled'. For instance, 'The Teenage Temptress', 'Purse-Snatcher', 'FA Cup' and 'Little Ginge' were just some of the characters monitored on the TV screens at this site. The security officers did use a number of derogatory terms to describe the local 'druggies', but the use of terms like 'Little Ginge' demonstrated a measure of familiarity, and even sympathy, with the local individuals under surveillance. It should come as no surprise, therefore, to find that when low paid, low status, mainly working-class security officers find themselves monitoring their own locales and neighbourhoods, they are not always willing to co-operate with the police. Thus, when 'wanted' persons were identified on the cameras at the Housing Estate Mall, the security officers decided not to contact the local beat officer. As one security officer said: 'I wouldn't grass (i.e. pass the name of a wanted suspect on to the local beat officer) on Tommo 'cause he's all right, he's never given me any bother. Anyway, he's off the smack now' (McCahill, 2002, p. 199). Thus, despite the shift towards a 'new penology' and 'actuarial justice', empirical research shows that ‘front-line’ practitioners in criminal justice agencies and security networks continue to be guided by the ‘old’ concerns of ‘welfare’ and ‘reform’ (Kemshall & Maguire, 2001). Research on the operation of CCTV systems in mass private property, for example, has shown that while some CCTV operators may use ‘economic’ forms of reasoning (Garland, 2001), choosing to monitor ‘vulnerable’ targets and ‘time periods’ rather than individual suspects, others continue to work with the old ‘welfare mentality’, empathising with the plight of local working-class youths (McCahill & Norris, 2003c). When asked by local police officers to remotely monitor a ‘black’ youth, for example, one CCTV operator in a South London shopping mall said, ‘this Nicky Green needs a job man. He’s gonna spoil himself, you know. If he had a job he’d be a nuisance to no one’ (McCahill & Norris, 2003c, p. 29). This, of course, was why the nineteenth-century police expert, Edwin Chadwick, suggested that policing should be carried out by anonymous state officials, ‘because it decontextualized policing from local, class, interpersonal and family conflicts’ (McMullan, 1998, p. 118).

CONCLUSION

Drawing upon Michel Foucault’s idea of ‘governmentality’, a number of writers have argued that policing and the ‘governance of security’ cannot be restricted to the study of the public police and allied specialist institutions of the state. Johnston and Shearing (2003) and Rose (1999), for example, have argued ‘that politics has fragmented and pluralized into a series of competing “security networks” or “nodes” in multifarious governable places in which no single actor is able to exercise control over a given territory’ (quoted in Hughes, 2007, p. 11). While recognising the important contribution made by these scholars to contemporary debates on ‘governance’ and ‘policing’, other writers argue that the ‘death’ of the sovereign state is very much exaggerated. Some writers, for example, have pointed to the continuing role of sovereign and coercive techniques of rule which has seen crime control strategies shift from welfare and reintegration towards punitive goals of incapacitation, the militarisation of public space and exclusionary practices (Stenson & Edwards, 2001, p. 73). For others, the introduction of ‘acceptable behaviour contracts’, ‘ASBOs’, ‘dispersal orders’ and ‘parenting orders’ show how the British state, far from withdrawing its role in the regulation of social behaviour, ‘is engaged in ambitious projects of social engineering’ (Crawford, 2006). Meanwhile, ‘critical’ accounts of the rise of CCTV have attempted to show how the construction of public–private surveillance networks is part of a 'social ordering strategy' which augments and extends 'the sovereignty of local and nation-state decision-making structures' (Coleman, Sim, & Whyte, 2002, p. 96). This view is supported by Webster (2004) who states that while

> The emergence of partnerships and co-regulation in the CCTV policy arena point to the growth of an independent self-governing network, these developments have been carefully steered and co-ordinated by key central government institutions ... to ensure that the diffusion of the technology and appropriate networks are in the interests of the state. (2004, pp. 24–25)

The account of 'plural' policing and CCTV surveillance presented in this chapter has attempted to throw some empirical light on these theoretical debates on the relationship between 'policing' and 'state power'. The chapter has attempted to avoid the tendency in some of the 'governmentality' literature to 'airbrush out the state ... in debates on governance' (Hughes, 2007, p. 184), whilst at the same time showing that the aims and intentions of dominant state forces and elites will not always be realised in practice. The chapter has also tried to avoid any simplistic notion of a shift in policing strategies from 'crime fighting' to 'risk management'. The aim instead has been to show how the construction of surveillance networks is blurring the boundaries of the 'public–private' divide along the 'sectoral', 'geographical', 'spatial', 'legal' and 'functional' dimensions (Jones & Newburn, 1998), giving rise to a *plural policing continuum*.

At the *state-power* end of the continuum, central state actors ('geographical'), publicly funded out of taxation (sectoral), with a legitimate monopoly of coercive powers (legal), use surveillance systems to monitor public space (spatial), with the aims of identifying, apprehending and punishing known offenders (functional). As we saw earlier, for example, in the first nine months of the Home Office-initiated ANPR scheme, 'intercept teams' comprising of 'state' police officers stopped over 130,000 vehicles which led to the arrest of over 10,000 people, mainly for non-driving related offences (Norris, 2006, p. 3). In contrast, at the *private-justice* end of the *plural policing continuum*, private security officers sell their services in the free market, armed only with the powers of ordinary citizens, and use surveillance systems in a 'pro-active' way to enforce the internal rules of the organisation, rather than state-based norms of apprehension and punishment. In the 'privately owned', but publicly accessible, space of the shopping mall, for example, private security officers use CCTV systems to enforce 'exclusionary' policies based on the informal rules of 'private justice', rather than the formal rules of the criminal justice system (McCahill, 2002). Within this continuum there is, of course, enormous scope for diversity and complexity in the policing of 'public' and 'private' space. For instance, policing activities in many cities across the UK now mirror those of the shopping mall with open-street public space CCTV systems being used by 'state' and 'non-state' actors in a 'pre-emptive' way to monitor and exclude undesirables from the new spaces of consumption.

However, while this chapter has tried to draw out some of the empirical complexity involved in 'plural' policing networks, and attempted to avoid giving the impression that one field ('state' or 'private' sector) marshals or dominates the other, it would, as Kevin Stenson (2005) argues, be 'naïve to present state agencies as simply operating on a common plane of security provision alongside the commercial sector' (p. 280). With their symbolic power, legitimacy claims, huge command over resources and information and key position as a 'back-up of last resort', it would be 'implausible to reduce state organizations ... to the status of equivalent nodes in the market for security' (Stenson, 2005, p. 273; Crawford, 2006, p. 459). We should not be surprised to find, therefore, that while the expansion of CCTV in 'semi-public' and 'private' spaces may be driven by 'beyond-the-state' interests of loss prevention and commercial image, these systems can be easily and routinely co-opted for traditional policing. Thus, while the 'public' police may increasingly find their actions shaped by the actuarial concerns of compliance-seeking institutions (Ericson & Haggerty, 1997, p. 49), it could equally be argued that in reacting to the 'crime control' requirements of central state actors, 'private security' officers have added the goals of 'crime fighting' and 'law enforcement' to their central concerns of 'private justice' and 'loss prevention'.

NOTES

1. Indeed it has been shown that in many national jurisdictions there are now more non-state policing agents than there are police officers with private security to state police ratios of five or six to one in South Africa, two or three to one in Canada and the US and two to one in the UK (Shearing & Wood, 2003; Jones & Newburn, 2002).

2. Research carried out as part of the Europen Union-funded URBANEYE study, for example, found that less than 1 in 20 CCTV systems in London recorded data in a digital form that would allow computerised analysis of the images (Norris & McCahill, 2006).

ACKNOWLEDGMENT

The author would like to thank Roy Coleman and Charlie Cooper for comments and suggestions made on an earlier draught of this chapter.

REFERENCES

Anon. (2005a). Cameras can make ASBO's work. *CCTV Image*, (May).

Anon. (2005b). Newark meets rural needs. *CCTV Image*, (Summer).

Anon. (2006a). London's new partnership model. *CCTV Image*, (May).

Anon. (2006b). The view of Lincoln city centre. *CCTV Image*, (May).

Anon. (2006c). Taking the bull by the horns. *CCTV Image*, (Spring).

Cohen, L. (2005). Naval review. *CCTV Image*, (Winter).

Coleman, R. (2004). *Reclaiming the streets: Surveillance, social control and the city*. Cullompton: Willan.

Coleman, R., Sim, J., & Whyte, D. (2002). Power, politics and partnerships: The state of crime prevention on merseyside. In: G. Hughes & A. Edwards (Eds), *Crime control and community: The new politics of public safety*. Cullompton: Willan.

Crawford, A. (2003). The pattern of policing in the UK: Policing beyond the police. In: T. Newburn (Ed.), *Handbook of policing*. Willan: Cullompton.

Crawford, A. (2006). Networked governance and the post-regulatory state? Steering, rowing and anchoring the provision of policing and security. *Theoretical Criminology*, *10*(4), 449–479.

Crawford, A., Lister, S., Blackburn, S., & Burnett, J. (2005). *Plural policing: The mixed economy of visible patrols in England and Wales*. Bristol: The Policy Press.

Ericson, R. V., & Haggerty, D. (1997). *Policing the risk society*. Oxford: Clarendon Press.

Fielding, J. (1758). *An account of the origin and effects of a police set on foot ... in the year 1753, upon a plan presented ... by Henry Fielding*. London: Millar.

Fussey, P. (2007). An interrupted transmission? Processes of CCTV implementation and the impact of human agency. *Surveillance and Society*, *4*, 229–256. Also available at http://www.surveillance-and-society.org

Garland, D. (1996). The limits of the sovereign state: Strategies of crime control in contemporary society. *The British Journal of Criminology*, *36*, 445–471.

Garland, D. (2001). *The culture of control: Crime and social order in contemporary society*. Oxford: Oxford University Press.

Gill, M. (2006). Police need more CCTV training. *CCTV Image*, (December).

Gill, M., & Spriggs, A. (2005). Assessing the impact of CCTV. *Home Office Research Study*, *292*, 1–176.

Goold, B. J. (2004). *CCTV and policing: Public area surveillance and police practices in Britain*. Oxford: Oxford University Press.

Graham, S. (1998). Spaces of surveillant simulation: New technologies, digital representations, and material geographies. *Environment and Planning D: Society and Space*, *6*, 483–504.

Henry, S. (1987). Private justice and the policing of labor. In: C. Shearing & P. Stenning (Eds), *Private policing* (pp. 45–71). Beverly Hills, CA: Sage.

Hillyard, P. (1987). The normalization of special powers: From Northern Ireland to Britain. In: P. Scraton (Ed.), *Law, order and the authoritarian state: Readings in critical criminology*. Milton Keynes: Open University Press.

Hughes, G. (2007). *The politics of crime and community*. Basingstoke: Palgrave Macmillan.

Johnston, L. (1992). *The rebirth of private policing*. London: Routledge.

Johnston, L., & Shearing, C. (2003). *Governing security: Explorations in policing and justice*. London: Routledge.

Jones, T., & Newburn, T. (1998). *Private security and public policing*. Oxford: Clarendon Press.

Jones, T., & Newburn, T. (2002). The transformation of policing: Understanding current trends in policing systems. *British Journal of Criminology*, *42*, 129–146.

Kelly, D. (2006). We've got your number. *CCTV Image* (Spring), 34–38.

Kemshall, H., & Maguire, M. (2001). Public protection, partnership and risk penalty: The multi-agency risk management of sexual and violent offenders. *Punishment and Society*, *3*, 237–264.

Lomell, H. M. (2004). Targeting the unwanted: Video surveillance and categorical exclusion in Oslo, Norway? *Surveillance & Society*, *2*, 347–361. Also available at http://www.surveillance-and-society.org/articles2(2)/unwanted.pdf

McCahill, M. (2002). *The surveillance web*. Cullompton: Willan.

McCahill, M., & Norris, C. (2002). *CCTV in London*. Urbaneye Working Paper no. 6. Centre for Technology and Society, Technical University of Berlin, Berlin. Also available at http://www.urbaneye.net/results/results.htm

McCahill, M., & Norris, C. (2003a). Estimating the extent, sophistication and legality of CCTV in London. In: M. Gill (Ed.), *CCTV*. Perpetuity Press.

McCahill, M., & Norris, C. (2003b). Victims of surveillance. In: P. Davis, V. Jupp & P. Francis (Eds), *Victimisation: Theory, research and policy*. Macmillan.

McCahill, M. & Norris, C. (2003c). *CCTV systems in London: Their structures and practices*. Working Paper no. 10. Also available at www.urbaneye.net

McGrail, B. (1998). Communication technology, local knowledges, and urban networks. Paper presented at Telecommunications and the City, University of Georgia, March.

McMullan, J. L. (1998). The arresting eye: Discourse, surveillance and disciplinary administration in early English police thinking. *Social and Legal Studies*, *7*, 97–128.

Morgan, R., & Newburn, T. (1997). *The future of policing*. Oxford: Clarendon Press.

Norris, C. (2003). From personal to digital: CCTV, the panopticon, and the technological mediation of suspicion and social control. In: D. Lyon (Ed.), *Surveillance as social sorting: Privacy, risk and digital discrimination* (pp. 249–281). New York: Routledge.

Norris, C. (2006). 'Criminal Justice' expert report in *A report on the Surveillance Society* for the Information Commissioners Office compiled by the Surveillance Studies Network, 1–9. Available at http://www.ico.gov.uk/upload/documents/library/data_protection/practical_application/surveillance_society_appendices_06.pdf

Norris, C., & Armstrong, G. (1999). *The maximum surveillance society*. Oxford: Berg.

Norris, C., & McCahill, M. (2006). CCTV: Beyond penal modernism? *British Journal of Criminology*, *46*, 97–118.

Pasquino, P. (1991). 'Theatrum politicum': The genealogy of capital-police and the state of prosperity. In: G. Burchell, C. Gordon & P. Miller (Eds), *The Foucault effect: Studies in governmentality*. Chicago: University of Chicago Press.

Reeve, A. (1996). *The private realm of the managed town centre*. Unpublished manuscript. Joint Centre for Urban Design, Oxford Brooks University, Oxford. A revised version also published in *Urban Design International, 1*, 61–80.

Rose, N. (1999). *Powers of freedom*. Cambridge: Cambridge University Press.

Shearing, C., & Wood, J. (2003). Nodal governance, democracy, and the new "denizens". *Journal of Law and Society, 30*, 400–419.

Shearing, C. D., & Stenning, P. C. (1987). Say "cheese!": The Disney order that is not so Mickey mouse. In: C. D. Shearing & P. C. Stenning (Eds), *Private policing*. California: Sage.

Smith, G. J. D. (2004). Behind the screens: Examining constructions of deviance and informal practices among CCTV control room operators in the UK. *Surveillance & Society, 2*, 377–396. Also available at http://www.surveillance-and-society.org

Stenson, K. (2005). Sovereignty, biopolitics and the local government of crime in Britain. *Theoretical Criminology, 9*, 265–287.

Stenson, K., & Edwards, A. (2001). Rethinking crime control in advanced liberal government: The "third way" and return to the local. In: K. Stenson & R. R. Sullivan (Eds), *Crime, risk and justice: The politics of crime control in liberal democracies*. Devon: Willan.

The Guardian. (2006). DNA of 37% of black men held by police. *The Guardian*, January 5.

Travis, A. (2007). Ground floor perfumery, stationary ... and cells, *The Guardian*, March 15.

Urbaneye. (2004). *On the threshold to urban Panopticon? Analysing the employment of CCTV in European cities and assessing its social and political impacts*.Final report to the European Union. Technical University of Berlin, Berlin.

Wakefield, A. (2003). *Selling security: The private policing of public space*. Cullompton: Willan.

Wakefield, A. (2005). The public surveillance functions of private security. *Surveillance and Society, 2*, 529–545. Also available at www.surveillance-and-society.org

Webster, W.R. (1998). Surveying the scene: Geographic and spatial aspects of the closed-circuit television surveillance revolution in the UK. Paper presented at the XII meeting of the Permanent Study Group on Informatization in Public Administration, European Group of Public Administration Annual Conference, Paris, France, September 14–17.

Webster, W.R. (2004). The evolving diffusion, regulation and governance of closed circuit television in the UK. Paper presented at *CCTV and Social Control: The Politics and Practice of Video Surveillance*, University of Sheffield, January 7–8.

Williams, C. A. (2003). Police surveillance and the emergence of CCTV in the 1960s. In: M. Gill (Ed.), *CCTV*. Leicester: Perpetuity Press.

Youth Justice Board. (2004). ISSP: The initial report. Also available at http://www.yjb.gov.uk/Publications/Resources/Downloads/ISSP%20Initial%20Report%20

Zedner, L. (2006). Policing before and after the police. *British Journal of Criminology, 46*, 78–96.

PART III: OBJECTIVES AND COUNTER-OBJECTIVES

THE NEW LATERAL SURVEILLANCE AND A CULTURE OF SUSPICION

Janet Chan

ABSTRACT

This chapter examines the 'new lateral surveillance', spearheaded by government anti-terrorism campaigns urging citizens to report any suspicious people and objects they encounter. Drawing a comparison between this and the community crime prevention (CCP) programmes of past decades, the chapter discusses the likely effectiveness of such campaigns in controlling crime and increasing security, suggests an alternative interpretation and discusses the consequences of the culture of suspicion generated by this form of surveillance. It concludes that the new lateral surveillance is a form of 'high policing' that is both political and dangerous in its vulnerability to errors.

INTRODUCTION

Much has been written about the spread of surveillance in modern society. It is truly remarkable that in the first decade of the 21st century, we have quietly, and without a great deal of resistance, taken for granted the

Surveillance and Governance: Crime Control and Beyond
Sociology of Crime, Law and Deviance, Volume 10, 223–239

ISSN: 1521-6136/doi:10.1016/S1521-6136(07)00210-2

ubiquitous presence of surveillance in everyday life: CCTV cameras in public and private spaces, scanners and sniffer dogs at airports and the web of digital dataveillance technologies (Ball & Webster, 2003; Marx, 2002). Once we may have been alarmed by the rise of the 'surveillance society' (Lyon, 2001; Norris & Armstrong, 1999), but the silent and apparently neutral presence of audiovisual, biometric and digital systems for monitoring identities, movements, possessions and communications is no longer a burning issue for citizens. Routine, intense and even intrusive surveillance technology has become an accepted part of the landscape in Western democracies. In effect, we have an assemblage of surveillant mechanisms that permeate and connect in a rhizomatic network of sensors and detectors, memories and logics, to prevent, pre-empt or control the eruption of disorder or catastrophe (Haggerty & Ericson, 2000).

There is, of course, a similar pervasiveness of non-technological surveillance by policing agents – private or public – whose visible presence is often utilised by organisations or demanded by citizens to provide security. In spite of the by now accepted fact that policing agents are not necessarily police or security officers, the phenomenon of citizen-to-citizen (or lateral, peer-to-peer) surveillance (Andrejevic, 2005) has received relatively little attention. Yet, there is an undeniable line of continuity between the informal watching of communities by their members before the institution of public police, through the more or less organised citizen patrols and neighbourhood watch programmes of the 1970s and 1980s, to contemporary government campaigns to recruit citizens to take part in the War on Terror.

This chapter is concerned with examining this neglected aspect of surveillance – lateral or peer-to-peer surveillance – in the current context of heightened threats and uncertainties. In particular, it compares the lateral surveillance of the current political climate with earlier movements to involve community members in surveillance and social control. It argues that like previous attempts, the new lateral surveillance is a conscious effort by governments to engage citizens to be the 'eyes and ears' of the police in their war against crime, disorder or terrorism. However, unlike community crime prevention (CCP) initiatives, the current campaign is global rather than local, general rather than specific in aims; it is more explicitly political and less concerned with outcomes. While CCP had been subjected to various degrees of monitoring and evaluation, the new lateral surveillance is open-ended and unaccountable. If CCP has been a move to 'responsiblise' communities against crime and disorder, the new lateral surveillance is an irresponsible campaign to stir up fear and paranoia against terrorism.

Where the former has died a natural death following its proven ineffectiveness and loss of interest among community members, the latter has lived on in the subconscious of the public, sustaining a culture of suspicion and a positive support for extreme precautionary and pre-emptive strategies. While the former is meant to build social capital, the latter is designed to destroy it. Rather than strengthening communities and building partnerships, the new lateral surveillance destroys trust and produces interpassivity (Zizek, 1998).

THE NEW LATERAL SURVEILLANCE

The terrorist attacks of the United States on 11 September 2001 (or 9/11 for short) have led to a host of responses from governments all over the world – these include various forms of special legislation to deal with terrorism, increased surveillance and various pre-emptive initiatives, including the Iraq War (Ericson, 2007). The new lateral surveillance of interest to the current discussion is spearheaded by government campaigns that tell ordinary citizens to be vigilant and report unusual or suspicious objects, people and circumstances to authorities (see Chan, 2007).

In Australia, a $15 million 'Be Alert, Not Alarmed' advertising and public relations campaign in 2003 has emphasised the need for individual vigilance and 'the right to suspicion' (Bossio, 2005, p. 3). The campaign distributed 'terror kits' to eight million Australians. The kits contain fridge magnets with a 24-hour National Security hotline number, a booklet on anti-terrorist and emergency procedures and a two-page letter from the Prime Minister John Howard (Tilley, 2004). In 2004, a new 'Help Protect Australia from Terrorism' advertising campaign was launched. Large posters placed in public places and television advertisements represented the War on Terror as an exercise in putting together a jigsaw puzzle: 'Small pieces of information from members of the public can help keep Australia safe from terrorism. Police and security agencies are working hard but you could help them complete the picture'. Under the heading 'Every piece of information helps', the poster asks citizens to look out for unusual activities such as surveillance or filming of official buildings, cars parked in public places for a long time, packages abandoned in public places, suspicious purchases of chemicals or explosives and unusual uses of garages.

Mythen and Walklate (2006) describe a national campaign in the United Kingdom in 2004: a 22-page booklet was sent to 25 million households giving advice on how to respond to terrorist attacks. An excerpt from the booklet shows the familiar plea to citizens to help to 'prevent a terrorist attack' by keeping alert, looking out for suspicious or unusual behaviours or objects, and call the police if 'you believe there is an immediate threat to life' (Mythen & Walklate, 2006, p. 135). Similar campaigns have been used by the London Metropolitan Police to warn citizens to look out for suspicious bags, unusual activities and seemingly ordinary situations. One series of posters with a heading 'terrorism: HELP US DEFEAT IT' shows images of unattended bags or briefcases on the street, in a mall or on public transport. They raise questions about the 'innocence' of these objects: 'Innocent bin bag or terrorist bomb? Innocent box or incendiary device? Innocent briefcase or terrorist bomb?' Citizens are urged not to touch the object, warn others, move away and call 999: 'If you suspect it, report it'. A more recent series put the responsibility for foiling terrorist attempts squarely on citizens: 'Terrorists won't succeed if someone reports suspicious bags, vehicles or behaviour. You are that someone. Call 999'. These posters, printed in deep red with black and white photographs cropped to narrow horizontal windows suggestive of someone peeking through a small opening, extend the net of vigilance to include other kinds of 'suspicious' people or activities:

- Terrorists need places to live. Are you suspicious of your tenants or neighbours?
- Terrorists need storage. Are you suspicious of anyone using garages, lock-ups or storage space?
- Terrorists need funding. Have any cheque or credit card transactions made you suspicious?
- Terrorists need transport. Has a vehicle sale or rental made you suspicious?

Another example is the New York City Metropolitan Transportation Authority's attempt to involve the public to be part of the terrorist watch. The slogan 'IF YOU SEE SOMETHING, SAY SOMETHING' was printed in big black letters against a bright yellow background, together with the warning 'BE SUSPICIOUS OF ANYTHING UNATTENDED' in smaller red letters. The picture above the slogan shows a row of empty seats on a train and a brown paper bag with heavy oily stains on the floor. Citizens are urged to 'Tell a cop, an MTA employee or call 1-888-NYC-SAFE'. Similar ideas are involved in a government call-in programme called TIPS which resembled the TV show *America's Most Wanted* and a CIA postcard campaign that was a type of 'neighbourhood watch program against terrorism' (Andrejevic, 2005, p. 486).

At first blush, these campaigns look very similar to the CCP initiatives in the 1970s and 1980s such as Neighbourhood Watch and Crime Stoppers. They operate under the same rhetoric of responsibilisation and self-management as CCP (Rose, 2000). As Ericson and Haggerty (1997, p. 156) note, watch programmes mobilise local residents 'to become reflexive with respect to risks', not only for crime prevention or arrest of criminals, 'but also to help residents confront their emotional responses to risk and to become responsible for policing their own territories'. The new lateral surveillance also emphasises citizen's local knowledge and how this can be put to use in a partnership with policing authorities (Mythen & Walklate, 2006). As Andrejevic demonstrates in citing the US Senate Majority Leader Bill Frist's book on bioterrorism:

> Frist suggests that one of the most important forms of civilian participation is to 'become the eyes and ears of our law enforcement agencies': 'You know your communities better than anyone else. You know when something looks out of place, whether it's a package left on the subway or someone acting in an unusual or suspicious manner in your neighbourhood'. (Andrejevic, 2005, p. 486)

Both types of surveillance are premised on the notion of suspiciousness – anything unusual, out of place or incongruent (Norris & Armstrong, 1999; Chan, 2007). Local knowledge helps define normality and abnormality. For example, real estate agents participating in a Realty Watch programme in Canada are told to 'watch and listen for suspicious activity wherever they go' and to report anything they find suspicious to the police (Ericson & Haggerty, 1997, pp. 164–165). Participants are given a 17-point 'Checklist of Suspicious Activity' to guide them in their surveillance; these include activities involving juveniles especially during school hours, persons running and carrying something valuable, parked cars with occupants at unusual places or times, and apparent business transactions conducted from a car (*ibid.*, p. 165). These indicators of suspicion are similar to those listed in the UK campaign posters.

A closer look at the new lateral surveillance, however, points to a clear departure from the old CCP. Where CCP is principally directed towards the prevention of specific categories of crime and disorder – minor street crime, vandalism, incivilities, trespassing, burglary, etc. – the new lateral surveillance is concerned with the detection of organised terrorist activities that are broad, ill-defined and not locally based. There is no way to ascertain the usefulness of a piece of information – it can be a crucial piece in the jigsaw puzzle, or it can be totally useless or even misleading. More often than not, a suspicious package found in a public place turns out to be something quite innocent or benign (Chan, 2007). But the point of vigilance is it *could have been* a bomb or a biological weapon. With terrorism, it is almost impossible to predict or even imagine when and how attacks will occur. As a result, the logic of risk management gives way to a 'precautionary logic' (Ewald, 2002; Ericson, 2007) that imagines the worst and prepares for it. Citizens are to 'accept the state of heightened risk' and to plan, prepare and get *ready* (Andrejevic, 2006, p. 442).

This orientation towards *preparedness* is, as Lakoff (2006, p. 265) argues, 'the primary strategic logic' United States' security experts have dealt with diverse forms of disasters and threats. As 'an ethos and a set of techniques for reflecting about and intervening in an uncertain, potentially catastrophic future', preparedness enjoys political support and legitimates a set of technical solutions, without questioning the assumptions and validity of current economic and social order (*ibid.*, p. 266).

The new lateral surveillance both shapes and is shaped by this new 'politics of uncertainty' (Ericson, 2007). The universality of terrorist threat is often reduced to the 'specter of a simple, cheap, but devastating attack on national security by a lone terrorist hunkered down in a suburban garage' (Andrejevic, 2006, p. 446). The need to responsibilise citizens in the War on Terror is a far more desperate and perhaps honest acknowledgement of the impotence of the state policing apparatus than found in the rhetoric of CCP:

> The recurring message is an admonitory invocation of self-reliance as the compensatory response to the disturbing recognition that the lumbering institutions of mass society – mass armies and their hyper-expensive equipment – aren't nimble enough to counter the flexible threat of terrorism. The work of defence has to be offloaded onto the civilian population. (Andrejevic, 2006, p. 447)

As Mythen and Walklate (2006, p. 134) point out, the UK emergency advice booklet is heavily 'skewed' towards individual responsibility (what you can do to protect yourself and your community against terrorism) and 'decidedly mute' about what the government is doing to protect citizens.

THE EFFECTIVENESS OF PEOPLE WATCHING PEOPLE

While CCP may have embodied the ideal of community-based policing, the new lateral surveillance is more driven by politics than ideas. One important rationale for the new lateral surveillance is to recruit the public into active supporters of the War on Terror. As reported in Chan (2007), John Rendon, an American 'public relations guru', has argued that the United States is in fact fighting two wars, one against 'real terrorists' and the other against 'potential terrorists', or more accurately, a war to win over potential allies:

> That means in reality, the threat comes not from the 12-million people, the 1%, the threat comes from the rest, if we don't get them engaged in the nature of this conflict. What we need to do is ... we need to turn the street into an active ally and away from being a passive observer. (John Rendon, quoted in ABC, 2006)

In line with the US President's famous statement that 'You're either with us or against us in the fight against terror' (CNN, 2001), citizens are conscripted figuratively into 'soldiers' who are 'ready' for the War on Terror (Andrejevic, 2006).

Tilley's (2004) analysis of the content of the Australian Prime Minister's letter in the government's 2003 'terror kit' suggests that 'particular propaganda devices' were used in 62% of the 267 phrases. Using a coding scheme adapted from Lee and Lee's (1939) categories, Tilley concludes that the letter consists mainly of 'positive' propaganda, such as 'glittering generality' (positive but unverified abstract words) and 'bandwagon' appeals (words that imply collective benefits) (Tilley, 2004, p. 40). Of particular relevance to the new lateral surveillance is the letter's use of 'plain folks' appeals:

> Pictures of the ubiquitous 'Aussie barbecue', beach scenes, family backyard cricket game, and a multicultural schoolyard strongly connoted normalised 'average' middle-class family values ... The plain folks concept can be an influential tool in demonising the 'enemy' as 'outside' what is 'normal', as external to the idealised and over-simplified 'community' (in which 'we' are all assumed to belong to families, live in neighbourhoods, go to work every day and hold the same ideas about what is 'ordinary' in order to 'spot ... things that are out of the ordinary' [Abetz, 2003, p. 6]). (Tilley, 2004, p. 39)

As Bossio suggests, 'The idea of an "ordinary" individual being able to thwart global terror is a powerful appeal to patriotic – and somewhat nostalgic – ideals of freedom'. More important, this ideal and the effectiveness of public vigilance through lateral surveillance are supported by 'real-life' success stories in London and Israel (Bossio, 2005, p. 8).

This patently political agenda may explain why these campaigns have not been subject to scientific scrutiny as to their effectiveness. The open-ended and unaccountable nature of the new lateral surveillance is in stark contrast with CCP initiatives which have virtually spawned a growth industry of programme evaluation. There is a wealth of research evidence that documents the mixed success of CCP programmes in a variety of settings. The dismal failure of programmes, such as Neighbourhood Watch, in reducing crime is well known and openly debated (Rosenbaum, 1987).

The fact that such programmes can actually heighten the fear of crime is an unintended consequence that is taken seriously in crime prevention policymaking (Norris & Kaniasty, 1992).

In the absence of evaluation research on the effectiveness of the new lateral surveillance or even a definition of what constitutes effectiveness, it is possible to at least speculate how it may actually 'work'.

Suppose that an 'alert' citizen Mr Blogg, while going on his morning walk, spots someone watching or filming the Sydney Opera House (one of the assumed terrorist targets). What makes this person, say a man, suspicious and worthy of reporting? Obviously, thousands of people visit the Opera House every week and millions of hours of video would have been taken. Something must have triggered Mr Blogg's suspicion. Perhaps it is a man of 'Middle Eastern appearance', wearing clothes that are different, filming for an excessively long time or focusing on places that are not normal for tourists. Perhaps he carries a big bag which conceivably may contain weapons or bombs. He may be taking notes in between filming. And so on. Mr Blogg, being a conscientious citizen who heeds the call 'If you suspect it, report it', decides to call the police. After a period of time, perhaps an hour, perhaps more, depending on how busy the police are, a police officer comes to Mr Blogg's apartment and asks questions about his observation. By this time the suspicious man is no longer near the Opera House, and Mr Blogg is asked for a detailed description of the man, or he is invited to go to the police station to look at some photographs of offenders to make a positive identification of the person.

Even if the government insists that 'Every piece of information helps', there is a very low likelihood that Mr Blogg's surveillance will result in any positive identification or even useful piece of intelligence for the police. First of all, the suspicious man may be totally innocent – for example, a documentary maker who has a lot of equipment in his bag – and even if he were 'caught', it is not going to help the anti-terrorism campaign at all. Second, the reliability of Mr Blogg's memory and his ability to provide a sufficiently detailed description or positive identification in a line-up may be questionable. Finally, the interviewing and investigation procedure can often compromise the quality of Mr Blogg's information.

There is a vast psychological literature that demonstrates the fallibility of eyewitness testimonies (Cutler & Penrod, 1995; Brewer, Weber, & Semmler, 2005). Cutler and Penrod's (1995) review of the literature estimated that 'there may be as many as 4,500 erroneous convictions each year arising from mistaken identification' in the United States (p. 8). A meta-analysis of four field experiments conducted in the late 1980s to early 1990s, involving

291 mock eyewitnesses who underwent 536 identification tests finds that the average rate of correct identification is 41.8%, while the average rate of false identification is 35.8%:

> What we learn from these experiments is that identification for persons seen briefly, in nonstressful conditions, and attempted after brief delays, are frequently inaccurate. In customer-present [or target-present] photoarrays that resemble the situation in which the suspect is guilty, only two out of five guilty persons were correctly identified. In customer-absent [or target-absent] photoarrays that represent the situation in which the suspect is innocent, one out of three innocent persons was falsely identified. In one of these studies (Pigott et al., 1990), the mock-eyewitnesses were bank tellers, 77% of whom reported that they had received training for eyewitness situations. (Cutler & Penrod, 1995, p. 12)

Cutler and Penrod (1995, p. 13) point out that field experiments probably 'overestimate the accuracy of eyewitness identifications' because eyewitnesses in real-life, especially violent situations, are likely to experience emotional stress. The external validity (generalisability) of experimental research is, of course, always a matter for concern (see Brewer et al., 2005, pp. 180–181), but this body of research does provide evidence that eyewitness identification is a human process and thus vulnerable to social, psychological and situational influences. For example, longer exposure time to the 'event' can improve face recognition, the presence of a weapon may reduce identification accuracy, a delay between the event and the identification test can lead to poorer identification and giving a verbal description of the offender before making an identification can lower the rate of accurate identification (the verbal overshadowing effect) (Brewer et al., 2005). The actual organisation of the line-up, including subtle cues about the presence or absence of the offender, line-up instructions, the composition of the line-up and its presentation mode (simultaneous or sequential) all affect the accuracy of identification (*ibid.*). What is more disturbing, but perhaps not too surprising, is the so-called cross-race effect:

> The basic finding is that attempts to recognize faces of the same race produce a higher proportion of hits, or correct identifications, from target-present line-ups than attempt to identify faces of another race. Conversely, a lower proportion of false alarms, or false identifications, from target-absent line-ups is made for same- compared with other-race face identifications. (Brewer et al., 2005, p. 190)

A dramatic instance of mistaken identity is one that led to the killing of an innocent man Jean Charles de Menezes by the London Metropolitan Police Service (MPS) in 2005. Although the Independent Police Complaints Commission's (IPCC) 2006 investigation report (*Stockwell One*) into the killing has still not been made public, the BBC's special report and the photographs of Menezes and Hussain Osman, the suspect police were looking for, suggest that eyewitness identification can be very unreliable, even among trained police officers (BBC News, n.d.). The IPCC's 2007 report on the MPS's handling of the public statements in fact blames some of the misinformation on civilian witnesses' accounts, which were subsequently repeated by police without checking:

> In the immediate aftermath of the shooting, civilian witnesses gave their account of what had happened to the media. These accounts mistakenly described Mr. de Menezes as wearing un-seasonal clothing, running away from the police, jumping the ticket barrier and acting suspiciously in other ways. It is apparent that some of the witnesses confused police officers with Mr. de Menezes. These accounts were inaccurate, but the IPCC accepts they resulted from genuine mistakes in a very stressful situation. Mr. de Menezes did nothing and wore nothing that could be considered suspicious. These mistaken civilian accounts that were given to and broadcast by the media became accepted and in some cases repeated, by the MPS. (IPCC, 2007, p. 26)

The effectiveness of surveillance technologies and practices is, as Haggerty and Ericson (2006, p. 13) point out, 'the net outcome of often contentious political struggles, including symbolic politics'. The new lateral surveillance is promoted by governments not necessarily because of its effectiveness, but there are 'more symbolic and emotional factors that contribute to the adoption of such tools' (*ibid.*, p. 12). In view of this emotional and political dimension, it is difficult to judge whether the fact that the Australian National Security Hotline was not inundated by callers – averaging only 16 calls per hour in its first week rather than the anticipated 2,000 per hour (Bossio, 2005) – should be interpreted as a failure of the campaign or a triumph of commonsense.

THE THREAT OF TERRORISM AND ITS CONSEQUENCES

The campaign for lateral surveillance is premised on the probability and imminence of terrorist attacks and serves to highlight this threat in the public mind. Research has consistently shown that real or perceived threat has a tendency to 'increase intolerance, prejudice, ethno-centrism, and xenophobia' (Huddy, Feldman, Taber, & Lahav, 2005, p. 594). Threat also increases support for 'punitive action' and even 'belligerent solutions' against threatening groups (*ibid.*, p. 594). Huddy et al.'s (2005) national survey of 1,549 adult US citizens between October 2001 and March 2002 not surprisingly found very high levels of perceived threat of future terrorist attacks on the United States, but they also found a significant link between perceived threat and negative stereotyping of Arabs and increased support for domestic anti-terrorism policies:

> Americans who perceived a high future threat of terrorism not only supported aggressive action against the enemy, they were also more likely to negatively stereotype Arabs and support restrictive immigration and intensified surveillance policies directed at Arabs and Arab-Americans … Perceived threat consistently increased support for domestic antiterrorism policies. Support for a national identification card and government monitoring of telephones and email rose significantly as the perceived threat of future terrorism increased. Similarly, threat was linked to a greater concern about the failure to enact strong antiterrorism measures than such laws would place undue restrictions on civil liberties. (Huddy et al, 2005, p. 602–603)

These findings help explain the results of Aly and Balnaves' (2007) survey of Australians (year of survey unknown, response rates unknown) which suggests that Muslim Australians perceive the threat of terrorism differently from non-Muslim Australians: while the latter were fearful of 'homegrown terrorism', the former were more concerned with backlash against their community. Focus group participants who expressed fear of terrorism linked their fear to media reports and the Australian government's media campaigns. Some have adopted behavioural changes, such as avoiding public transport and air travel, but 'the most frequently cited change to behaviour was an increase in suspicion of others, described by one focus group as "paranoia", particularly suspicion of people who conformed to the media image of terrorists' (Aly & Balnaves, 2007, p. 117). Participants were also critical of government and media discourse about terrorism:

> The focus group findings indicate that both Australian Muslims and those in the broader community believe that the government and media discourse is constructed with the overriding aim of instilling fear with comments such as: 'I think the messages that are coming in to us are definitely "Be Afraid", "Watch your Back", "Be Very Careful", "Don't trust anyone"'; 'this message of fear is being jammed down our throats' and 'There are these evil terrorists around … like trying to create unease, suspicions'. Muslim participants expressed an acute awareness of themselves as the objects of fear and suspicion: 'the government love to create a monster you know to feed and to frighten people – now we are the monster'. (Aly & Balnaves, 2007, p. 118)

These findings echo reports of 'racially motivated' attacks, discrimination and harassment, threats, property damage and verbal assault in public against Arab, Muslim and Sikh Australians during the two months following 9/11 (Dreher, 2005). Similar findings of discrimination, abuse and violence have been reported in Poynting and Noble's (2004) study of Muslim Australians and Sheridan's (2006) of British Muslims. The apparently pervasive and routine nature of such incidents suggests that the culture of suspicion has in fact developed into a culture of hatred:

> One of the most disturbing findings of the study was the frequency of these incidents of racism – these were often everyday experiences and accepted as commonplace by the interviewees. The pervasive nature of these incidents suggest that for many citizens of Arab and Muslim background racism, abuse and violence form part of an everyday landscape of fear and incivility'. (Poynting & Noble, 2004, p. 17)

The new lateral surveillance, with its emphasis on the unusual, the incongruent and the out-of-place, seems to complement the racial stereotyping of technological surveillance, such as CCTV, which in the United Kingdom targeted young black men for special scrutiny (Norris & Armstrong, 1999).

What, then, are the consequences of this new culture of suspicion? If crime control is ultimately to engender social order and physical security, then a culture of suspicion is the anti-thesis of order and security because it undermines the ontological security of social interactions (Giddens, 1984). Claus Offe (1999) who has written about 'horizontal trust' has a lot to say about lateral surveillance and the culture of suspicion. Offe postulates that the social order of modern society is ideally reproduced by three coordination resources: money, legitimate state authority and knowledge (Offe, 1999, p. 42). However, the 'ideal of an intelligently regulated market economy' neglects the role of informal social, cultural and moral coordination resources such as trust, which cannot be bought by money, manipulated by strategic action or taught by gaining knowledge (p. 43). Most of the literature on trust has been concerned with the vertical dimension of citizens' trust of political or other elites (Tyler, 1990), but it is the horizontal trust between citizens that is most relevant to lateral surveillance. A culture of suspicion can be simply conceptualised as a climate lacking in trust, often indicated by a reluctance of people to enter into any relations with others without intensely monitoring the risks of failure (Offe, 1999, p. 47).

Traditionally, trust is built on long-term experience between the truster and the trustee, with the stability of the trust relationship being premised on the latter developing a 'sense of moral *obligation* to honor trust' or the presence of strong incentives, such as reputational or social capital, to continue the trust relationship (Offe, 1999, p. 50, emphasis in original). In the case of horizontal trust between citizens who do not have a history of direct interaction – as is the case with trusting strangers – there are simply no opportunities to build trust, test its validity, or establish appropriate incentives for sustaining it (*ibid.*, p. 55). In effect, the more complex society has become and the more mobile its citizens, the more difficult it is for horizontal trust to be established.

Nevertheless, trustworthiness can be assessed in a non-experiential way through substituting 'categorical trust' for personal experience: by '[a]ttaching positive or negative trust value to entire social categories' based on stereotypes of trustworthiness, for example, in the form of visual markers:

> Trust-*inviting* visual markers include the wearing of licensed uniforms (e.g. military, religious, professional) or chosen visual signs such as haircut, style of dress, and other such bodily signifiers of identity and belonging. They have all the advantage of being easily perceived from some distance and even in the absence of and prior to any actual communication and interaction taking place. (Offe, 1999, p. 64, emphasis in original)

It is easy to see how 'categorical trust' is simply the flipside of 'categorical suspicion' (Lyon, 2001). The danger of categorical trustworthiness has been anticipated by Offe:

> The more trust is based upon group identity and group belonging (as a substitute for direct interaction), the more limited it is and the more likely trust is to be withheld from anybody outside of the boundaries of the group, with the result of massive discrimination and aggressive distrust directed at the outside world of those who do not clearly belong to 'us'. (Offe, 1999, p. 65)

Some would argue that 9/11 has in fact eliminated categorical trust/suspicion as 'we are all suspects now' (Ericson, 2007, p. 214). In practice, however, a hierarchy of suspicion still operates: while security checks are becoming routine in hot-spots such as airports and government buildings, certain visual markers still set off more attention from agents of surveillance than others (Norris & Armstrong, 1999). In fact, the most pressing problem for law enforcement and security is how to use technology to identify indicators of suspicion and hence automate the detection of 'suspicious activity patterns'.

If the lack of horizontal trust is the mark of a culture of suspicion, then it is not surprising that this is more than made up for by an increased reliance on vertical trust – the trust of political and security elites by citizens. What is intriguing about the current situation is the apparently high trust citizens have been willing to place on Draconian laws and pre-emptive strategies, coupled with an almost reckless distrust of established institutions such as the rule of law and the protection of basic human rights. As Offe has argued, this lack of horizontal trust comes at a price:

> This difficulty of trust-building is paid for in terms of transaction costs. In the absence of trust, actors must rely on formal monitoring and enforcement. Also, low trust conditions

> involve opportunity costs, as potential gains from cooperation are being missed due to the prevailing (but actually untested) anticipation of non-cooperative disposition of others. (Offe, 1999, p. 61)

How, then, can horizontal trust be rebuilt in the wake of 9/11 and the culture of suspicion that has been generated? Offe nominates two strategies: a top-down and a bottom-up approach. The former aims to generate horizontal trust by inducing citizen compliance through legitimate and competent governance: ‘institutions and procedures generate an impeccable record in terms of truth-telling, promise-keeping, fairness and solidarity – and thus reasons for suspicion and cynicism are virtually nullified’ (Offe, 1999, p. 85). The bottom-up approach, on the other hand, relies on citizens learning to trust each other by ‘getting involved, together with strangers, in voluntary and open associative communities’ (*ibid.*). It is unfortunate that under the current paranoia about terrorism, there is every indication that governments have no intention of moving towards any form of truth-telling, let alone being concerned with fairness or legitimacy. The bottom-up task of building horizontal trust is therefore all the more pressing, albeit made difficult by the political climate. Perhaps we can turn lateral surveillance on its head rather than using it as an ineffective tool of control, we could try morphing it into an effective mechanism for mutual care and protection. By looking out for each other, instead of spying on each other, we may come close to the original ideal of building strong and resilient communities.

As I have pointed out elsewhere (Chan, 2007), far from creating a community of active citizens that look out and care for each other, our current obsession with security may have in fact transformed the ideal of the public sphere with interactive citizenship into what van Oenen calls, following Robert Pfaller (2003) and Slavoj Zizek (1998), an ‘interpassive securityscape’ (van Oenen, 2004, p. 7). As citizens, we have been made to feel responsible for our own security, and yet we are aware of our limited ability to change anything. Increasingly, we are relying on others – the government, police, private security and technology – to do the work for us:

> Involvement or engagement is delegated, outsourced. We would like to get involved, but we no longer believe we can; therefore we ask others to get involved, on our behalf… What the modern citizen wants is not to commit, but merely to ‘join’. He does not want to act, but merely to appear. Whatever happens, he is neither for nor against, but simply ‘present’. The contemporary citizen is, in short, a one-person flash mob. He seeks protection and security, but in doing so only makes the public domain more unstable and more insecure. (van Oenen, 2004, pp. 7, 16)

Lakoff contrasts two types of security – population security and nation-state security – that have co-existed in the past century. Where the former is concerned with improving the health and well-being of citizens as a collective, the latter seeks to defend territory and protect defence and industrial infrastructure to 'guarantee the continuity of political and economic order' (Lakoff, 2006, p. 272). The new lateral surveillance, as part of the logic of preparedness, is not interested in sustaining trust and building healthy communities. Rather, the call for eternal vigilance and constant suspicion is at best a feeble attempt to build support for the War on Terror, at worst it is a means to turn populations into atomised interpassive individuals who watch the War on Terror on television.

CONCLUSION

The new lateral policing has all the hallmarks of a down-market, low-tech version of 'high policing' which is characterised by the extensive use of informants 'that infiltrated all walks of society', the 'hoarding [of] all-encompassing intelligence', and an explicit goal of preserving the political regime 'and not the protection of civil society' (Brodeur & Leman-Langlois, 2006, p. 171). Like high policing, the new lateral surveillance is 'absorbent' and 'quiet': it is *absorbent* in its ambition to accumulate intelligence from the general public everywhere, it is *quiet* in its low emphasis on prosecution success, but it manages to instil 'dread in the whole population through the stealthy character of its operation' (*ibid.*, p. 182). As discussed above, the new lateral surveillance offers very little value to community protection, and may in fact escalate fear and distrust in civil society, but it may have been effective in recruiting allies in the War on Terror and hence in preserving the political survival of the government. The danger, as Brodeur and Leman-Langlois (2006) point out, is that the new 'craving for intelligence' may have created 'an irrational sellers' market', which can lead to 'the maximization of the possibilities of making mistakes' (p. 195). In this climate of fear and paranoia, it is not the Big Brother, but the Big Bungler, that ought to concern us (*ibid.*, p. 196). The killing of Jean Charles de Menezes is a chilling reminder of the price we pay for a culture of suspicion that gives the Big Bungler 'too much power and too much speed' (*ibid.*, p. 196).

ACKNOWLEDGMENT

The author would like to thank Noreen Metcalfe for her research assistance.

REFERENCES

Aly, A., & Balnaves, M. (2007). 'They want us to be afraid': Developing a metric for the fear of terrorism. *International Journal of the Diversity*, *6*(2), 113–122.

Andrejevic, M. (2005). The work of watching one another: Lateral surveillance, risk, and governance. *Surveillance and Society*, *2*(4), 479–497.

Andrejevic, M. (2006). Interactive (in) security: The participatory promise of ready.gov. *Cultural Studies*, *20*(4–5), 441–458.

Ball, K., & Webster, F. E. (2003). *The intensification of surveillance: Crime, terrorism and warfare in the information age*. London: Pluto Press.

BBC News (n.d.) 'The Menezes Killing' Special Report. http://news.bbc.co.uk/2/shared/spl/hi/uk/05/london_blasts/tube_shooting/html/default.stm. Accessed 20 August 2007.

Bossio, D. (2005). *Be Alert, Not Alarmed: Governmental communication of risk in an era of insecurity*. The Annual Meeting of the Australian and New Zealand Communication Association Christchurch, New Zealand. http://www.mang.canterbury.ac.nz/ANZCA/FullPapers/07PeacePolCommfinal.pdf. Accessed 24 August 2007.

Brewer, N., Weber, N., & Semmler, C. (2005). Eyewitness identification. In: N. Brewer & K. D. Williams (Eds), *Psychology and law: An empirical perspective* (pp. 177–221). New York: Guilford Press.

Brodeur, J.-P., & Leman-Langlois, S. (2006). Surveillance fiction or higher policing. In: K. D. Haggerty & R. V. Ericson (Eds), *The new politics of surveillance and visibility* (pp. 171–198). Toronto: University of Toronto Press.

Chan, J. (2007). Dangerous art and suspicious packages. *Law Text Culture*, *11*, 51–69.

CNN. (2001). 'Bush says it is time for action', 6 November 2001. http://archives.cnn.com/2001/US/11/06/ret.bush.coalition/index.html. Accessed 7 January 2008.

Cutler, B. L., & Penrod, S. D. (1995). *Mistaken identification: The eyewitness, psychology the law*. Cambridge: Cambridge University Press.

Dreher, T. (2005). Targeted: Experiences with racism in NSW after September 11, 2001. *UTS Shopfront Monograph Series, No. 2*. Sydney: University of Technology Sydney.

Ericson, R. V. (2007). *Crime in an insecure world*. Cambridge: Polity Press.

Ericson, R. V., & Haggerty, K. D. (1997). *Policing the risk society*. Toronto: University of Toronto Press.

Ewald, F. (2002). The return of Descartes's malicious demon: An outline of a philosophy of precaution. In: T. Baker & J. Simon (Eds), *Embracing risk: The changing culture of insurance and responsibility* (pp. 273–301). Chicago: University of Chicago Press.

Giddens, A. (1984). *The constitution of society*. Cambridge: Polity Press.

Haggerty, K. D., & Ericson, R. V. (2000). The surveillant assemblage. *British Journal of Sociology*, *51*, 605–622.

Haggerty, K. D., & Ericson, R. V. (Eds). (2006). *The new politics of surveillance and visibility*. Toronto: University of Toronto Press.

Huddy, L., Feldman, S., Taber, C., & Lahav, G. (2005). Threat, anxiety, and support of antiterrorism policies. *American Journal of Political Science*, *49*(3), 593–608.

Independent Police Complaints Commission. (2007). *Stockwell two: An investigation into complaints about the metropolitan police service's handling of public statements following the shooting of Jean Charles de Menezes on 22 July 2005*. London: IPCC.

Lakoff, A. (2006). Techniques of preparedness. In: T. Monahan (Ed.), *Surveillance and security: Technological politics and power in everyday life*. New York: Routledge.

Lee, A. M., & Lee, E. B. (1939). *The fine art of propaganda: A study of Father Coughlin's speeches*. New York: Harcourt Brace.

Lyon, D. (2001). *Surveillance society: monitoring everyday life*. Buckingham: Open University Press.

Marx, G. T. (2002). What's new about the 'new surveillance'? Classifying for change and continuity. *Surveillance and Society*, *1*(1), 9–29.

Mythen, G., & Walklate, S. (2006). Communicating the terrorist risk: Harnessing a culture of fear? *Crime Media Culture*, *2*(2), 123–142.

Norris, C., & Armstrong, G. (1999). *The maximum surveillance society: The rise of CCTV*. Oxford: Berg Publishers.

Norris, F. H., & Kaniasty, K. (1992). A longitudinal study of the effects of various crime prevention strategies on criminal victimization, fear of crime, and psychological distress. *American Journal of Community Psychology*, *20*(5), 625–648.

Offe, C. (1999). How can we trust our fellow citizens? In: M. E. Warren (Ed.), *Democracy and trust* (pp. 42–87). Cambridge: Cambridge University Press.

Pfaller, R. (2003). Little gestures of disappearance: Interpassivity and the theory of ritual. *Journal of European Psychoanalysis 16*. http://www.psychomedia.it/jep/number16/pfaller.htm. Accessed 25 August 2007.

Poynting, S. & Noble, G. (2004). Living with Racism: The experience and reporting by Arab and Muslim Australians of discrimination, abuse and violence since 11 September 2001. *Report to The Human Rights and Equal Opportunity Commission*, Human Rights and Equal Opportunity Commission, Sydney.

Rose, N. (2000). Government and control. *British Journal of Criminology*, *40*, 321–339.

Rosenbaum, D. P. (1987). The theory and research behind neighbourhood watch: Is it a sound fear and crime reduction strategy? *Crime and Delinquency*, *33*(1), 103–134.

Sheridan, L. P. (2006). Islamaphobia pre- and post-September 11th, 2001. *Journal of Interpersonal Violence*, *4*(3), 203–214.

Tilley, E. (2004). Propaganda – who, us? The Australian government terror kit. *Media International Australia*, *113*, 30–43.

Tyler, T. R. (1990). *Why people obey the law*. New Haven: Yale University Press.

Van Oenen, G. (2004). Languishing in securityscape: The interpassive transformation of the public space. *Open: Cahier on Art and the Public Domain – (In)Security*, *6*, 6–16.

Zizek, S. (1998). The interpassive subject. *Traverses*. http://www.egs.edu/faculty/zizek/zizek-the-interpassive-subject.html. Accessed 12 August 2007.

CITIZENSHIP, HYPER-SURVEILLANCE, AND DOUBLE-CONSCIOUSNESS: RACIAL PROFILING AS PANOPTIC GOVERNANCE

Karen S. Glover

ABSTRACT

Incorporating DuBois's concept of a racial "double-consciousness" and extending Foucault's work on the Panopticon, I examine current day racial profiling processes and the effects of hyper-surveillance on communities of color. DuBois suggests that the citizen of color has a sense of duality based upon minority status and being an American. This duality offers insight into the way race "works" that few Whites comprehend. Foucault argues that the permanent visibility of those subjected to the Panopticon generates awareness of the power differential between individuals and the state. The current examination is a contextualization of narratives from people of color who experience governance and surveillance via racial profiling.

Surveillance and Governance: Crime Control and Beyond
Sociology of Crime, Law and Deviance, Volume 10, 241–256

ISSN: 1521-6136/doi:10.1016/S1521-6136(07)00211-4

The racial construction and governance of citizenship in the United States has taken various forms over the centuries. Surveillance and governance of citizenship identity was and continues to be guided by both formal and informal practices of the state. Racial profiling processes, I argue, offer an example of how informal historical and contemporary racial paradigms and practices of society and the state affect citizenship. Racial profiling processes are defined here as the use of racial and/or ethnic status as the determinant factor in decisions to stop motorists either in the absence of indications of criminality or in determining whom to enforce law against.

THEORETICAL FRAMEWORKS

In this chapter, I engage the work of W.E.B. DuBois (1986) and Michel Foucault (1977) to theoretically contextualize contemporary processes of surveillance and governance that manifest in the racialized traffic stop. DuBois's concept of double-consciousness provides a race theory component, with Foucault's conception of the Panopticon underpinning a hyper-surveillance and state social control analysis.

The contextualization within these frameworks provides a much needed theoretical grounding perspective in the literature on racial profiling. This chapter posits racial profiling as an issue of racial oppression and less an issue of "crime control." To illustrate how less overt forms of governance and surveillance such as racial profiling restrict citizenship *racially* in the Post-Civil Rights Era, I analyze narratives from people of color who have been racially profiled in hopes of extending our understanding of the contexts and effects of surveillance and governance on the citizenry, particularly on citizens of color.

Surveillance and governance is about identity. Notions of "who belongs where" have long organized policies in racialized states such as the United States, through institutionalized slavery and Native American dissemination experiences to historical and contemporary immigration debates (Omi & Winant, 1994). Identity imposed from the state during racial profiling processes involves the criminalization of citizens.

While DuBois was at the fore of contextualizing issues of crime and social control within the bounds of a racial state, his work has not been engaged to explain these matters, with notable exceptions largely from African-American scholars. DuBois's concept of double-consciousness refers to an awareness one has and cultivates that recognizes the importance of status and identity concerns in a racially ordered society. Fundamentally, it is a concept to understand how racialization processes influence those who are subjected to racial and other oppressions. What are the mechanisms that emerge to deal with, contextualize, and resist racial oppression? DuBois's work provides theoretical grounding for understanding a neglected aspect of racial profiling studies – examination of the effects of constant surveillance and governance on the social actors who experience it.

As described by DuBois (1986, p. 364), "One ever feels his two-ness, – an American, a Negro; two souls, two thoughts, two unreconciled strivings; two warring ideals in one dark body, whose dogged strength alone keeps it from being torn asunder." The sense of "two-ness" described by DuBois is a direct reference to the limits of citizenship in practice for people of color when they confront, with varying modes of resistance, the social fact of racialized citizenship emergent through surveillance and governance processes of the state. Minority identity, often hyper-visible through racialization and criminalization processes like racial profiling, is not reconciled with citizenship identity in the eyes of many Whites. This contradiction between identifying as a person of color yet having full citizenship denied creates awareness of racial power dynamics.

In addition to using DuBois's work, I extend Foucault's concept of the Panopticon to contextualize racial profiling processes and effects as surveillance and governance systems. The Panopticon is a social control mechanism introduced by Jeremy Bentham in the 1800s as part of his prison reform efforts. The basic design is a circular prison complex with a tower in the center of the structure. The inmates reside in the cells that face outward to the central tower and, because of architectural/interior design that includes backlighting of each cell, monitoring of inmates is facilitated. A key aspect of the Panopticon experience, however, is that the inmates need not *always* be under surveillance in order for governance to occur. Consequently, according to Foucault, a self-monitoring and self-conscious effect occurs with the inmates. In essence, these individuals experience being under suspicion by authorities at all times. Foucault explains:

> … the major effect of the Panopticon: to induce in the inmate a sense of conscious and permanent visibility that assures the automatic functioning of power … the surveillance is permanent in its effects, even if it is discontinuous in its action; that the perfection of power should tend to render its actual exercise unnecessary. (Foucault, 1977, p. 201)

The "permanent visibility" of race as a functioning marker of criminality and the permanent visibility induced by the Panopticon – always being under suspicion – are similar. An "automatic" functioning of power develops as the individual – here, the citizen of color – is reminded of the power relationship they are in with the state.

Foucault saw the move to what he termed "discipline power" of the state as a shift away from overt and more conventional forms of social control. He suggests mental stress and coercion may be more socially harmful than traditional social control efforts. He states:

> ... there is a shift in the point of application of this power: it is no longer the body, with the ritual play of excessive pains, spectacular brandings in the ritual of the public execution; it is the mind or rather a play of representations and signs circulating discreetly but necessarily and evidently in the minds of all. It is no longer the body, but the soul ... (Foucault, 1977, p. 101)

Surveillance processes by the state such as racial profiling provide a contemporary example. No longer are race-based restrictions about "who belongs where" promulgated through official state-sanctioned policies. Racial profiling, with its somewhat elusive character, emerges in the contemporary era as a way to govern racially without reliance on more overt forms of racial governance. The move from "the body" to the "mind/soul" is representative of a move from more overt forms of racial oppression to more covert formations in the contemporary era, which is in line with other examinations of racialization processes in the Post-Civil Rights Era (Bonilla-Silva, 2003).

It is useful to engage the works of DuBois and Foucault on the issue of racial profiling because each expresses a concern with the effects of constant surveillance and evaluation on the individual, in addition to acknowledging how these surveillance and criminalization processes are tools of the (racial) state.

For DuBois, this concern manifests in the double-consciousness concept where individuals of color recognize the ramifications of racial/ethnic status in determining how far citizenship rights and protections are extended, both for themselves and for Whites. The recognition that full citizenship is generally excluded from their experience in a White-dominated social realm, particularly in the face of public discourse that may fervently suggest otherwise, offers people of color unique insight into the racial state that Whites do not possess (Feagin, 2006). Experiential knowledge that the use of racial status is employed as a tool of the state when making assumptions of criminality provides an example of the limitations of "equal protection under the law" that is fundamental to double-consciousness awareness.

At a basic level, Foucault's panopticonism concerns social control. The Panopticon effect – where individuals have an awareness of constant monitoring, patrolling, and surveillance by the state – induces an awareness of power relations with the state. Thus the panoptic effect also involves an awareness of oppression, similar in process to DuBois's double-consciousness. For instance, when one of my respondents comments on the regularity of racialization and criminalization processes evident in the racial profiling phenomenon, I consider this as part of the Panopticon effect.

In the examination to follow, I provide excerpts from a series of in-depth interviews I conducted with people of color about their experiences with racial profiling. These excerpts and the surrounding discourse show how the respondents framed racialized traffic stops as involving broad legal and political realms. These encounters were not bounded by the police-minority relationship, as is the orientation of much of the research on race and law enforcement in criminology. Indeed, the racialized and criminalized encounters with law enforcement were viewed as reflections of much greater contestation than a "mere" traffic stop may suggest and point to the larger realm of citizenship.

CITIZENSHIP

Citizenship emerged as a dominant narrative with my respondents. It is characterized by inclusion and exclusion, by formal and informal means that either promote or deny "full social, economic, and political rights" (Cook, 1999, p. 136). For my respondents, citizenship is reflected discursively by their embracement of a liberty and justice framework of democratic ideals. The ideology is characterized in part by freedom from unwarranted state intervention typified in racialized surveillance and governance practices of the state. For example, one respondent comments that individuals "belong" in the United States when they get treated fairly under the Constitution. The liberty and justice framework also serves as a ruling ideology posited by dominant Whites to bolster the "white racial frame" that organizes society (Feagin, 2006). Thus, those who get served well under the protections of citizenship invoke the rhetoric of "equality under the law" to explain a justice system that privileges them, while people of color invoke similar narratives to demonstrate how the substantive ideals of citizenship that frame our public discourses are not extended to them in practice.

Marshall (1964) classically defined citizenship as concerned with the rule of law and its application to the liberty of individuals and their right to justice. The dimensions of limited citizenship via panoptic processes include struggles when invoking knowledge of due rights and a lack of protection and accountability from the state. The use of citizenship discourse reflects my respondents' own position that they are due the rights and protections of citizenship and thus resist the subordinate identity imposed by the state.

NARRATIVES AND DISCUSSION

The first narrative to illustrate the deep sense of citizenship expectations and awareness of racial oppression inherent to racial profiling is Darrell, a Black male in his 20s with some college education. Throughout Darrell's interview, a running theme was his knowledge of due rights and protections:

> I have that right to say what I need to say in order to defend myself. To make him [an officer] aware that I know what I'm talking about and that I know what the laws are. To let him know that I'm not going to just sit up here and let him do what he's doing without knowing that I'm aware that he's doing it to me.

Evidence of DuBois's double-consciousness is clear: Darrell identifies as a citizen as he articulates protections under full citizenship, while recognizing the limitations he is subjected to as a young man of color. At another point, Darrell introduces the concepts of "mental freedom" and "psychological warfare" in relation to racial profiling. I asked him to expand on these ideas. Darrell says:

> just to be free in general is to be unbound. To be unshackled. It doesn't matter whether that means metaphorically speaking, mentally, or physically. Just to be free means not to be bound by anyone or anything, basically. When you put that stress on individuals with certain types of laws or hidden rules that hinders their freedom. It makes their freedom even harder to be enjoyed. Racial profiling definitely does it.
>
> Oh because when you're expecting to be stopped just because of your ethnicity, of your age demographic or just the way you look. Mentally you have that in the back of your mind – it can happen at any time. So you sit up and think, well man! ... So it's a psychological thing because here you are a human being and that's taking away a large piece of your mental freedom because you know you're going to get stopped. You don't know when. You don't under what circumstances. Even when it does go down, you don't know how far they're going to go with the stop ... It cheats an individual out of some of their mental freedom because you have to worry about that.

Metaphorically, his choice of "unbound" and "unshackled" frame the racial profiling experience within the context of slavery. Foucault views the Panopticon experience as one released from the body but one that effectively binds the mind. While racial profiling processes interact with both the body and the mind, the mental coercion that surrounds the encounters create unique alienating relations with the state that go beyond the physical. Foucault suggests that while both the mind and the body were bound in early times, the effect of the Panopticon ushered in an emphasis on the mind over the body.

The consistency of the Panopticon effect is evident in Darrell's several references to the mental/psychological burden of constant surveillance. By engaging citizenship ideals such as freedom from unwarranted state intervention and recognizing the limitations placed on him, Darrell reflects the two-ness DuBois argues is both a burdensome, negative "condition" of people of color and, in the same instance, an insight that allows people of color to see the world "as it really is." Few Whites experience and thus comprehend the implication of targeted law enforcement, which explains much about the differential in views of the police across racial lines (Hagan & Albonetti, 1982; Parker, Onyekwuluje, & Murty, 1995; Pate & Fridell, 1993; Russell, 1998; Tuch & Weitzer, 1997). Darrell recognizes the effects of the Panopticon as a way of restricting freedom, thus reflecting Foucault's discussion of how the Panopticon is a function of power. Darrell's remarks also indicate the "conscious and permanent visibility" Foucault sees as resulting from panopticonism when he comments on his awareness that "... it can happen at any time."

Another respondent, Javier, a Latino in his late teens with some college education, contrasts his lived experiences with the experiences of Whites in similar situations with law enforcement. By doing so, Javier demonstrates a fundamental "truth" of racial profiling for people of color: awareness of differential enforcement of the law in the case of actual violations, or differential allocation of law enforcement resources (who is being surveilled, patrolled, monitored?) is key in making determinations of whether racialization is occurring in a given encounter with the state. DuBois's description of how being both African-American and American were conflicting identities on the streets of White America is articulated in the following quote where Javier compares his minority status experiences with those of similarly-situated Whites. This complicates the current discourse in the racial profiling literature and the more general public discourse on race in the United States that suggests people of color are (too) quick to assume minority status shapes their experiences in important ways. On the contrary, Javier's judgments on the racialization of his encounter with law enforcement is contextualized within the larger legal and political realm of the justice and liberty frame that promises equal treatment under the law.

Javier explains:

> I had gotten an MIP (minor in possession). Me and four of my friends ... it was kind of late and it was dark out. We were parked near the shore. We had Budweiser on us and we were drinking. We obviously got stopped. After we got the tickets and everything, we had stayed there. Of course we had to spill out all of the beers. We went walking down the beach and there were a couple of people ... they were white. They had a bonfire and all that. They had like a little party. We asked them – 'did the cops stop by here?' They said 'yeah.' I noticed that some of them didn't look too old. They said they didn't even get questioned or anything. I was pretty mad about that. I thought that these cops – I thought it was some kind of a racial thing. I mean it was clear over here in my little junkie car, we don't have a fire or lights or anything and it is only four of us. Then, the other people have the big bonfire and everything out and had a big radio making lots of noise. I was thinking why didn't they get stopped? Isn't that more of probable cause?

Javier's sense is that his experience was racialized through the actions of the state when similar actions by a group of young White males were not sanctioned. A fundamental component of racial profiling is the targeted application of law enforcement resources, such as surveillance, to communities of color when Whites engage in similar behaviors but do not receive such a level of scrutiny. Javier's immediacy in contrasting his experience with the group of White youths points to his awareness that the rule of law inherent to the citizen-state relationship is differentially imposed. He directly engages the citizenship discourse through his use of the expression "probable cause," – a significant and formal legal element of citizen protections against state intervention, thus constructing a Constitutionally bounded sense of identity for himself. As noted earlier, surveillance is an issue of identity. In the case of racial profiling, that identity is criminalized through the actions of the state.

Jesse, a Latino in his early 20s with some college education, discusses how not receiving a ticket from law enforcement may support initial notions that a traffic stop is race-based. He invokes the "out of place" doctrine inherent to racial profiling processes by recognizing limitations on his freedom in certain social spaces.

> J: I've been, all over this city, pulled over. I've kept count too – I know where not to drive. I've been pulled over eight times and I've only been given a ticket that one time. I already know the routine. They pull me over, they grab my license, I guess they run my license for warrants, and they just let me on my way. It's pretty blatant.
>
> K: So you're not ticketed for anything – they're not saying you're speeding?
>
> J: No, I'm not ticketed at all.

K: So this happened about eight times over the course of ...

J: One year. I've only been here a year. I haven't even been here an entire year.

In the following narrative, it is clear that the state need not hand down official documentation of an actual violation for the traffic stop to become a defining moment in a citizen's relationship with the state. Carlos, a Latino in his 30s with a graduate degree, discusses an array of identity concerns that emerge in the wake of a racialized and criminalized traffic stop after which he did not receive a ticket. Carlos engages in credentialing – making note of the various identities he holds – to position himself as someone who is showing commitment to society and holding up the ideals of citizenship.

> It upsets me because you know as a veteran and a college student and whatever I think I am, um, that someone could look at me and the way I'm driving or the way I'm dressed and instigate a situation by pulling me over and stopping me from what I'm doing. I consider myself busy and with a purpose. I'm stopped for no apparent reason and not given a ticket. It does upset me. I know what they're looking for ... I know what they want and I am not that person. Yet, when they profile me, they think I fit this mold. I will pull him over and I will find something. It gets very upsetting because it's harassment. That's what it is. If you've never been in that situation, then you don't know what it feels like. And if you're not angry about it, it's because you've never been in that situation. This repeated situation.

The criminal identity forced upon him due to his ethnic status is not easily reconciled with own self-identification, clearly pointing to the two-ness theorized by DuBois. The criminal identity imposed by the state via racial profiling is rejected ("I know what they want and I am not that person"), yet the toll of being criminalized is not avoided. Later, he refers to his experiences as "... like constantly being under a microscope."

Carlos envisions a scenario of one of these criminalizing encounters on a young person of color and its potential as entrée into the criminal justice system:

> He's a good kid ... who got pulled over at the wrong time and just snaps – he's branded as a juvenile or a trouble-maker. Or a cop doing this or that or whatever just because of that single act. That's it. He's branded for life. I mean the significance of that – of racial profiling – is just the fact that one incident can really affect a person for the rest of his life. Maybe not bringing it out in the open like we are but constantly day after day when you see that cop car. We don't see that's our friendly neighbor.

> It's almost like we've been given this fairy tale that cops are the good guys. The cop standing there, you know, helping old ladies – whatever. This is reality. This is the way it really is. Sure, you may be educated, you may hear about it on TV, you may read about it but until you experience it – you're like "Oh shoot." ... That's the feeling you get in

your gut. It's a loss of innocence. This is reality. This is the way your life is going to be. You're a Hispanic male – get used to it.

In the above excerpt, Carlos expands on a particularly troubling aspect of racialized social control practices: because of the differential in where state resources are directed and what types of behavior ("street'" crime versus "white collar" crime, for example) are targeted, young men of color are at a greater risk of getting "initiated" into the criminal justice system compared to dominant group members who may be engaging in similar behaviors. He understands that "one incident" can be a determinant factor in the life chances of young men of color. Carlos also dissects the public discourse about the law. His experiential knowledge of police-minority relations on the street contradicts the ideology of justice that characterizes citizenship discourse in the United States. From DuBois, the experience of people of color in a racial state includes a struggle to reconcile citizenship ideology with experiential knowledge.

Edgar, a Latino in his 20s with a bachelor's degree, discusses the explicit nature of criminalization found in racial profiling processes. His framing of the role of the state and its citizenry includes a theoretical neutrality of the state to which Edgar's experience is contrasted.

they admit it ... "We are looking for Hispanic males driving this type of car ... driving a nice car." They just flat out admit it ... If you've never felt it before, you don't know what it feels like. For me, the first time it was almost like, I was confused. I'm like, man ... is what happened ... did what happen just happen? It was one of those things that, you know, this has never happened to me before. It happened a second time and I kind of realized it. Wow, this is pretty messed up.

You get that tense feeling ... That's not the way it should be. We should be able to look at them like they are my friendly neighborhood cop, you know, or Officer Friendly.

Edgar has a sense of how his relationship with the police is theorized to operate, yet understands the relationship is different in practice, thus exhibiting the double-consciousness experience of many people of color. Dominant group members welcome state efforts at social control in part because of their belief in the restrictions on state powers (Thompson & Lee, 2004; Vila & Morris, 1999; Websdale, 2001). However, people of color, through the experiential knowledge that both DuBois and Foucault argue shapes individuals' sense of the social world, have a greater context in which to frame the ideology of liberty and justice. Edgar also expresses incredulity about the overt way criminalization of communities of color occurs. His comments underscore the reliance on the experiential quality ("If you've never felt it before ...") of racial profiling for people of color in determining if and when racialization of their everyday encounters with the state occurs. The struggle with reconciling his underlying belief in citizenship with what he has come to experience himself is evident.

Angela, a Black female in her 40s (no education given), provides an example of the omnipresence of the Panopticon effect in her community:

> Sometimes it might just be a reaction they see you with. When they pull up they might see you with your hands in your pockets at the wrong time. There's really not a right way or a wrong way to do that over here. If you see an officer come around this corner and say I'm coming out of this parking lot, I might just so happen to put my hand in my pocket walking to my door. They quickly assume and pull over and pull up in here and say "I saw you stick your hands in your pocket." But when they get here, you don't have nothing but your keys and change. They don't even apologize. They just assume. There's always an assumption!
>
> K: So they're thinking that you're putting something away – like a bag [of dope]?
>
> Yeah. All the time. When they come, it's like there's never no good person over here. They want to say everybody over here is the same. We all sell, we all, you know. It's like everybody's judged by the same color.

Angela's reflections concern her knowledge of the everyday quality of racialization and criminalization processes in her community. Her experiences suggest that behavioral indicators to signal criminality are not necessary components of law enforcement surveillance: "... there's really not a right way or a wrong way" to maneuver in the everyday because there is "... always an assumption [of guilt]." The Panopticon effect – this assumption of guilt that Angela and her community must negotiate daily – creates alienating relations with the state. Dominant group members are largely protected from developing similar alienating relations with the state (Feagin, 2006).

That whole groups of people are under suspicion and governed collectively is mirrored in the excerpt below from Francis, a Black woman in her 50s with some college education. She responds to a general inquiry about instances where racial and ethnic status figured into an encounter with law enforcement:

> I've seen incidents where policemen have searched, looking for one person, and have really searched the whole group of Black boys when they claim they are looking for one ... that they are not looking for this one certain person yet still they are ... having them all against fences and things which kind of upset the Blacks because we figure they wouldn't be doing nobody else's child like that. I've had incidents where they've all, the children, have told us about how police treats them, talks to them, kind of demanding and making them say things they didn't want to say ... which is also against us because mostly we feel like ... they're picking on our Black kids.

Francis's remark that the young boys are subjected to state intervention through a physical search demonstrates the criminalization of Black male youth as a collective. This practice is sometimes referred to as "the round-up" in critical analysis of social control practices by the state (Covington, 2001). Francis directly frames the citizenship issue in the context of unequal treatment by the state – "we figure they wouldn't be doing nobody else's child like that." She views this behavior as unequal treatment under the law and as a direct reflection of the state's broader treatment of communities of color. Clearly, Francis possesses the double-consciousness and insight DuBois describes as characteristic of communities of color.

Hector, a Latino in his 40s with a high school education, brings up identity concerns within an ultra-patriotic discourse to signal dissatisfaction with the citizen-state relationship he experiences:

> I know I'm an American because I was born here. I went to school here. I was educated here. I'm as American as anybody else. Red, white, and blue – that's all I bleed. For them to come up and question me as far as that ... I'm not a citizen of the United States. I'm disenfranchised!

His invocation of the term "disenfranchised," defined as a loss of rights and protections, is particularly meaningful and reinforces a later comment from him that he is "not a part of this country" when criminalization and racialization processes occur via panoptic processes of the state. As with other respondents, he engaged in credentialing to signal his conformity as a citizen, and contrasted his lived experience as a person of color with the ideology of citizenship in the United States.

Another respondent, Nathan, a Black male in his 30s with a graduate degree, after relaying a particularly traumatic story about having a gun placed to his head by a police officer shortly after purchasing a soda and exiting a convenience store, discussed the Panopticon effect in the following way when asked why he considered the incident racialized.

> Simply because it was not an isolated experience. I've experienced similar stuff over the years. Different name, different place or location. Just unusual stuff for no reason ... It's sort of in the subconscious that the combination of male, Black, and young looking – put the three together and the image that the public, whether laity or professionals, has is a negative image. The three combined are just automatic prime suspects for just anything.
>
> At this stage in my life, I'm not surprised by it. Not that I agree with it but when it happens, I accept it as part of the Black experience. I agree that there is something seriously wrong with it ... I've been dealing with that part of stuff for ages. It is a part of the experience.

Nathan's narrative illustrates how the regularity of profiling practices – a product of surveillance and governance by the state – becomes normative for many people of color. His expression "automatic prime suspects" is informative as it indicates how the criminalization of young men of color manifests under panopticonism. These instances also point to DuBoisian double-consciousness as it reflects awareness and struggle over identity.

Jason, a Black male in his mid-30s with an associate's degree, offered two intricately detailed accounts of racial profiling incidents he experienced. As with Nathan and others, he engages the idea that the constant surveillance and patrolling of the Panopticon effect is part of his experience as a Black male. It is contrasted with the justice and liberty ideals of American citizenship that nonetheless comprise his self-identity. The following is Jason's response to my question about how he deals with these encounters by the state.

> As a race I think that is the idea or the feeling that we have from law enforcement – that we're suspicious because we're African American. We have to be up to something. If they stop us then it's going to lead to some type of bust. It is aggravating. It can be very aggravating. Sad to say that it is a part of life that as a race we teach our young men that "Hey – this is going to happen to you and you better be prepared for it. It's going to happen. Not that it might happen, but that it is going to happen sooner or later. You're going to get stopped and you won't have to be doing anything but crossing the street. You're going to get stopped and you're going to get questioned." That's just life. We just learn to accept it as part of our interaction with law enforcement.

Jason's comments point to the everydayness of racial oppression for minority communities, particularly young men of color. Jason also makes reference to what has been described as "the lesson" (Russell, 1998). In communities of color, discussions about how to negotiate in a racial state become a part of community discourse. For young males of color in particular, a specific discourse about expectations from law enforcement is circulated as very purposeful communication for basic group survival concerns (Collins, 2000). Some argue that because communities of color maintain this dialogue across generations, the communities perpetuate a racial divide with law enforcement. A related point in the racial profiling debate is that people of color should not be concerned about being stopped by law enforcement on occasion if they have nothing to hide. A similar rationalization occurred in the public discourse following the events of September 11 to specifically address the increased attention given those who were judged to be of Middle Eastern descent. Placing "national security" concerns above individual Constitutional protections became a mainstay of pro-racial profiling arguments.

However, as Jason's comments illustrate, the regularity of surveillance, monitoring, and patrolling of his community by the state demands communal knowledge about how to negotiate these realities.

CONCLUSION

Any examination of surveillance and governance concerns in a racial state such as the United States must include the contemporary phenomenon of racial profiling. This chapter examines how the personal experiences of people of color in racialized encounters with law enforcement go well beyond the local, micro-level association focused on in the current racial profiling literature. My respondents clearly reflect on these encounters as racializing and criminalizing experiences with the state that are experienced as watershed moments in their lives. The overarching theme that emerges from their narratives is one of a break from citizenship and the liberty and justice rights frame that encompasses it. In other examinations of how race operates in the criminal justice system, this process is referred to as an "attenuation" of citizenship (Pettit & Western, 2004). Western (2006, p. 193), in his examination of the role and effects of status differentials in regard to incarceration, views the effects of race and socioeconomic status as an "evolutionary" aspect of African-American citizenship because of the *retrenchment* of citizenship that results from disenfranchisement resulting from incarceration (of which racial profiling is a potential precursor to). Yet, as shown by the active and frequent engagement of the justice and liberty rights frame that many of my respondents engage, my respondents continue to make claim to the citizenship realm and resist denial of full citizenship by the racial state. Writing in the late 1800s, DuBois (1986, p. 364) described this same struggle to reconcile the "warring ideals" of minority identity and citizen identity imposed by the racial state with the self-identified sense of being a full citizen.

In this regard, the current study finds that for people of color, the racialized traffic stop is deeply contextualized within a well-developed base of knowledge about how race operates in the United States. This goes to the DuBoisian perspective that people of color possess insight into the inner workings of the social world – they know the "souls of white folks" to a greater extent than Whites understand the experiences of people of color. My respondents indicate that they assess their encounters with the state by comparing similarly-situated conditions with Whites or through a process of elimination that racial status is the motivating factor in their being stopped, among other things. Indeed, many, though not all, respondents articulated initial reluctance to view a traffic stop (and the criminal justice system generally) as *raced*, having adopted and believed in much of the liberty and justice frame that orients our national, and specifically criminal justice, discourse. This classic construction of living in both worlds is very much a part of the contemporary experience of people of color. Strides that have been made in the years following DuBois's 1897 treatise may have dulled the demarcation of citizenship that existed in DuBois's day, but those strides remain outweighed by the restricted *substance* of citizenship for communities of color.

As Foucault argues, surveillance as a tool of governance by the state is a form of "disciplinary" power that is:

> exercised through its invisibility; at the same time it imposes on those whom it subjects a principle of compulsory visibility. In discipline, it is the subjects who have to be seen. Their visibility assures the hold of the power that is exercised over them. It is the fact of being constantly seen, of being able always to be seen, that maintains the disciplined individual in his subjection. And the examination is the technique by which power, instead of emitting the signs of its potency, instead of imposing its mark on its subjects, holds them in a mechanism of objectification. (Foucault, 1977, p. 187)

Foucault's conception of panopticonism, as argued earlier, is an appropriate backdrop for contextualizing racialized traffic stops and the more general idea of the criminalization of communities of color. The obvious connection concerns the foundation of racial profiling processes: the omnipresent eye of the state on communities of color, especially young minority males. The Panopticon effect, as discussed, also includes a permanent change in the individual under its effects to where views of the state (as embodied by the panoptic processes) become alienated from previously neutral or even positive standpoints.

The racial surveillance and governance that manifests in racial profiling practices complicates the notion that surveillance as a tool of the state is primarily about crime control. These processes are fundamentally about racialized social control that exploit society's emphasis on particularly forms of behavior in order to maintain racial ordering spatially, ideologically, and politically. Ultimately, while my respondents acknowledge that racial governance via panoptic surveillance processes limits full citizenship, they still engage the promise of citizenship by self-identifying as such and resisting the criminal identity imposed upon them by the state.

REFERENCES

Bonilla-Silva, E. (2003). *Racism without racists: Color-blind racism and the persistence of racial inequality in the United States*. Lanham, MD: Rowman & Littlefield Publishers.

Collins, P. H. (2000). *Black feminist thought: Knowledge, consciousness, and the politics of empowerment*. New York: Routledge.

Cook, D. (1999). Racism, citizenship and exclusion. In: D. Cook & B. Hudson (Eds), *Racism & criminology* (pp. 136–157). London: Sage Publications.

Covington, J. (2001). Round up the usual suspects: Racial profiling and the war on drugs. In: D. Milovanovic & K. Russell (Eds), *Petit apartheid in the U.S. criminal justice system: The dark figure of racism* (pp. 27–42). Durham: Carolina Academic Press.

DuBois, W. E. B. (1986). *Writings: The suppression of the slave trade; The souls of black folks; Dusk of dawn; Essays and articles*. New York: Literary Classics of the United States-Library of America.

Feagin, J. R. (2006). *Systemic racism: A theory of oppression*. New York: Routledge.

Foucault, M. (1977). *Discipline & punish: The birth of the prison* (A. Sheridan Trans.). New York: Random House.

Hagan, J., & Albonetti, C. (1982). Race, class, and the perception of criminal injustice in America. *American Journal of Sociology*, *88*, 329–355.

Marshall, T. H. (1964). *Class, citizenship, and social development: Essays*. Garden City, NY: Doubleday.

Omi, M., & Winant, H. (1994). *Racial formation in the United States from the 1960s to the 1990s*. New York: Rutledge.

Parker, K. D., Onyekwuluje, A. B., & Murty, K. S. (1995). African-Americans' attitudes toward the police: A multivariate study. *Journal of Black Studies*, *25*, 396–409.

Pate, A. M., & Fridell, L. A. (1993). *Police use of force: Official reports, citizen complaints, and legal consequences*. Washington, DC: Police Foundation.

Pettit, B., & Western, B. (2004). Mass imprisonment and the life course: Race and class inequality in U.S. incarceration. *American Sociological Review*, *69*, 151–169.

Russell, K. K. (1998). *The color of crime: Racial hoaxes, white fear, black protectionism, police harassment, and other macroaggressions*. New York: New York University Press.

Thompson, B. L., & Lee, J. D. (2004). Who cares if police become violent? Explaining approval of police use of force using a national sample. *Sociological Inquiry*, *74*, 381–410.

Tuch, S. A., & Weitzer, R. (1997). The polls-trends: Racial differences in attitudes toward the police. *Public Opinion Quarterly*, *61*, 642–663.

Vila, B., & Morris, C. (Eds). (1999). *The role of police in American society: A documentary history*. Westport, CT: Greenwood Press.

Websdale, N. (2001). *Policing the poor: From slave plantation to public housing*. Chicago: Northeastern University Press.

Western, B. (2006). *Punishment and inequality in America*. New York: Russell Sage Foundation.

HACKING THE PANOPTICON: DISTRIBUTED ONLINE SURVEILLANCE AND RESISTANCE

Benoît Dupont

ABSTRACT

Surveillance studies scholars have embraced Foucault's panopticon as a central metaphor in their analysis of online monitoring technologies, despite several architectural incompatibilities between eighteenth and nineteenth century prisons and twenty-first century computer networks. I highlight a number of Internet features that highlight the limits of the electronic panopticon. I examine two trends that have been considerably underestimated by surveillance scholars: (1) the democratization of surveillance, where the distributed structure of the Internet and the availability of observation technologies has blurred the distinction between those who watch and those who are being watched, allowing individuals or marginalized groups to deploy sophisticated surveillance technologies against the state or large corporations; and (2) the resistance strategies that Internet users are adopting to curb the surveillance of their online activities, through blocking moves such as the use of cryptography, or masking moves that are designed to feed meaningless data to monitoring tools. I conclude that these two trends are neglected by a majority of surveillance scholars because of biases that make them dismiss the initiative displayed by ordinary users, assess

Surveillance and Governance: Crime Control and Beyond
Sociology of Crime, Law and Deviance, Volume 10, 257–278

ISSN: 1521-6136/doi:10.1016/S1521-6136(07)00212-6

positive and negative outcomes differently, and confuse what is possible and what is probable.

The panopticon concept occupies a pivotal position in the field of surveillance studies. Michel Foucault's (1977) analysis of Bentham's total surveillance architecture has become a ubiquitous reference in the literature (Haggerty, 2006; Lyon, 2006), despite Foucault's deliberate lack of interest for the emerging technologies of his time (Haggerty & Ericson, 2000). A few years later, Thomas Mathiesen (1997) highlighted the limits of relying exclusively on the panopticon's metaphor in a "viewer society" where television lets the many see what the few are up to. Although these two major contributions still partly resonate with the current state of surveillance and continue to provide useful theoretical insights, I will argue in this chapter that their hegemonic influence (Haggerty, 2006) is becoming counterproductive to understand two trends related to surveillance in the online environment. The first trend can be defined (for lack of a better term) as the "democratization of surveillance", where cheap surveillance software and hardware is marketed to individual customers so that they can monitor the activities of their family, coworkers, neighbours, and even their favourite celebrity or their most despised politician. The second trend concerns the resistance to surveillance, where efforts are deployed by the subjects of surveillance to understand, reveal, mock, evade, and neutralize surveillance technologies through the collaborative power of socio-technical networks. Because of their incompatibility with the dominant panoptic and synoptic conceptual frameworks, these two trends have been underestimated and sometimes even ignored by surveillance scholars.

These two facets of contemporary surveillance will be examined in a very specific context: the omnipresent network of computers, servers, software, and services that make up the Internet. The Internet is now routinely used to exchange information of personal and public interest, to conduct financial transactions, to acquire goods and services of all kinds, and to spend time (or waste it, depending on the perspective) by playing online games, downloading music and movies, and managing social networks of friends and acquaintances. Its architecture is decentralized and distributed, making it at the same time very exposed and very resilient to failures and malfeasances. Its invention is recent, and when *Discipline and punish* was first published in French in 1975, ARPANET (the ancestor of the Internet) was still in its infancy (Mowery & Simcoe, 2002). At first sight, the Internet seems to embody the worst fears of a panoptic world: total surveillance can be achieved at very low cost, making all exchanges traceable and significantly altering the notion of privacy (Lessig, 2006). As the Internet penetrates every aspect of our lives and the boundaries between the physical world and virtual world become irremediably blurred, we should be quite worried by these flows of digitized information that are used to create "data doubles" whose slightest alterations are constantly scrutinized (Haggerty & Ericson, 2000, p. 611). If one tool could manage to leverage the power of knowledge to govern the behaviour of a population, the Internet should figure among the top contenders (Graham & Wood, 2003). However, no matter how great the dystopian potential of the Internet is, it seems that it has not yet delivered its disciplinary promise. To be entirely fair, it has not liberated people from autocratic regimes either, as some of its most naïve promoters initially believed. One of the reasons for this lies in the "openness" paradox: while the technical protocols that underpin the Internet are public and standardized, therefore making surveillance relatively easy to carry out, the very same openness empowers application writers (programmers), who are free to design and distribute new tools of surveillance and resistance. For these reasons, the Internet seems like the perfect case study to assess the contemporary relevance of the panoptic and synoptic conceptual frameworks.

I do not contest the existence and growth of pervasive surveillance programmes run by governments that seek to unmask terrorist suspects before they strike or political opponents who criticize the abuses of authoritarian regimes. Nor do I want to minimize the impact of similar efforts by corporations that want to profile their customers better in order to increase their profit margins (Gandy, 1993; O'Harrow, 2005) or ensure the compliance of their employees (Associated Press, 2007). Recent developments in the United States – where the executive branch has authorized massive antiterrorist datamining initiatives despite their dubious constitutional legality (Eggen, 2007) – and elsewhere would make such a position untenable because of its complete disconnection from reality. However, a simple transfer of the panoptic model, so eloquently delineated by Foucault and refined by Mathiesen, does not provide a more accurate description of the reality of contemporary Internet surveillance. In the following sections, I will first explain why the panoptic and synoptic approaches provide an incomplete set of conceptual tools to analyze the proliferation of surveillance capacities in the online world, before examining how these capacities have become available to a broad range of social actors and are also increasingly resisted with a certain degree of success by a growing body of activists and ordinary users. Finally, in the conclusion, I offer a non-exhaustive list of biases that have, in my opinion, prevented a significant number of surveillance scholars from integrating the trends mentioned above in their existing work.

THE PANOPTICON: AN EXHAUSTED SURVEILLANCE METAPHOR?

Although this question might seem unnecessarily provocative, I would like to show in this section the perils of extending eighteenth century thinking, no matter how innovative it was at the time, to twenty-first century technologies. Foucault's work assumes a certain linearity in the development and refinement of surveillance techniques, "from a schemata of exceptional discipline to one of a generalized surveillance, [which] rests on a historical transformation: the gradual extension of the mechanisms of discipline throughout the seventeenth and eighteenth centuries" (Foucault, 1977, p. 209), ending in the formation of "the disciplinary society". This unrelenting expansion of the disciplines does not consider the possibility of disruptive technologies that would redefine how people watch each others and resist various efforts to monitor their activities.

Panoptic Features

Foucault's analysis of Bentham's panoptic prison emphasizes a number of features. The first innovation consists in the physical ordering of the cells in a ring, in the middle of which a focal point – the observation tower – affords a perfect view of all the inmates. Such a "hub-and-spoke" architecture allows a single warden to watch a large number of cells and creates a new economy of surveillance. The asymmetrical power relation created by this circular architecture is reinforced by the lighting arrangements that induce total and permanent visibility for the inmates, while the guardians are shielded behind blinds that make them invisible to the surveillance subjects. A third feature consists in the partition between cells. The solitude it creates seeks to make the inmate "a subject of information, never a subject in communication" (Foucault, 1977, p. 200), to remove the opportunities for coordination that could lead to a "collective effect". The expected result is a more effective institution, where the concentration of power facilitates the observation, classification, comparison, and ultimately, management of subjects.

Beyond an erudite description of Bentham's model, Foucault's main argument resides in the idea that the panopticon "must be understood as a generalizable model of functioning; a way of defining power relations in terms of the everyday life of men" (Foucault, 1977, p. 205). It is an ideal-type, "the diagram of a mechanism of power reduced to its ideal form; *its functioning, abstracted from any obstacle, resistance or friction*, must be represented as a pure architectural and optical system: it is in fact a figure of political technology that may and must be detached from any specific use" (Foucault, 1977, p. 205, my emphasis). Hospitals, military units, schools, or workshops were other places where Foucault identified panoptic mechanisms at work, in a trend that he predicted would result in the emergence of a disciplinary society. This total theory of surveillance and discipline proved very appealing and was embraced by a number of scholars, who extended its application to public spaces – where CCTV systems have become ubiquitous, in the workplace or on the Internet, just to name a few. While their interpretation of panopticism varies greatly (Lyon, 2006; Simon, 2005), they all implicitly subscribe to the idea of a power asymmetry between a small group of elite supervisors exercising a monopoly on surveillance tools, and a large mass of unsuspecting or passive individuals whose interests seem to rarely transcend their obsession for consumption (Bauman, 2000).

This hierarchical model of surveillance was famously challenged by Thomas Mathiesen, who introduced the concept of synopticism in his article on the "viewer society" (Mathiesen, 1997). Mathiesen reminds Foucault's readers that a significant piece of the contemporary surveillance puzzle is missing from the master's account:

> We have seen the development of a unique and enormously extensive system enabling *the many to see and contemplate the few*, so that the tendency for the few to see and supervise the many is contextualized by a highly significant counterpart. I am thinking, of course, of the development of the total system of the modern mass media. (Mathiesen, 1997, p. 219)

However, far from disagreeing with Foucault's conclusions, Mathiesen insists on the reciprocal functions of the panopticon and the synopticon, which are to control and discipline the "soul", ending his article on a very pessimistic note. Although he calls for political resistance as a moral imperative, his prognosis is very gloomy, and the Internet is merely seen as another media reproducing a familiar pattern of domination and oppression through surveillance and preformatted choices.

What is striking in this very severe judgement, which also resonates in many panoptic studies that extend Foucault's reasoning to computer technologies (Poster, 1990; Sewell & Wilkinson, 1992; Gandy, 1993), is that it transposes the rock and mortar architecture of the prison to the structure of the Internet, built on wires and bits. A more careful examination of the Internet's structural features should however introduce a dose of relativism and open up new avenues of enquiry for the study of contemporary surveillance practices. In that respect, Yochai Benkler's book on "the wealth of networks" (2006) offers one of the most detailed accounts of the Internet's structural and institutional features, as well as a consideration of their impact on political and cultural freedoms.

The Internet as an Anti-Panopticon

Where the panopticon and synopticon adopt the "one-way, hub-and-spoke structure, with unidirectional links to its ends" (the periphery in the case of the former, the centre for the latter), the Internet is built as a decentralized and "distributed architecture with multidirectional connections among all nodes in the networked information environment" (Benkler, 2006, p. 212). This distribution of ties allows members of the network (machines and individuals) to access and communicate with other members through a large number of simultaneously available paths that very rarely transit through a single central node. This is due to the fact that the concept of centrality is by definition excluded from the architecture of the Internet to increase its resilience in case of a major failure of the central node. In this model of information management, it is much harder for a central authority to control the flow of data than in a panoptic environment, while at the same time, it becomes much easier for a myriad of actors to observe and monitor their peers, since the distribution of ties also creates a hyper-connectivity conducive to the multilateralization of surveillance. So, while the panoptic and synoptic models placed the emphasis on "the fact that the disciplines use procedures of partitioning and verticality, that they introduce, between the different elements at the same level, as solid separations as possible, that they define compact hierarchical networks, in short, that they oppose to the intrinsic, adverse force of multiplicity the technique of the continuous, individualizing pyramid" (Foucault, 1977, p. 220), the Internet functions under entirely different premises. It connects people and let them form horizontal networks – largely independent from governments – that moderate the distribution of power instead of reinforcing its concentration (Lessig, 2006, p. 274).

This is not to say that the Internet is devoid of architectures of control: governments and businesses around the world spend considerable amounts of money to design surveillance systems able to tell them who is doing what, with whom, and from where on the Internet (Lessig, 2006, p. 38). But these technologies are not exclusive to a restricted group of supervisors. They are becoming increasingly accessible to individual users and fulfill a number of functions that range from the noble to the mundane, and the disciplinary to the playful. They must also contend with a number of resistance technologies and behaviours that thrive in the Internet environment because of its very un-panoptic architecture.

THE DEMOCRATIZATION OF SURVEILLANCE

The term democratization refers to the broadening accessibility of online surveillance through a plurality of tools and services that could previously only be afforded by governments and large companies. This trend reverberates both in the private and public spheres, and corresponds to a wide range of rationalities sustained by business-oriented ventures, non-governmental organizations (NGOs), and social units such as families and groups of friends. Low barriers of entry to the world of online surveillance are responsible for this democratization. Contrary to other mass media such as television or newspapers, the marginal costs for the distribution of information on the Internet are very low, because expensive proprietary infrastructure such as satellites, fibre-optic cables, printing presses, and delivery routes are not required (Benkler, 2006). All providers of Internet services share the same infrastructure and the same data transfer protocols, also known as TCP/IP (Lessig, 2006, pp. 143–146). Therefore, large investments in capital assets are not required to start disseminating information, as millions of bloggers have found out. Most of the costs incurred by new service providers are associated with the collection and sorting of data, or the development of new methods to collect and sort data more effectively or more efficiently. For example, the success of the very popular Google search engine can be attributed to the superior quality of its ranking algorithm, making the results it displays at the top of its page more relevant than those of its competitors. Once data or information has been processed, it can be distributed or accessed on a large-scale at little or no additional cost.

This combination of openness and cheap means of distribution constitutes a powerful incentive to innovations fuelled by entrepreneurs and social activists alike. These innovations can be categorized in two groups. The first group merges off-line observation technologies with online dissemination tools, while the second group is entirely made up of online technologies that are used to collect and distribute data. Among the observation technologies mobilized by the first group, we find digital photography and video recording, remote sensing, geographical information systems, human input, and social engineering. The following examples will provide a better idea of the democratization processes at work.

Online Diffusion of Content Collected by Off-Line Observation

YouTube[1] is probably the best-known video-sharing website, with an estimated monthly audience of 20 million people and 100 million video downloads per day. The company, whose slogan is "broadcast yourself", adds more than 65,000 videos every day to its library. Users of the site directly post these short segments with very limited interference from YouTube employees, whose number does not exceed 30 people (Reuters, 2006). Thousands of contributors find there a platform to share contents produced by the explosion of video-capable consumer devices such as camcorders, computer webcams, or mobile phones. Although YouTube and other less successful video-sharing websites are primarily promoting the entertainment aspect of their services, many videos uploaded on their servers have a distinctive surveillance flavour: shopkeepers or homeowners are routinely making surveillance tapes of burglars breaking into their property available in the hope that it will increase their chances of being arrested (Rodriguez, 2007), grainy videos capturing police brutality incidents or blatant instances of corruption are uploaded at regular intervals,[2] and politicians uttering racial slurs or contradicting themselves shamelessly in semi-private functions are also bound to find their duplicity exposed to an audience of millions within hours, with very limited opportunities for damage control.[3] The miniaturization of video recording devices and the ubiquity of Internet access points, even in conflict zones, also allow anyone with a connected computer to remotely experience the ferocity and confusion of close quarter combat: Iraqi insurgents and US troops alike profusely post uncensored videos of their deadly encounters, providing far bleaker pictures of the conflict than the sanitized versions offered by the main television networks. YouTube and its edgier competitors LiveLeak and Dailymotion return thousands of results for search terms such as "Iraq war", "insurgency", "sniper", or "IED" (improvized explosive devices).

At the other end of the spectrum, macro-observation technologies such as remote sensing and geographical information systems applied to the Internet information economy can foil the efforts deployed by governments and large corporations to conceal some of their most questionable activities. Google and Microsoft offer through their Google Earth and Virtual Earth services high resolution geocoded satellite pictures of the planet that can been used for surveillance purposes, despite the fact that the data provided is usually a few weeks to three years old.[4] These very popular tools are free to use, and Google claims that more than 100 million people have downloaded the software needed to access its imagery (Meyer, 2006). The primary use of these tools involves the first-hand observation of what past official maps deliberately omitted (Monmonier, 1991), hidden behind high walls, or too remote to be accessed by any other means. The Cryptome website offers, for example, a series of detailed "eyeball" pictures[5] that expose sensitive infrastructures such as military bases, intelligence agencies' headquarters, politicians', and company executives' residences, in an effort to dispel the myths surrounding these secretive places. Anyone with a connection to the Internet can comb the millions of satellite pictures available online in order to satisfy their idiosyncratic curiosity. Some people use this capacity to track the latest nuclear submarine launched by the Chinese navy[6] while others are just as happy having a peek at the houses of the rich and famous[7] or the places they will visit during their next vacation. NGOs are also enlisting Google Earth to call attention to civil wars and humanitarian disasters such as Darfur. Amnesty International has launched a campaign called "eyes on Darfur" that uses satellite imagery to present the extent of violence committed in this inhospitable part of the world and let Internet users "monitor [12] high risk villages [to] protect them from further attack" in what the NGO describes as the "global neighbourhood watch".[8] The United States Holocaust Memorial Museum offers a similar experience on its website, but on a much larger scale. It plans to use these satellite pictures to build an online "global crisis map" of emerging genocides or crimes against humanity, which would allow activists, journalists, and citizens to access and share information more quickly.[9] At the illegal end of the spectrum, some terrorists have even embraced these surveillance tools to identify possible targets and their vulnerabilities (Harding, 2007), an approach explicitly acknowledged by Google on its website when it describes how homeland security agencies can leverage the power of Google Earth to conduct "critical infrastructure vulnerability assessment" and "pattern visualization of surveillance data" for $ 400 a year.[10]

A more significant outcome of these online technologies derives from the capacity to combine satellite pictures and maps with other types of digital data provided by sensors such as mobile phones, or generated by users themselves. These new applications are known as "mashups" and are made possible by open and easy-to-use programming formats and tools (Eisenberg, 2007) that fuse layers of information into a single file, adding value to the original pool of diverse data. Some businesses incorporate mashups to the affordable surveillance tools they market, such as mobile phone companies that offer handsets equipped with global positioning systems and let their customers (usually parents) track online the movements of the person carrying the phone (usually a child) (Pogue, 2006). Beyond the rise of Big Mother and Big Father, mashups also assist citizens in their efforts to gain a more detailed awareness of their immediate environment. While interactive crime maps that let online users create personalized outputs based on criteria such as type of crime, zip code, location, or even transport route,[11] are popular in the United States, Europeans seem more interested in monitoring the location of speed and red light cameras. The SCDB website[12] claims to maintain a database of 18,000 cameras scattered all over Europe, whose coordinates are updated by road users (Big Driver?).

Online Surveillance of Online Activities

In the previous examples, the Internet was used as a mediator by millions of connected supervisors who access dispersed real-world data, then modify, aggregate, and disseminate it for their own benefit, for altruistic motives, or in some instance for criminal gain. The same process applies to the surveillance of online activities, which cannot structurally be monopolized by governments or large corporations. As the underlying rationale is fairly similar, I will only use three examples (two lawful, the last one criminal) to show how this works. The first example demonstrates how travellers who book their trips online can harness the power of self-surveillance to extract cheaper airfare and hotel room rates from companies that have developed predatory pricing systems based on consumers' surveillance. This practice is known in the tourism industry and in other sectors that deal in perishable items as "yield pricing" or "yield management" (Desiraju & Shugan, 1999) and involves the dynamic allocation of discounts so that revenues are maximized for each flight or room sold (Borenstein & Rose, 1994, p. 655). The complexity of this pricing system can only be managed by computers that constantly adjust prices to encourage purchases when sales are going slowly and maximize profits when the demand is strong, sometimes resulting in airfares that vary from one minute to another. Obviously, it creates a form of discrimination between consumers who pay fares that vary substantially for the same service, since they do not have access to the same data and tools on which to base their decision. The Internet resolved this informational asymmetry by creating a forecasting market that monitors the highs and lows of airfares or hotels rates. Services such as Farecast[13] or Kayak[14] use datamining techniques to comb the wild fluctuation of thousands of airfares over long periods of time and advise customers on the best purchasing strategy (wait or buy). Although they are applied to a fairly mundane activity, these tools should be understood as highly disruptive by nature. They bring meta-surveillance capacities to individuals who can deploy their own sophisticated technologies to uncover the routine surveillance to which they are submitted by large corporations.

The second example also illustrates how the democratization of surveillance can be used to expose the online activities of powerful interests. Whether it represents an improvement or not, the online collaborative encyclopedia Wikipedia[15] has become in a matter of years a source of reference material for millions of Internet users who also contribute to its entries. Content accuracy is a major issue (Giles, 2005), especially for controversial issues where conflicting interpretations of an event or someone's actions can lead to defamatory or plainly dishonest comments (Kolbitsch & Maurer, 2006). Government agencies that seek to defend their record on contested policy decisions or want to obscure their mistakes are tempted, in that context of openness, to edit entries that refer to them. Large corporations and NGOs might also use Wikipedia as a public relations tool to downplay their responsibility in embarrassing scandals or inflate their contributions to society. Unfortunately for them, the same surveillance tools that are used to undermine privacy and authenticate the identity of every Internet user can also be used to identify (to a certain extent) who has made changes on any given Wikipedia entry. This capacity has always been available to computer-savvy users through what is known as an IP tracer or IP locator. The term IP stands for Internet Protocol and refers to the addressing system that allows data to be sent to the right machine on the network. IP addresses are unique identifiers, and although they are not allocated on a geographical basis, it is still fairly easy to locate a user based on publicly available IP address tables (Lessig, 2006, p. 59). Hence, talented programmers can develop an IP mapping application that integrates seamlessly with another web application. Virgil Griffith, the designer of

WikiScanner,[16] is one of those talented programmers. His online search engine lets users find out which organizations are the most active Wikipedia editors. Thousands of changes made by people working for government agencies such as the US Department of Homeland Security, the Pentagon, or the CIA; companies such as Wal-Mart or Exxon; NGOs such as the American Civil Liberties Union (ACLU) or the Electronic Frontier Foundation or even religious entities such as the Vatican or the Church of Scientology are retrievable. While some of them are the results of bored employees taking a break to update a page that relates to their personal interests (in itself a form of resistance), many others are linked directly to attempts by these organizations to anonymously shape their image. The openness that characterizes the Internet's architecture renders these clandestine efforts much easier to detect, providing sufficient incentives exist for someone to provide monitoring tools and for users to take advantage of them.

The surveillance tools described above are not isolated or exceptional, but the democratization trend is not synonymous with equal access to surveillance resources either. The barriers to the deployment of highly intrusive online surveillance technologies are not financial resources, but instead technical skills. While governments have rapidly expanded their online surveillance capacities since 9/11, criminal actors have also been busy deploying their own elaborate webs of surveillance. Botnets (the contraction of software robot and network) are computer networks made up of compromised machines (called zombies) that have been infected by viruses or other malicious software and that can, as a result, be monitored and controlled remotely without the knowledge of their rightful owners. These botnets are used by hackers (called botmasters in this instance) to send spam, commit click fraud,[17] or launch large-scale attacks against websites in order to shut them down or extort money from their operators to stop the attacks.[18] Botnets are routinely used to perform scans of their host machines. With some of them including more than a million compromised computers (Gaudin, 2007) and conservative studies evaluating botnet infection at 11% of all computers connected to the Internet (Abu Rajab, Zarfoss, Monrose, & Terzis, 2006), their mass surveillance potential is not hard to imagine. In this last example, surveillance is no more horizontal and democratic than it is vertical or centralized, and the panoptic model can only be of limited assistance to analyze the distributed structure of supervision, and its disconnect from any disciplinary and social sorting project (Haggerty & Ericson, 2000; Lyon, 2006; Haggerty, 2006). Social and technical factors such as the plurality of functions associated with the monitoring of others' online activities, regulatory frameworks, new business models, computer skills of Internet users, and the open or faulty code of communication protocols all play an important role in the adoption of online surveillance technologies. Unfortunately, we have barely begun examining these variables' empirical architecture, which also influence the numerous resistance strategies employed by those who want to defend their privacy from the omnipresent surveillance of the state, their family and friends, or computer hackers.

RESISTANCE TO ONLINE SURVEILLANCE

In line with Foucault's lack of interest for resistance as a counteracting force to the oppressive panoptic gaze, many modern surveillance scholars have dismissed the possibility of collective neutralization and sabotage efforts or have been ambivalent about them, at best (Gandy, 1993, p. 147; Campbell & Carlson, 2002, p. 603), despite clear signs that they are not isolated occurrences (Bain & Taylor, 2000; Timmons, 2003; Lyon, 2004, Poster, 2005; Bogard, 2006, p. 101). Acts of resistance in surveillance studies are often presented as individual and localized efforts (Haggerty & Ericson, 2006, p. 18) that produce partial and temporary victories (Gilliom, 2006, p. 115) and merely reinforce the effectiveness of surveillance through an escalation process. There are, however, many ways for the subjects of surveillance to reclaim their privacy and autonomy, as Gary Marx (2003) so compellingly demonstrated. Although the eleven resistance strategies he describes in his article apply more or less to online surveillance, two of them will be considered in greater detail, and from a collective rather than an individual perspective. These strategies are: blocking moves and masking moves.

Cryptography as a Blocking Move

Blocking moves refer to the process that seeks "to physically block access to the communication" (Marx, 2003, p. 379). Blocking moves are inconceivable in the panoptic world, since partitions prevent subjects from contacting each others, whereas on the Internet, where messages transit through multiple paths, they become an essential tool to ensure the safety of communications. Cryptography is perhaps one of the oldest blocking moves. It can be defined as:

> A transformation of a message that makes the message incomprehensible to anyone who is not in possession of secret information that is needed to restore the message to its normal *plaintext* or *cleartext* form. The secret information is called the *key,* and its function is very similar to the function of a door key in a lock: it unlocks the message so that the recipient can read it. (Diffie & Landau, 1998, p. 13)

Cryptography has a long history that dates back to the invention of writing and played an instrumental role in several military conflicts (Singh, 1999; Pincock, 2006). Yet, its impact on Internet surveillance is rarely considered, despite the fact that the need to safeguard online financial transactions makes it one of the most widely used online privacy tools. If encryption procedures were mainly used by spies and diplomats before the advent of the Internet, the computing power available in each PC today is sufficient to produce scrambled messages that would foil the most determined code breakers. Since Philip Zimmermann made his Pretty Good Privacy (PGP) encryption software available on the Internet in 1990 and won his legal battle with the US Department of Justice, anyone who is not a mathematician or programmer can still enjoy the benefits of unbreakable encryption and defeat the most sophisticated surveillance technologies (Diffie & Landau, 1998). For example, terrorist organizations, pedophiles, and computer hackers have been known to use off-the-shelf or homemade encryption tools to conceal their unlawful activities (Denning & Baugh, 2000). Encryption is sometimes used by human rights organizations who want to protect their correspondents in authoritarian regimes. Although most popular e-mail programs such as Outlook or Thunderbird can send and receive encrypted emails, very few people actually use this facility. An Internet user survey conducted by Garfinkel, Margrave, Schiller, Nordlander, and Miller (2005) shows that 68% of people in their sample ($N=417$) were either unaware that encryption was available on their e-mail client or did not know what cryptography was. Hence, despite the fact that cryptography is widely available at virtually no charge to Internet users, resistance to online surveillance is informed by other factors than purely technical considerations. A study of political activists opposing US administration policies in the post-9/11 environment shows that users balance the need for secrecy with a reluctance to fall into what they perceive as a paranoid or abnormal state of mind (Gaw, Felten, & Fernandez-Kelly, 2006). Systematic resistance that applies indiscriminately to mundane and highly sensitive content is experienced as a mental burden denoting an unbalanced personality, while selective resistance is described by one respondent as similar to healthy eating and exercise: people know it is the right thing to do, but they are not always doing it themselves (p. 594). What these informed users tell us is that they resort to blocking moves with parsimony, maintaining a much more complex rapport to resistance than initially assumed by surveillance scholars.

Distributed Masking Moves

Masking moves that allow users to surf the web anonymously are more widespread than blocking moves. One reason that might explain this difference is that the former take full advantage of the distributed architecture of the Internet by establishing virtual networks of trust (Tilly, 2005). These resistance networks thwart surveillance attempts by randomly routing the information their members want to send or receive through other members of the network, thereby making it impossible for supervisors to know who is effectively communicating with whom and about what. TOR (The Onion Router), Freenet, and Psiphon[19] are examples of popular masking tools that are freely available for download and use on the Internet. Freenet's homepage claims that its software was downloaded more than two million times, and TOR's user base is said to reach hundreds of thousands, mainly from the United States, Europe, and China (Zetter, 2007). Although these programs differ slightly at the technical level, their overall approach is similar. Once people have installed them on their computer, a portion of their hard drive is automatically encrypted and secure connections are established with other computers that run the same software when the user logs on the Internet. All communications transit seamlessly through other nodes of the trust network before they are allowed into the more open and easily monitored part of the Internet. Attributing a particular online behaviour to a specific machine, and hence to its owner or operator, becomes a fruitless endeavour since complex algorithms are used to blur the patterns of data that enter and exit the trust network. What makes this type of trust network different from the more traditional ones described by Tilly (2005) is that it is scalable and does not require its members to share the same objectives. It is scalable in the sense that the more members these masking tools can enlist, the more effective they will be, while traditional trust networks expose themselves to failure and malfeasance when their membership becomes too large and difficult to manage. The second feature of these virtual trust networks is that credentials are allocated on a technological basis (the willingness to encrypt and relay encrypted communications with no control over the contents being transmitted) more than on ethno-religious traits or shared social or political interests, making strange bedfellows in the process. Even though they are primarily destined to privacy and anti-censorship activists, diplomatic missions, intelligence agencies, and armed forces – including from authoritarian regimes such as Iran – also make intensive use of these free masking tools (Zetter, 2007), a good indicator of the trust these surveillance organizations place in them to protect their sensitive information against their counterparts.

Less drastic masking moves involve the manipulation by consumers of registration and search data in order to minimize the generation of profiles based on viewing patterns and datamatching techniques. The free online service BugMeNot[20] (BMN) offers to bypass the registration process that is compulsory to enter many websites by providing its users access to a database made up of active accounts (usernames and passwords) obtained by submitting fake socio-demographic details. BMN also provides disposable e-mail addresses that can be used for twenty-four hours as an alternative to disclosing real e-mail address to online merchants and data-brokers. Because the online interface allows users to directly submit new accounts and retrieve passwords from the database, there is a positive correlation between the number of users and the utility they derive from this service. As of September 2007, BMN provided accounts to more than 175,000 websites. Another interesting initiative is TrackMeNot[21] (TMN), a little program written by two New York University professors.[22] This application is used whenever the Firefox browser[23] accesses Internet search engines such as Google, AOL, Yahoo, and MSN. These websites keep track of all the searches performed by individual users in order to return context or location-relevant advertisements to accompany search results (Barbaro & Zeller, 2006). TMN uses an obfuscation strategy to drown real search queries in a cloud of randomly generated queries that makes profiling considerably more difficult and much less accurate, if not totally meaningless. The inventors of TMN actually acknowledge on their webpage that Gary Marx's article "A tack in the shoe" (2003) partly inspired their application.

CONCLUSION

The reified panoptic metaphor that dominates the field of surveillance studies appears increasingly detached from the complex reality of online monitoring (Boyne, 2000; Haggerty, 2006). Through a detailed analysis of several diverse meta-surveillance and resistance technologies, I have attempted to expand the register of legitimate research questions on this issue. For example, how do some disruptive technologies concretely modify the underlying distribution of knowledge and power in the surveillant assemblage (Haggerty & Ericson, 2000)? How are expanding monitoring technologies appropriated by people and institutions for unexpected uses? What are the individual, social, political, economical, and technological factors that impact on resistance or constrain the effectiveness of surveillance? Can resistance be integrated to the study of surveillance, or should it be treated as a separate subject? These questions challenge the panoptic framework, but they also have the potential to make it more relevant to twenty-first century technological conditions. To be answered, they require a more grounded knowledge of the actual interactions between those who watch, the machines and infrastructure they design and use to carry out their surveillance, the people being watched and the flows of data that are generated as a result. These connections involving humans, machines, and places are easier to map in high-technology environments, because they leave behind a profusion of traces or markers, but it cannot be done without first abandoning the paranoid and megalomaniac tendencies the panopticon so often fuels (Latour, 2005).

While compiling example upon example of distributed surveillance and widespread resistance, I could not help wonder why so many surveillance scholars had carefully avoided this less travelled path. In an important contribution, Kevin Haggerty (2006) offers some interesting hypothesis to explain this reluctance, such as the critical thinking tradition of surveillance scholars, their simplified understanding of Foucault's integral intellectual legacy, a focus on human surveillance that neglects human/technological hybrids, and a methodological approach that overemphasizes discourse and document analysis to the detriment of more grounded empirical data. This last trait makes surveillance scholars overly dependent on the public transcripts that explain power relations between subjects and supervisors. Unfortunately, the official story is rarely the whole story, and hidden transcripts that can be defined as "offstage speeches, gestures, and practices that confirm, contradict, or inflect what appears in the public transcripts" (Scott, 1990, p. 4) should also be studied. However, the critical posture or methodological choices made by surveillance scholars cannot entirely explain the lack of interest for the "arts of resistance" and their impact on the governance of surveillance.

I offer an additional interpretation inspired by Gary Marx's (2007) techno-fallacies article and the heuristics' theory of Tversky and Kahneman

(1982). Just like technophiles often succumb to the false belief that there is a technological fix for every security problem, surveillance scholars (as an epistemic community, not as individuals) are not immune to biases that lead them to assume that the monitoring technologies embedded in virtually every aspect of our lives are a clear indicator of our inexorable fall into a 1984 reality. Three biases are particularly salient in this belief system. The first bias is the initiative bias, which leads people to attribute less initiative and less imagination to others than to themselves (Kahneman & Tversky, 1993, p. 3), especially if they belong to a lower socio-economic group. While surveillance scholars are able to offer elaborate narratives of the hidden power of the electronic panopticon and its significance, they frequently discount the interpretive capacities and agency of surveillance subjects and the resistance strategies that ensue. The loss aversion bias refers to the asymmetrical evaluation of positive and negative outcomes, where losses are systematically overestimated and gains are underestimated. This bias seems particularly pronounced "when the reference point is the status quo, and when retention of the status quo is an option" (Kahneman & Tversky, 1993, p. 14). This bias corresponds in surveillance studies to the reticence manifested toward the study of positive developments (Haggerty, 2006, p. 35) such as the accountability produced by meta-surveillance applications or the independence afforded to elderly patients by monitoring systems that let them stay at home. The tendency to predict widespread erosions of freedom has also been a prominent feature of surveillance studies, despite the lack of empirical and historical data to support this claim. Democracies have not crumbled since advanced monitoring technologies have invaded our lives, and the lack of sophisticated surveillance tools has never prevented authoritarian states to enroll thousands of informers to control internal dissent (Pfaff, 2001). Finally, the third heuristic is the probability bias whereby a confusion is made between what is possible and what is probable (Ohm, 2007). This bias is very closely connected with the previous one, because on contentious subjects such as surveillance and privacy, people tend to focus on disastrous outcomes and neglect the role played by randomness (Taleb, 2004), complexity, and contested rationalities (Espeland, 1998) among supervisors. Surveillance scholars frequently present what may happen as what will happen, obscuring the mechanisms that so often derail the best plans. Perhaps, the fact that Bentham's panopticon was actually never built and that the British government preferred instead to deport its prisoners to Australia, an open-air prison where convict supervision was deliberately kept at a minimum (Kerr, 1989;

Jackson, 1998), should serve as a reminder that dystopias are about as likely to materialize as utopias.

NOTES

1. http://www.youtube.com, accessed September 4, 2007.
2. See for example the string of videos showing Moroccan police officers receiving cash payments from truck drivers at http://www.youtube.com/watch?v=Afed8wvYwmc, accessed September 11, 2007.
3. Former US Republican senator George Allen (with presidential aspirations) lost his bid in the 2006 election after a video in which he called an aide to his opponent a «macaca» was made available on YouTube at http://www.youtube.com/watch?v=r90z0PMnKwI, accessed September 11, 2007.
4. See Google Earth help centre at http://earth.google.com/support/, accessed September 15, 2007.
5. http://eyeball-series.org/, accessed September 15, 2007.
6. http://www.fas.org/blog/ssp/2007/07/new_chinese_ballistic_missile.php, accessed September 16, 2007.
7. http://www.gearthhacks.com/dlcat25/Famous-Homes.htm, accessed September 16, 2007.
8. http://www.eyesondarfur.org/, accessed September 16, 2007.
9. http://www.ushmm.org/googleearth/projects/darfur/, accessed September 16, 2007.
10. http://earth.google.com/security.html, accessed September 16, 2007.
11. http://www.chicagocrime.org/; http://www.latimes.com/news/local/crime/homicidemap/; http://www.mapufacture.com/feeds/1000398-Oakland-Crime-Feed, all accessed September 16, 2007.
12. http://www.scdb.info/. It is one among others: see for example http://www.speedcameramap.co.uk/ and http://www.spod.cx/speedcameras.shtml for the United Kingdom, all accessed September 16, 2007.
13. http://www.farecast.com, accessed September 22, 2007.
14. http://www.kayak.com, accessed September 22, 2007.
15. http://www.wikipedia.org, accessed September 22, 2007.
16. http://wikiscanner.virgil.gr/, accessed September 22, 2007.
17. A practice where online advertisers are charged for clicks on banners that originate from computer software and not legitimate users interested in their product.
18. They are known as DDoS or distributed denial of service attacks.
19. http://tor.eff.org/, http://freenetproject.org, and http://psiphon.civisec.org/, all accessed September 25, 2007.
20. http://www.bugmenot.com, accessed September 25, 2007.
21. http://mrl.nyu.edu/~dhowe/trackmenot/, accessed September 25, 2007.
22. Daniel C. Howe, from the Media Research Lab and Helen Nissenbaum from the Culture and Communication department.
23. Unfortunately, the program is not available with the most popular Microsoft Explorer browser.

REFERENCES

Abu Rajab, M., Zarfoss, J., Monrose, F., & Terzis, A. (2006). A multifaceted approach to understand the botnet phenomenon. In: P. Barford (Ed.), *Proceedings of the 6th ACM SIGCOMM on Internet measurement* (pp. 41–52). New York, NY: ACM Press.

Associated Press. (2007). New breed of 'compliance software' makes office computer monitoring more sophisticated. *Technology Review*, published August 20, 2007, retrieved August 21, 2007, from http://www.technologyreview.com/Wire/19271/page1/

Bain, P., & Taylor, P. (2000). Entraped by the 'electronic panopticon'? Worker resistance in the call centre. *New Technology, Work and Employment*, *15*(1), 2–18.

Barbaro, M., & Zeller, T. (2006). A face is exposed for AOL searcher no. 4417749. *The New York Times*, published August 9, 2006, retrieved September 25, 2007, from http://www.nytimes.com/2006/08/09/technology/09aol.html

Bauman, Z. (2000). *Liquid modernity*. Cambridge: Polity Press.

Benkler, Y. (2006). *The wealth of networks*. New Haven, CT: Yale University Press.

Bogard, W. (2006). Surveillance assemblages and lines of flight. In: D. Lyon (Ed.), *Theorizing surveillance: The panopticon and beyond* (pp. 97–122). Cullompton: Willan Publishing.

Borenstein, S., & Rose, N. (1994). Competition and price dispersion in the U.S. airline industry. *The Journal of Political Economy*, *102*(4), 653–683.

Boyne, R. (2000). Post-panopticism. *Economy and Society*, *29*(2), 285–307.

Campbell, E., & Carlson, M. (2002). Panopticon.com: Online surveillance and the commodification of privacy. *Journal of Broadcasting & Electronic Media*, *46*(4), 586–606.

Denning, D. E., & Baugh, W. E. (2000). Hiding crimes in cyberspace. In: D. Thomas & B. D. Loader (Eds), *Cybercrime: Law enforcement, security and surveillance in the information age* (pp. 105–131). London: Routledge.

Desiraju, R., & Shugan, S. (1999). Strategic service pricing and yield management. *Journal of Marketing*, *63*(1), 44–56.

Diffie, W., & Landau, S. (1998). *Privacy on the line: The politics of wiretapping and encryption*. Cambridge, MA.: MIT Press.

Eggen, D. (2007). Lawsuits may illuminate methods of spy program. *The Washington Post*, August 14, p. A01.

Eisenberg, A. (2007). Do the mash (even if you don't know all the steps). *The New York Times*, September 2, p. 5.

Espeland, W. (1998). *The struggle for water: Politics, rationality and identity in the American Southwest*. Chicago, IL: The University of Chicago Press.

Foucault, M. (1977). *Discipline & punish: The birth of the prison*. New York, NY: Pantheon Books.

Gandy, O. H. (1993). *The panoptic sort: A political economy of personal information*. Boulder, CO: Westview Press.

Garfinkel, S., Margrave, D., Schiller, J., Nordlander, E., & Miller, R. (2005). How to make secure email easier to use. In: W. Kellogg & S. Zhai (Eds), *Proceedings of the SIGCHI conference on human factors in computing systems* (pp. 701–710). New York, NY: ACM Press.

Gaudin, S. (2007). Storm worm botnet more powerful than top supercomputers. *Information Week*, published September 6, 2007, retrieved September 23, 2007, from http://www.informationweek.com/story/showArticle.jhtml?articleID=201804528

Gaw, S., Felten, E., & Fernandez-Kelly, P. (2006). Secrecy, flagging and paranoia: Adoption criteria in encrypted e-mail. In: R. Grinter, T. Rodden, P. Aoki, E. Cutrell, R. Jeffries & G. Olson (Eds), *Proceedings of the SIGCHI conference on human factors in computing systems* (pp. 591–600). New York, NY: ACM Press.

Giles, J. (2005). Internet encyclopedias go head to head. *Nature*, *438*(7070), 900–901.

Gilliom, J. (2006). Struggling with surveillance: Resistance, consciousness, and identity. In: K.D. Haggerty & R.V. Ericson (Eds), *The new politics of surveillance and visibility* (pp. 111–129). Toronto: University of Toronto Press.

Graham, S., & Wood, D. (2003). Digitizing surveillance: Categorization, space, inequality. *Critical Social Policy*, *23*(2), 227–248.

Haggerty, K. D. (2006). Tear down the walls: On demolishing the panopticon. In: D. Lyon (Ed.), *Theorizing surveillance: The panopticon and beyond* (pp. 23–45). Cullompton: Willan Publishing.

Haggerty, K. D., & Ericson, R. V. (2000). The surveillant assemblage. *The British Journal of Sociology*, *51*(4), 605–622.

Haggerty, K. D., & Ericson, R. V. (2006). The new politics of surveillance and visibility. In: K. D. Haggerty & R. V. Ericson (Eds), *The new politics of surveillance and visibility* (pp. 3–25). Toronto: University of Toronto Press.

Harding, T. (2007). Terrorists 'use Google maps to hit UK troops'. *The Telegraph*, published January 13, 2007, retrieved September 16, 2007, from http://www.telegraph.co.uk/news/

Jackson, R. V. (1998). Jeremy Bentham and the New South Wales convicts. *International Journal of Social Economics*, *25*(2/3/4), 370–379.

Kahneman, D., & Tversky, A. (1993). *Conflict resolution: A cognitive perspective*. Toronto: University of Toronto.

Kerr, J. (1989). Panopticon versus New South Wales. *Fabrications : The Journal of the Society of Architectural Historians, Australia and New Zealand*, *1*, 4–32.

Kolbitsch, J., & Maurer, H. (2006). The transformation of the web: How emerging comm unities shape the information we consume. *Journal of Universal Computer Science*, *12*(2), 186–213.

Latour, B. (2005). *Reassembling the social: An introduction to actor-network-theory*. Oxford: Oxford University Press.

Lessig, L. (2006). *Code version 2.0*. New York: Basic Books.

Lyon, D. (2004). Globalizing surveillance: Comparative and sociological perspectives. *International Sociology*, *19*(2), 135–149.

Lyon, D. (2006). 9/11, synopticon and scopophilia: Watching and being watched. In: K. D. Haggerty & R. V. Ericson (Eds), *The new politics of surveillance and visibility*, (pp. 35–54). Toronto: University of Toronto Press.

Marx, G. T. (2003). A tack in the shoe: Neutralizing and resisting the new surveillance. *Journal of Social Issues*, *59*(2), 369–390.

Marx, G. T. (2007). Rocky Bottoms: Techno-fallacies of an age of information. *International Political Sociology*, *1*(1), 83–110.

Mathiesen, T. (1997). The viewer society: Michel Foucault's 'Panopticon' revisited. *Theoretical Criminology*, *1*(2), 215–234.

Meyer, D. (2006). Google, Microsoft vie for earth domination. *CNET News.com*, published September 12, 2006, retrieved September 15, 2007, from http://www.news.com/

Monmonier, M. (1991). *How to lie with maps*. Chicago, IL: The University of Chicago Press.

Mowery, D. C., & Simcoe, T. (2002). Is the Internet a US invention? An economic and technological history of computer networking. *Research Policy*, *31*(8–9), 1369–1387.
O'Harrow, R. (2005). *No place to hide*. New York: Free Press.
Ohm, P. (2007). The myth of the superuser: Fear, risk, and harm online. *University of Colorado Law Legal Studies Research Paper No. 07-14*, retrieved September 28, 2007, from http://www.ssrn.com/abstract=967372
Pfaff, S. (2001). The limits of coercive surveillance: Social and penal control in the German Democratic Republic. *Punishment & Society*, *3*(3).
Pincock, S. (2006). *Codebreaker: The history of codes and ciphers, from the ancient pharaohs to quantum cryptography*. New York, NY: Walker & Company.
Pogue, D. (2006). Cellphones that track the kids. *The New York Times*, December 21, 2006, C1.
Poster, M. (1990). *The mode of information: Poststructuralism and social context*. Chicago, IL: The University of Chicago Press.
Poster, M. (2005). Hardt and Negri's information empire: A critical response. *Cultural Politics*, *1*(1), 101–118.
Reuters. (2006). YouTube serves up 100 million videos a day online. *USA Today*, published July 16, 2006, retrieved on September 3, 2007, from http://www.usatoday.com/tech/news/2006-07-16-youtube-views_x.htm
Rodriguez, G. (2007). YouTube vigilantes. *Los Angeles Times*, published August 6, 2007, retrieved September 11, 2007, from http://www.latimes.com/news/opinion/
Scott, J. C. (1990). *Domination and the arts of resistance: Hidden transcripts*. New Haven, CT: Yale University Press.
Sewell, G., & Wilkinson, B. (1992). 'Someone to watch over me': Surveillance, discipline and the just-in-time labour process. *Sociology*, *26*(2), 271–286.
Simon, B. (2005). The return of panopticism: Supervision, subjection and the new surveillance. *Surveillance and Society*, *3*(3), 1–20.
Singh, S. (1999). *The code book: The science of secrecy from ancient Egypt to quantum cryptography*. Toronto: Random House.
Taleb, N. N. (2004). *Fooled by randomness: The hidden role of chance in life and the markets*. New York, NY: Texere.
Tilly, C. (2005). *Trust and rule*. Cambridge: Cambridge University Press.
Timmons, S. (2003). A failed panopticon: Surveillance and nursing practices via new technology. *New Technology, Work and Employment*, *18*(2), 143–153.
Tversky, A., & Kahneman, D. (1982). Judgement under uncertainty: Heuristics and biases. In: D. Kahneman, P. Slovic & A. Tversky (Eds), *Judgement under uncertainty: Heuristics and biases* (pp. 3–20). Cambridge: Cambridge University Press.
Zetter, K. (2007). Rogue nodes turn to anonymiser's into eavesdropper's paradise. *Wired*, published September 10, 2007, retrieved September 25, 2007, from http://www.wired.com/politics/security/news/2007/09/embassy_hacks

SURVEILLANCE AND SOVEREIGNTY

Kevin Stenson

ABSTRACT

An industry of description and interpretation has developed around the growth of surveillance, accelerated by: the development of the internet; volatile international relations since the collapse of communism; demographic mobility, segregation by class and ethnicity in the rich and poor worlds, sharpening inequalities, and post 9/11 fears of terrorism. Influential narratives have emphasised the diminishing power of sovereign nation-states in a marketised and globalised world. This chapter challenges the notion that coercive, sovereign modes of rule are a monarchical survival in decline. Rather, sovereign technologies of rule, in which surveillance is central involves strategies of governance from below as well as from above. They combine coercive with rhetorical, metaphorical communication and other 'soft' modes of rule. These make thinkable the nation-state as a discrete, defensible entity. Political communication translates between the complex technical expertise of evolving surveillance and security technologies and language intelligible to the public. Though surveillance technologies and information can be produced by commercial and other non-state sites of governance, metaphorically, much surveillance can be viewed as the extension of the eye of the sovereign. Although we are all targets of surveillance, those seen as threatening to the majority help to constitute and reproduce the social collectivity.

Surveillance and Governance: Crime Control and Beyond
Sociology of Crime, Law and Deviance, Volume 10, 279–301

ISSN: 1521-6136/doi:10.1016/S1521-6136(07)00213-8

The eye is the first circle; the horizon which forms is the second; and throughout nature this primary figure is repeated without end. It is the highest emblem in the cipher of the world.

– Ralph Waldo Emerson

All along the watchtower, princes kept their view.

– Bob Dylan

I would not make windows into men's souls.

– Queen Elizabeth I

INTRODUCTION: GLOBAL THREATS AND SURVEILLANCE

The collapse of communism in 1989 and the rapid growth of increasingly interdependent global markets have escalated anxieties about security and danger, and the growth of their counterpart in multiple, interacting forms of surveillance, the systematic gathering of data through multiple modes of oversight, using technologies of increasing sophistication (Haggerty & Gazso, 2005; Haggerty & Ericson, 2006). The predictable threat of mutually assured destruction of the bi-polar cold war era has given way to a more volatile, multi-polar world in which previously poor states like China, India, Pakistan, Libya, Venezuela, and Brazil, have become industrialised, richer, increasingly militarised, and key players on the international stage. Just as many states in the Horn of Africa and elsewhere have disintegrated into competition between warlords and militias. From the perspective of the US Pentagon, the latter areas are deteriorating into 'rogue' or 'failed' states – the 17th-century philosopher Thomas Hobbes' vision of a war of all against all, which can only be averted with strong central sovereign state authority. These risk becoming the breeding ground for anti-Western 'terrorist' movements claiming to represent the wretched of the earth. Hence, movements like Al Qaeda and Hamas that claim to be liberationist but are defined by Western governments as terrorist have emerged as centres of governance in their own right, able to conduct their own forms of surveillance and intelligence gathering (Burke, 2004). Al Qaeda, though centred in remote areas like the Afghan/Pakistan borders, has networks that operate across the globe through shadowy alliances and both direct and indirect association. The attacks of 9/11/2001 in New York and Washington, in Madrid in 2004, and London in 2005, involve asymmetrical warfare. Groups with limited resources are able to use secrecy, careful intelligence gathering, cell phones, and other low-cost new technologies to inflict massive damage. This exposes the vulnerability of even the richest states, no matter how sophisticated their surveillance and security systems. The New Wars in unstable regions like the Balkans and the Middle East, that have engaged the major Western powers since the early 1990s must be understood in this context (Kaldor, 1999).

This has been linked with an exponential global growth in state security systems, transnational innovations in policing, and commercial security, in attempting to coordinate responses to terrorism and cross-border crime. This is in addition to perceived escalating domestic threats from organised crime to juvenile offenders, domestic violence against women, violent gangs, and anti-social behaviour. The opening up of domestic and international economic activity creates myriad new opportunities for offending. They range from people, drugs, and weapons trafficking to the laundering of stolen money to finance terrorist and other criminal activity. They also include downloading child pornography, stealing from on-line bank accounts, identity fraud, and other internet-based crime that can be organised thousands of miles away from the victims (Sheptycki, 1997; Wall, 2007; Metcalf, 2007).

Managing these issues draws crime control from the political periphery to the centre, reinforcing the increasing global power of the commercial corrections and security industry, as well as the state agencies. This is in combination with powerful state lobbies of police, security, criminal justice personnel, all with privileged access to the media, politicians, and civil servants in order to advance their claims for funding and extra legislation (De Lint & Virta, 2004). In the three years to 2006, the top 100 US surveillance companies had doubled in value to $400 billion (Caulkin, 2007). This has spurred huge investment in security technologies. These range from expanded state and commercial surveillance of citizens, from the compiling of consumption data from credit cards by supermarkets, and by credit rating agencies to combat identity fraud, to iris recognition technologies and use of fingerprint records to protect airlines and national geographic and cyber borders from terrorists, cross-border criminals, and illegal immigrants. This facilitates growing demands to exchange data within and across jurisdictions, even when the data may encode error and falsehood, easily accessed by organised criminals. These are often one jump ahead in the cat and mouse game of cracking encryption security systems (Lyon, 2003; Wall, 2007).

Despite global trends, countries have developed unevenly depending on resources and their proximity to the front line of threats. Just as Northern Ireland during the Troubles was a laboratory for new security technologies and systems later used on mainland Britain, the UK now provides a global benchmark, inspiring even liberal Canada in the development of street-level close circuit TV systems (CCTV) (Lippert, 2007). The UK's citizens are the most watched in the world, with about 4.2 million CCTV cameras; urban citizens may be caught on camera around 300 times a day (Murakami Wood, 2007). Many protect and control government, military, and commercial spaces. The latter include the mass, private property spaces of shopping malls. This peaks in the City of London, surrounded by a 'ring of steel' following a devastating Irish Republican Army (IRA) bomb. It involves sophisticated technologies, including automatic car number-plate recognition, and other technologies progressively faster in operation and less-dependent on human operatives, and often tied into automatic, criminal database checks. These new systems have been brought together and further honed in 'lock down' security exercises that involve the suspension of normal mobility by citizens during, for example, gatherings of heads of state, major sporting events or the aftermath of major terrorist events, natural disasters, or other states of emergency (Murakami Wood, Coaffee with Blareau, 2007). These innovations facilitate global exchange and learning by security agencies, and blur the boundaries between public and private spaces. This bolsters the power of state and commercial personnel to gather information about individuals and social groups profiled to be suspicious, criminal, or a security threat. So often they are the 'usual suspects,' the poor, homeless, mentally ill, addicted, and illegal immigrants, as well as young people from the major ethnic minorities (Spalek & Lambert, 2007; McCahill, 2007).

There are parallels between the Catholic population of Britain and Ireland, for centuries considered a suspect, disloyal minority, and Britain's two million Muslims. Princess Diana complained that the second Elizabeth's courtiers spied on her, but in an era of Catholic Inquisitions in Spain and elsewhere, England's first Queen Elizabeth expressed caution about her new protestant state prying on the inner thoughts of her subjects – making windows into their souls. This included those suspected of harbouring treasonable sympathies for the Pope and encircling Catholic powers. A small seed of restraint was planted, which perhaps helped to blossom eventually into English liberal conceptions of the relation between the sovereign, the state, and the citizen. Yet, those genteel qualms did not prevent the hounding and barbaric execution of generations of English and Irish Catholics, often living a fugitive existence. For there were tensions between protestant Christian notions of freedom of conscience and a personal relationship with God, and the need for any sovereign to see threats to her/his power. Even now the monarch cannot marry a Catholic and in Northern

Ireland and Scotland, up until the Good Friday Agreement in 1999, for many servants of the state, the Catholic population – and not just those who supported Glasgow's green clad Celtic football team – remained 'papists.' As such they, and the families and communities that nurtured them, were seen as a treasonable threat to British sovereignty, and hence a warrantable target for surveillance by the police and security services (English, 2006).

Britain's Muslims include significant minorities who identify with the 1.6 billion global Muslim community, or Ummah, and want a global Caliphate, governed by theocratic Sharia law that would replace the secular law of the nation-state. This, the involvement of British-born Jihadists in the London tube bombings of July 2005, and a series of bomb attacks narrowly foiled by effective surveillance and detective work, has heightened concerns – at times unfair and indiscriminate – about the loyalty of Muslims to the sovereign British state. These deep concerns, shared, for example, in France, Germany, Denmark, and the Netherlands, that they may be an enemy within, have been used by police and government ministers to justify selective risk profiling. This involves police stop and search, trawls of phone and e-mail messages, bank records, DNA matching, the use of electronic tagging, and other surveillance and policing interventions. Some commentators fear this exacerbates the disaffection of young Muslims in Europe, often living, like generations of Irish Catholics and their descendants in British cities, in the poorest neighbourhoods, driving them further into the arms of Jihadists, just as sections of the Irish populations harboured sympathies for the IRA (Spalek & Lambert, 2007).

While many commentators would endorse the description of these trends in surveillance there is less agreement about how to interpret them theoretically and the implications this may have for policy. Our focus here is sovereignty, the struggle to monopolise the use of coercive force in controlling populations and geographical territory through the institutions of state and law. These include the apparatuses of policing, security, criminal justice, and military defence against internal and external threats. Defence of the realm inevitably casts some individuals and groups into the role of criminals, moral deviants, traitors, or enemies of the state (Stenson, 2005). Some emphasise that with new forms of market-oriented governance after the decline of the welfare states, and increasing global interdependence, there are multiple sites of governance beyond the state (Rose & Miller, 1992). These are said to diminish the discretionary scope for action by particular sovereign nation-states. There are multiple agents of surveillance which do not necessarily connect into a unitary 'Big Brother' surveillance society, viewing the population from one visionary vantage point.

In similar vein, it is claimed that nation-states become market states, subordinate to the interests of international commercial corporations. They have diminishing ability to protect domestic industry and populations, try to suppress the power of organised labour and create business-friendly conditions to attract inward investment (Bobbitt, 2002). Others, in Marxist and neo-Marxist schools, and some on the political right, deny that nation-states are declining in power and significance. Even if the nation-state is not a simple unitary actor, its institutions and huge resources deployed in the effort to monopolise the use of coercion to control populations and geographical territory, and build national solidarity, remain at the heart of government (Coleman & Sim, 2002; Fukuyama, 2004). The argument in the remainder of this chapter, building on the author's work over many years, is that sovereignty is best viewed as a set of *processes*, rather than simply a function of state power representing the interests of the dominant groups in society. It involves sites of governance operating from below as well as the official agencies of state government and national law. These can include, for example, youth groups in the street, residence associations, organised crime and revolutionary, militarised networks, religious communities, to a range of social movements (Stenson, 2005). In addition to coercive practices, it also includes cultural and social processes, in which surveillance plays a key role. Complex surveillance technologies are intelligible as extensions of governing and transgressing gazes. Our technologically complex modes of rule have to be translated for the public into the metaphorical language of sovereign political leadership, which in turn is fed by the fruits of surveillance. Other perspectives which view sovereignty in terms of coercion and violence, leftovers from our monarchical past, have neglected this interaction between wider cultural and coercive processes. They are deployed in trying to build community and national solidarity at every spatial scale and maintain sovereign control in the face of multiple rival and alternative attempts to control territory and populations emanating from below.

SOVEREIGNTY, METAPHOR, AND THE SOCIAL BODY

Academics in this field are preoccupied with examining the evolution of new technologies of surveillance in the digital age, the possibilities they open up for new forms of power, the constraint on them and the possibility of our sleepwalking into a surveillance society, as George Orwell and other dystopian, literary futurologists have warned (Murakami Wood, 2007).

Yet, for all that new technological sophistication, there are features of political communication, especially the emotive rhetoric of power and the relations between the leaders and the led, that demonstrate continuity with past regimes (Stenson, 2007). It is unwise to exaggerate the role of rational ideas in policy formation. This is particularly so given that, increasingly over the last two decades, in the advanced democracies as well as in poorer countries, social problems and proposed solutions to them have been coded in the highly emotive language of crime, risk, fear, and insecurity. As in centuries past, in wrapping up rational and emotional themes, political leaders deploy figurative language. The exercise of sovereign coercive power to govern populations and territories needs not just 'power' in the form of police, security agencies, and military personnel, it also needs 'knowledge,' much drawn from the systematic surveillance of populations and those seen as potential or actual enemies, and the use of rhetorical techniques and emotive, figurative language.

Metaphors have powerful emotional force as calls to action and operate in clusters to structure our thoughts and understanding of social relationships (Lakoff & Johnson, 1980, p. 5). Politicians use personalised and simple images of power based on surveillance, trawls of public opinion, and feedback from campaigning, to mobilise popular support and resistance. This is manifest in key metaphors of power, ancient and modern, that disproportionately focus on the human body, especially the head and the eye. Such metaphors have ancient provenance in human cultures and are at the heart of ways to think into being the human collectivity, including images of the nation and law. Historically, this rhetoric bridged between the individual and the social collectivity by presenting the latter as a social body (Stenson, 1993a, p. 49). While direct references to the social body are now less common, our modern images and concepts of collectivity though notions, for example, of society, community, and nation, still bear traces of these old metaphors. While broad images of the social body can provide a metaphorical language through which political communication can work, the construction of nation-state sovereignty involves ongoing rhetorical work at every spatial scale from national to local. Many politicians picture the social body in terms of moral polarities. This fuels the desire for tough punishment and demonises deviants as alien, even 'monsters,' threatening the common interest and well being of the law abiding, respectable and hard working (Wilkins, 1991). It also enables politicians to mobilise, largely through the media, public sentiment, and support, creating collective solidarity in the face of these perceived common threats. The sources of these mediatised images have deep oral cultural roots, and in literature

(Reiner, 2007, pp. 141–151). Leaders across the spectrum can use these techniques, from George W. Bush to Osama Bin Laden (Redlawsk, 2006).

For Ralph Waldo Emerson, the American poet quoted at the beginning of this chapter, poets name the world we inhabit. The most complex of concepts begin life as images. The eye, one of the most pervasive images in human culture and history, is the author shaping the images we have of the world and a metaphor for Emerson's vision of thought and language. Surveillance is essential to sovereign power and so is the eye that observes the objects of power, linked to ears attuned to dissent. How are images of sovereign order constituted? Between the end of the middle-ages and the formation in Europe of the absolutist regimes of the 17th and the 18th centuries, the proliferation of governmental programmes and practices was accompanied by what liberals may term dystopian visions. Some of these were ancient in origin, in which a sovereign attempted to exercise a totalising will over the whole of the territory under his/her command. For example, one vision with currency among rulers involved a revival of the old Roman republican discourses and the use of metaphors, in which the whole field of human association within a territory was conceived of as a social body.

This social body was coterminous with the state, indistinguishable from civil society. The prince or King is the head, the warriors the hands, the servants the feet and so on. This vision drove ambitions to regulate every corner of social life and helped to crystallise *Polizeiwissenchaft* (police science) in the German principalities of the 18th century (Foucault, 1977, p. 213; Stenson, 1993, p. 376). Furthermore, in the 17th century, the English philosopher Thomas Hobbes conceived of his notion of the social body through two metaphors. First, it was a contract between sovereign and people of mutual benefit. However, second, the parties could not easily separate. They formed part of the image of a sea monster, the Leviathan, which encapsulated all its living beings. The head represented the organising source of the social body. The very term capital city, from the Latin for head, expresses this vision of the sovereign's head as the source of vision and command. Note also that from ancient times the image of the sovereign's head (from Kings to Presidents) has been used on coins, statues, and pictures to represent his/her law and reign.

SOVEREIGNTY AND THE JURIDICAL SUBJECT

These visions of omniscient total control by the sovereign were far removed from the brutal realities of life and represented wishful thinking on the part of would be absolute rulers. In much of the Europe for centuries after the collapse of the Roman Empire, central authority remained temporary and fragile in the ongoing Shakespearean intrigues of plotting, murder, and rebellion by rivals. Hence, sovereign modes of rule develop and are sustained in opposition to rival powers that may resist them. Even the most advanced states have to cope with and gain sight of transgression in the shape of organised and other modes of crime, rebellious political movements, and other dissidents. In this respect, we should note that the eye was also a key metaphor in the work during the 1930s of the French Nietzschian and surrealist philosopher of transgression George Bataille, an influence on Michel Foucault (Macey, 1994, p. 137). For Bataille, sovereignty represented law, proscription, and the enchainment of the life force. Transgression involved breaking the limits and chains of sovereign authority and subverting its central images. His notorious erotic story *L'Histoire de l'oeil* included a series of vignettes involving eyes and metaphorical transformation into testicles, eggs and so on, used for pleasure. It also involved voyeurism and exhibitionism in which the transgressors challenge and return the censorious gaze of authority and convention. In turn, this transgressing gaze is a metaphor for the ways that agents and sites of governance that are indifferent or resistant to sovereign power have their own 'eyes,' that is modes of surveillance, information gathering, and processing that are deployed in their own strategies and tactics of governance.

Hence, surveillance and sovereign state authority go hand in glove (Foucault, 1977). It is hard to envisage even the simplest state that does not – whatever their nomenclature or official status – use the familiar cast of police, spies, agents provocateurs, and turncoats, to keep an eye on those who may be fomenting disorder and seditious plots, in castles, markets, ale houses, or anywhere the jealous and disaffected may gather. As Michel Foucault argued with graphic horror, while the sovereigns of old may have tried to keep an eye on the population through their spies, largely, their rule was buttressed by the exercise of draconian, cruel punishment in public spectacles, theatres of punishment. For those who transgressed, the sovereign's representative inscribed the power of the sovereign literally onto the body of the offender through burning, dismemberment, hanging, and other ritualised bloody torments, often linked precisely to the part of the body held responsible for the crime. The mutilated body is, hence, a metaphor for the terrible power of the sovereign. This signalled to the offender the folly of challenging sovereign power, and, more significantly, pour encourager les autres, the assembled awe-struck throng, and to whom they relayed their tales (Foucault, 1977). In this mode of rule, in addition to the gaze from above by agents of the sovereign, the gaze from below is even more significant. This involves the crowd gazing up at the majestic demonstration of sovereign, coercive power. In this sense, in Foucault's influential view, under sovereign rule the individual is seen principally as a juridical subject who can, perhaps, be ruled by constraint of law.

BIOPOLITICS AND THE WATCHTOWER

The core principles of sovereignty were strengthened in the 17th century. The Treaty of Westphalia 1648 brought to a close a long period of debilitating warfare in mainland Europe, in which Scottish and Irish mercenary troops (the commercial security personnel of their day) played a major role, despite or because of their compatriots' subjugation to English sovereignty. This treaty provided an initial loosening of the bonds between church and state, and the basis of a mutual (if asymmetrical and capricious) guarantee between states of their rights to jurisdiction over their own populations and territories (Hunter, 1998). This also guaranteed the surveillance and coercive juridical power over domestic populations and the movements of goods and people over borders. It made possible the eventual alignment of nation and democracy (and more authoritarian polities) and a broadening agenda for public government to include fostering the health, wealth, and military capacities of populations. This enhanced governmental agenda operated through the apparatus of law, security, and national and local administration, crystallised in 19th- and 20th-century nation-building and modes of civic governance (Foucault, 1991; Stenson, 2007). Hence, in Foucault's view particularly, sovereign power, as centralised, coercive force, was identified principally with the age of Kings and Queens, the mediaeval states of justice, and destined to decline as the world modernised.

For Foucault, the new modes of government, which first emerged in philanthropic and other sites beyond the state in the 19th century, created new possibilities for ruling human bodies, which would transcend crude, coercive sovereign methods. In the new vision humans are not simply juridical subjects, but beings with economic, emotional, physical, and other capacities that may be enhanced or inhibited, and about which information can be gathered. This opened the way to broader, new mentalities, or ways of thinking about government, first developed outside the state by philanthropists, physicians, and other social reformers. In the 19th and 20th centuries they developed forms of professional data and expertise from demography to medicine, psychology, and criminology. The new visions involved *biopower* or *biopolitics*, in the sense that they made the human body a topic for a disciplinary form of governing: a *biopolitics* of the individual and wider social body (Foucault, 1979). The accent shifts towards the use of disciplinary powers fostering a docile conformity to command and what were considered respectable lifestyles. They operated through systematic surveillance, judgment, and education in factories, armies, schools, mental hospitals, prisons, and other large institutions, social work and other interventions into the domestic lives of the poor. The emphasis was on deeper modes of rule that made broadly liberal democratic societies possible, yet, it was clear that similar techniques operated in non-democratic and un-liberal soviet societies. Foucault's key metaphor in representing this was Jeremy Bentham's early 19th-century Panopticon design for the new rational prison combining punishment with the reshaping of behaviour. This enabled the warder to have a near total, all seeing view of prisoners from a central watchtower. Watchtowers have ancient provenance and extend the gaze of the sovereign to distant outposts, and featured, for example, in Hadrian's Roman Empire wall to watch for the savage Picts in Scotland, to the survey posts of the ancient Chinese Wall.

Yet the watchtower is a metaphor for, and feature of, more recent social orders, which try to unify the commanding gaze of those who function as the sovereign, and his/her agents in both liberal and authoritarian polities. The stark watchtowers overlooking the Irish republican controlled 'bad-lands' of Northern Ireland, where the writ of British government and law barely ran and along which agents of princes kept their view, are being dismantled. If the image of the watchtower represents old technologies of surveillance, we should recall their extraordinary effectiveness in governing populations, in conjunction with street patrols and other direct surveillance technologies and practices. In recent history, in sharp contrast with the American and British occupation, Saddam Hussein's Baathist Party was an effective surveillance apparatus that identified and suppressed dissent with ruthless cruelty, holding together Iraq's incipiently warring tribes and religious communities. In the Muslim world, the most powerful challenge to secular, totalitarian military dictatorships like Saddam's comes not from liberal democracy but from Islamist movements wanting a theocratic state, organised under the authority of versions of Sharia law. In Saudi Arabia and the Afghani, Taliban state in the late 1990s, governing techniques, backed up by body-centred punishments, have involved religious police on street patrol. These have powers of surveillance and intervention, to police dress codes, gender relations, and the most intimate aspects of life. Whether or not this makes windows in men and women's souls, such governance leaves scant room for freedom of conscience, self expression, dissent, and the distinction between public and private life (Burke, 2004).

The story of the Stasi, the communist East German secret police, is only gradually yielding its secrets. The recent German film *Lives of Others* provides a touching account of the ubiquity of surveillance into every corner of life, the paper records of which were compiled with meticulous detail. With, by today's standards, primitive surveillance technology, this penetration into what in Western liberal democracies would be termed the private spheres of life, included friendship, sexual, and family relations, with the betrayal of friends and lovers to the state stretching over decades. The museums devoted to the terror of the Nazi and Soviet states near Berlin's erstwhile 'Checkpoint Charlie,' a residue of the boundary wall between East and West, provide a grim epitaph for the savagery and rationality of state surveillance and power. Long lists of people killed trying to escape the soviet watchtowers and armed guards remind us that the latter had to make terrifying choices. Even they were under scrutiny, expecting ruin for failing to fulfil their ghoulish duty. Trust was scarce and nobody, even at the apex of the system, could know for sure if they were being watched and betrayed.

In the narratives of the dominant (what I have termed) *discourse* governmentality theorists (Stenson, 2005), inspired by the work of Foucault, like sovereign modes of rule, the older disciplinary, surveillance-based modes of power that tried to turn people into compliant, docile subjects, did not suit fast changing, social and economic conditions and have, since the 1960s, been on the wane. Their principal concerns have been with transformations from 'social' rationalities, or mentalities, of government associated with Roosevelt's New Deal and social democratic welfare states, to post-social rationalities or mentalities of government associated with New Right intellectuals and administrations. The emphasis has shifted, in market-oriented societies towards the fostering of self regulatory, flexible forms of subjectivity and self-governance (Rose, 1989). In addition, they have followed Foucault's view that Marxists and other radicals tend to exaggerate the malevolent, omniscient, panoptic, power of the state, considered as a unitary, collective actor and cold monster. In this version of governmentality theory, Marxists are said to underestimate the extent to which there are, particularly in liberal societies, multiple domains, or nodes, of governance both within and beyond the formal boundaries of the state.

Hence, the post-Foucault vision involves not a single panopticon of surveillance but multiple surveillant eyes with multiple agendas. These can include professional, commercial, philanthropic, and other relatively autonomous networks and institutions, each with their own systems of surveillance, data gathering, and brokerage, and agendas of governance. This view is maintained by a number of Foucault's influential followers, who, if they recognise centralised sovereign power, see it as an atavist survival from the past (Rose & Miller, 1992; Garland, 1996; Wood & Shearing, 2007; Rose, O'Malley, & Valverde, 2006). This approach chimes with neo-liberal arguments that, with globalisation, states are now less nation than 'market states,' servants to the markets and major corporations, desperate to attract and retain footloose capital, and with diminishing scope for independent action to protect and govern domestic populations and industries (Bobbitt, 2002).

RECOGNISING SOVEREIGNTY: FROM MARXISM TO REALIST GOVERNMENTALITY THEORY

There are problems in the accounts of the discourse governmentality theorists. This author has argued for a number of years for an alternative version of Governmentality theory, more compatible with mainstream, realist social science. From this alternative, 'realist Governmentality' perspective, Foucault's view of sovereignty as the government of juridical subjects, associated with the mediaeval states of justice, is the Achilles heel in his accounts of changing forms of liberal government (Stenson, 1998, 2005; Lea & Stenson, 2007). He (and his followers) failed to recognise that sovereign modes of rule have not atrophied or died. They developed in creative interaction with cultural processes. This involves disciplinary and other modes of rule, and both civic and ethnic forms of nation building. The intellectual confusion of discourse Governmentality theorists results from their narrow focus on discourse. While paying lip service to the need to study material, human practices, they detach the study of liberal mentalities of government from the varied, complex historical settings in which they operate. Moreover, discourse Governmentality theorists' principal sources of evidence for theorising are archives and policy texts. Rather, we need to engage with a rich range of both oral and textual data sources, using the range of realist methods of research, including the direct observation of human practices. For example, policy texts often disguise the context of political struggles and compromises that underlie their production, contexts, and tensions that remain hidden to the analyst confined to tidy texts (Stenson, 1998; Stenson & Watt, 1999). Before elaborating the realist version of Governmentality theory account of surveillance and sovereignty, building on the foregoing analysis of the role of metaphor, let us acknowledge that other writers have also highlighted sovereignty.

Commentators from the Marxist and social democratic left, and the right, have argued that reports of the demise of the sovereign, nation-state are premature. It remains, in a dangerous age with fearful populations attracted to strong leadership, a robust survivor. It is the best arena for the political and emotional mobilisation of increasingly diverse populations, the organisation of security, the more equitable delivery of services, use of resources and enforcement of law and justice (Hirst & Thompson, 1996; Fukuyama, 2004; Stenson, 1998, 2005; Loader, 2000; Lea & Stenson, 2007). Neo-Marxists are still convinced by Lenin's description of the state in capitalist society as the executive committee of the bourgeoisie. From this perspective, capitalism generates a need for political instruments to protect capitalist interests in the poor but resource-rich global south, through military colonial conquest and other means (Hardt & Negri, 2000; Hall & Winlow, 2003). In addition, they point out that the embrace of the global markets in rich countries involves economic restructuring, and the outsourcing of production to where labour is cheap and compliant.

This creates enormous inequalities, social dislocation, disaffection, and growing geographical, economic, and social segregation, conflict by class and race, the growth of illegal economies and a growing population of incarcerated poor. This expanding incarceration and media publicity about it, accompanied by 'get tough' punitive political rhetoric can be seen as a contemporary way to make a spectacle of the demonised offender population (Hallsworth, 2002). In alliance with burgeoning commercial security, but orchestrated by state agencies at local, national, and international levels, this creates expanding police and criminal justice apparatuses, and increasingly restrictive CCTV and other surveillance and control systems, to contain and manage the growing underclass. These people are seen to threaten shopping malls and other major sites of consumption (Coleman & Sim, 2002; McCahill, 2002). Despite force in these arguments, these theorists underestimate the cultural dynamics and political pressure welling up from ordinary citizens as well as the tabloid media they consume for politicians to enforce tough policing and criminal justice to contain groups they fear (Stenson, 2000).

In addition, Giorgio Agamben (2005) has emphasised the growing significance of extreme sovereign rule by reference to 'states of emergency,' especially in the volatile, war-torn years since the collapse of communism. In time-honoured fashion, as Amnesty International reminds us, kleptocrat despots in nominally democratic countries from Zimbabwe to the central

Asian republics use the fig leaf of states of emergency to justify the suspension of habeas corpus and other liberal civil rights and due process of law, and/or attempts to destroy or exile all opposition with extreme coercion. Usually, such constitutional rights had never had much currency in their countries in the first place. So far, so familiar, but Agamben's concern is that this logic is increasingly being used also in the advanced liberal democracies, fuelled particularly by the rhetoric of endless 'wars' against terrorism, drugs, and illegal immigration. This helps to blur the boundaries between the role of law and government in peace and war. It warrants the accumulation of additional sovereign legal powers, and, we may add, vast increases in expenditure on public and commercial security and surveillance technology and personnel. These institutions then have an interest in stoking public fears, further enhancing their funding and power in a self-reproducing spiral.

The watershed was the passage in haste after 9/11 of the US Patriot Act in October 2001, extending state powers and creating a major new state security framework. This was followed by similar legislation in a range of democracies and a commitment to inter-state partnerships to combat common enemies. For example, against opposition from sections of the judiciary and civil liberties lobbies, the UK has echoed the US Patriot Act in constructing a sovereign legal armoury, for example, with the Terrorism Act 2000, the Anti-Terrorism, Crime and Security Act 2001, the Prevention of Terrorism Act 2005, and the Terrorism Act 2006. They build upon the provisions against anti-social behaviour in the Crime and Disorder Act 1998 and later legislation introducing Anti-Social Behaviour Orders, acceptable behaviour contracts, curfews, dispersal orders, long-term detention without charge or trial, and other constraining measures. Richard Ericson sees this international trend, linked with the new surveillance technologies, as the development of 'counter law,' eroding the foundations of liberal conceptions of justice and due process (Ericson, 2007).

Carl Schmitt, a Nazi sympathiser and German legal theorist at the time of the Weimar Republic, anticipated this. In his view, the committees, corruption, and prevarications of representative democracy undermine the capacity of leaders to make and implement decisions that will serve the truly democratic purpose of sovereign government, to protect and advance the higher needs of the nation (Schmitt, 1996). The sovereign is precisely the one who can decide what is the state of exception, or emergency, and hence suspend the encumbrance of due process and civil rights. Adolph Hitler employed this logic to suspend law and rule by decree for 12 years after the Reichstag fire in 1933.

For Agamben, it is naïve to see extreme measures as a temporary over reaction to threats, and to see law as a check on violence. For centuries, violence has underpinned the word of sovereign law, which has the capacity to distinguish between those who have a 'qualified life,' recognised as having the full range of citizenship rights (*bios*), and those people who are seen only as having the 'bare life' (*zoe*) of animal bodies. The inmates of Guantanamo Bay, Cuba, incarcerated by the Americans after the invasion of Afghanistan, under conditions of super surveillance, are seen as enemy combatants. They do not have the rights to due legal process of the citizen, nor even the rights of prisoners of war. For Agamben (1998) they illustrate the reduction of people to the status of 'bare life.' Furthermore, this may not be confined to Guantanamo. There may be a form of legal mission or function 'creep' in the extension of the use of this principle in what are seen as conditions of emergency. The Australian, British, and Italian states have confined asylum seekers in camps, in which their rights are extremely restricted. These are zones of exception in which large categories of people are placed under conditions of permanent surveillance and suspicion.

Agamben's work is a salutary warning about where democracy may be heading. However, there are large gaps between legal principle and everyday practice and conditions vary considerably between countries. Even in the US, these developments may not survive the Bush administration. Not all democracies have created such zones of exceptions, and there remain powerful liberal sentinels who resist such moves. It is not yet clear how far his theories can explain these variations. However, this work reminds us that sovereign nation-states are intrinsically particularistic. In order to define and differentiate themselves it may be essential to construct scapegoats and folk devils, enemies within and outside which constitute a forge for the construction of identity and solidarity. Sporting rivalry can provide a relatively tamed and benign forum for this but there are much darker expressions (Young, 1999). Let us now turn to an alternative perspective that also takes issues of sovereignty seriously, but in less negative terms.

REALIST GOVERNMENTALITY THEORY AND SOVEREIGNTY

In the author's realist governmentality theory, sovereignty is not simply the operation of coercive powers on behalf of the state (Stenson, 2005). Whatever the claims of the discourse governmentality and neo-liberal theorists of governance (and these two categories overlap), in the long history of liberal democratic polities, the imagery of sovereignty has stubbornly refused to expire. Filtered through and shaped by media images, it is manifest in the organisation of the United Nations as a framework of (albeit unequal) nation-states led by powerful, named leaders, with whom it is possible to negotiate and who are in command over military and other resources. It is also manifest in the rituals of summits between powerful leaders, both democrats and dictators, in the dignity and authority of the judge in court and the deep grammar of jurisprudence exalting the authority and majesty of law. This is despite faltering attempts by social movements led by liberal criminologists to create less hierarchical, more republican, and restorative-based alternatives to criminal justice (Braithwaite & Pettit, 1992). Hence, sovereignty is best seen as a set of processes that aim to monopolise the government of populations and territories with coercive controls in combination with building collective cultural solidarity and identity, including fostering a sense of nationhood (Loader & Mulcahy, 2003). It is important to recognise that there would be no attempt to use force and other means to create sovereign control if there were not continuous, primordial challenges from criminal, political, religious, and myriad other networks engaged in the struggle to govern their own separate domains and operating with their own forms of folk *biopolitical* rhetoric and knowledge (Stenson, 2005). As witnessed by struggles between warlords in the Horn of Africa and the Afghan–Pakistan borders, this deeper process of territorialisation operates whether or not a sovereign state exists (Burke, 2004).

Sovereign technologies involve hard and soft powers: warfare and policing, but also education, the mass media, citizenship training, and political campaigning. Moreover, in advanced liberal democracies, particularly, sovereignty also embodies pressures from and governmental agendas of ethnic minority, religious and other social movements, and lobbies operating as sites of governance from below. Aside from all those criminal or subversive political groupings that operate outside or in opposition to the law and state, these are groups wishing to align their sectional agendas with those of the state and law, creating policy and legal changes that will incorporate their interests and values. For them, this can harvest a host of cultural, legal, and material powers and resources under the auspices of the state, greatly strengthening their position within the body of society (Stenson, 1998, 2005; Stenson & Edwards, 2001). In other words, in the advanced liberal democracies, privileged elites and state servants do not always impose sovereign rule, coercively, from above without enrolling groups from below.

There are deep underpinnings to these new, cultural dimensions of sovereignty in liberal democracies. Notwithstanding the ornamental monarchies of Europe and awesome power of the Saudi and other Middle Eastern Arab dynasties, the age of powerful monarchs, ruling by divine right, has passed. Modern liberal democratic government builds on the visions of intellectuals of the 18th, 19th, and 20th centuries who formulated democratic images of rule. Montesquieu in France, Locke in England, Adams and Jefferson in the USA, and other architects of liberal democracy devised principles, laws, and constitutional measures to provide checks and balances to concentrated power, traditionally understood in terms of the (separate powers) of the executive, legislative, and judicial branches of authority. Rousseau, presaging the political experiments of Robespierre and the French revolutionaries made thinkable the notion of a shared sovereignty in a constitutionally framed social contract: the people as sovereign, either directly, in his imagination, or in that of Locke and other Anglo-Saxon thinkers, through representative democracy (Bertram, 2003). In these visions, the gaze of citizens high and low need be less hierarchical and vertical, less 'top down' or 'bottom up.' Rather, as in Alexis de Toqueville's conceptions of early American democracy in the prairies, government and law could be seen as horizontal: people survey, police, and support each other through community organisation (Putnam, 2000). Yet, the conception of the 'representative' of the people could be a Trojan horse within which notions of personalised sovereign power are smuggled back into the democratic polity. How different is the powerful figure of the elected Prime Minister or President from the sovereign?

Let us build on our earlier discussion of the role of metaphor and rhetoric. Sovereign rule involves the technical administration of surveillance and other technologically sophisticated technologies, but in conjunction with rhetorical work by political leaders that offers people frameworks of interpretations to make sense of their experience and perceptions. Modern surveillance involves cameras and other complex technologies that most citizens struggle to understand in scientific detail. Yet they can be rendered intelligible through simple images. New surveillance technologies are presented to the public, implicitly or explicitly, as metaphorical extensions of the authoritative, censuring, and guarding sovereign eye. A paradigm of this is the eye that symbolises the Iris recognition system, enabling enrolled citizens to fast track through immigration at British airports. Civil liberties lobbies struggle to present these tools as agents of oppression, a danger to freedom and universal, individual human rights. However, to the extent that sovereign authorities are publicly defined as looking with a protective guardian's eye, these anxieties are, using collectivist, communitarian and nationalist logic and rhetoric, trumped by the notion that surveillant technologies are needed to defend the over-arching human right to security of life and limb (Edwardes, Hosein, & Whitley, 2007).

These messages are reinforced by political rhetoric that provides metaphors of sovereign leadership within the social body. These include, for example, shepherd to a flock, captain of the ship of state sailing through treacherous seas to his crew, or valiant general in the vanguard of his troops. In the USA, the Political Right has invested hugely in media campaigns depicting the desirable leader as a strong father of unruly children, while liberal and left leaders are presented as unmanly and ineffectual, like over-indulgent mothers. In the terms of California's muscle-clad Governor Arnold Shwarzenegger, these leaders are 'girly men.' Powerful lobbies frame issues in such a way as to foreclose debate and downplay political and ethical alternatives through the medium of simple metaphors for the wider public. This leaves governing to the wise patriarch, served by capable experts (Lakoff, 2004). Mayor Rudi Giuliani successfully deployed a strong father frame in legitimating his New York policing strategies, based on sophisticated surveillance of precincts and tough management systems, and drew heavily on these images in his presidential bid in 2007–2008. In the UK, operating with similar rhetoric and strategies so did Ray Mallon in Hartlepool and Middlesbrough, as both police chief and later, popular mayor since the mid-1990s. He explains crime, anti-social behaviour, and other social ills as the product of denigrated, ineffective, or non-existent fathering. As with Giuliani, this has been undertaken with numerous media images of his strong, penetrating, authoritative, sovereign like, gaze. Strong leadership and policing then serve as surrogates for a refurbished fatherhood, presented as battered by the breakdown of the family, the rise of feminism, and political correctness (Stenson, 2000).

Figurative language is used to resolve or defuse the tensions between maintaining a recognition of, on the one hand, individualist, universalistic, liberal rights, liberties, and due legal process and, on the other hand, a communitarian and nationalist recognition of the collective rights to security for the fearful majority, realised through sovereign controls. In the post-9/11 era, this tension is sharpening, reinforcing the creation of secure urban enclaves and a creeping segregation of populations deemed troublesome. Historically, 'social' modes of government developed in close interaction with nationalism. In nationalism, there is often an uneasy mix of communitarian values emphasising the particularistic needs and identities of the majority and the collectivity, and, on the other hand, universalistic liberal values that focus more on the individual, universal ethical values, and rights of minorities. Nationalism can involve uneasy hybrid mixtures of both civic nationalism, open to newcomers to a society, and ethnic nationalism that speaks to those with centuries old attachments to blood and soil (Dench, 1986; Stenson, 1998). These are people – often the solid middle mass – who identify a nation with the culture of its dominant ethnic groups, religious faiths, and heroic history. There is a continuum from progressive to conservative versions, but nationalist and communitarian visions tilt away from emphasising rights towards responsibility, duty, and the well being of social collectivities, from the family, to nation and higher levels of human association (Hughes, 2007). In this perspective, the relatively stable rule of populations in geographical areas requires not just hard coercive power, but also 'soft' power. It needs domestic political strategies to persuade people, including waves of immigrants from very different cultures, to identify with and be loyal to an over-arching organic unity of the nation-state.

Such modes of nationalism accompany the struggle to find new forms of surveillance. These include, for example, ID cards and passports with encoded biometric and other information, to identify and record citizens and their rights to public services, and to police borders, both geographical and virtual (Goodhart, 2004). This view can be deeply unpalatable to liberal and left intellectuals (including progressive criminologists) and the social movements and constituencies they seek to represent. By contrast, they are often committed to a cosmopolitan, human rights ethic. This value position may be sympathetic to immigrant populations identifying mainly with their global, ethnic, or religious diaspora communities, and distrustful of nationalism and racism (often seen as an indissoluble couple), and restrictions on demographic mobility (Wood & Shearing, 2007, p. 95). As social scientists, we need to reflect on how our personal and shared value positions affect our theories and how this impacts on how we see the fellow humans we study.

CONCLUSION

The great complexity of forms of surveillance, the global nature of the internet, and the ability of criminal offenders to target their prey from anywhere in the globe can encourage the view that the sovereign, nation-state is in decline, lacking effective knowledge of its populations and territories and control of its geographic and e-borders. There may be some merit in this view, as natural and social disasters demonstrate, chaos lurks beyond the semblance of order. However, as yet, international forms of jurisdiction and law enforcement remain undeveloped, nation-states remain the principal units of political and legal organisation (Hirst & Thompson, 1996), and in financial crises commercial corporations still rush to the fiscal skirts of nation-state governments. They are as dependent on the state as any single mother on welfare. In these conditions, amidst increasing anxiety about terrorism and global flows of populations, we should not underestimate the extent to which people still look for protection to sovereign modes of power. These embody both coercive and 'soft' cultural forms of nation-building both domestically and abroad. As in the past, politicians have to mediate between the worlds of expertise and the public with emotionally attuned, metaphorical discourse.

In metaphorical terms, surveillance technologies operate as extensions of the eye of commercial and other sites of governance beyond the state. Yet state sovereign modes struggle to orchestrate, combine, incorporate, or suppress these visions through new legislation and other means in extending the eye of the sovereign, the directing intelligence of the social body. Constructing this communitarian and nationalist image of the social body involves focusing on threats by publicly demonised, predatory, 'anti-social' groups and individuals to weak and vulnerable citizens. These can range from noisy 'neighbours from hell,' paedophiles, aggressive beggars, street drug dealers and robbers, to those seen as terrorists, and growing numbers of both legal and illegal immigrants. In turn, this can generate ever-evolving modes of targeted surveillance, from 24/7 satellite-based electronic tagging to the use of pilot-less drone aircraft flying over increasingly segregated troubled areas. This very struggle against perceived enemies, and the social exclusion it entails, helps to constitute the contours of the nation it aims to protect.

ACKNOWLEDGMENT

Thanks to Nicola Madge and Simon Hallsworth for their comments.

REFERENCES

Agamben, G. (1998). *Homo sacer: Sovereign power and bare life*. Stanford University Press.

Agamben, G. (2005). *State of exception*. Chicago: University of Chicago Press.

Bertram, C. (2003). *Rousseau and the social contract*. London: Routledge.

Bobbitt, P. (2002). *The shield of Achilles: War, peace, and the course of history*. London: Allen Lane.

Braithwaite, J., & Pettit, P. (1992). *Not just deserts*. Oxford: Clarendon Press.

Burke, J. (2004). *Al Qaeda: The true story of radical Islam.* London: I.B. Tauris.
Caulkin, S. (2007). Watch it, or surveillance will take over our lives. *The Observer*, August 7.
Coleman, R., & Sim, J. (2002). Power, politics and partnerships: The state of crime prevention on Merseyside. In: G. Hughes & A. Edwards (Eds), *Crime control and community, the new politics of public safety.* Cullompton: Willan.
De Lint, W., & Virta, S. (2004). Security in ambiguity. Towards a radical security politics. *Theoretical Criminology, 8*(4), 456–489.
Dench, G. (1986). *Minorities in the open society: Prisoners of ambivalence.* London: Routledge & Kegan Paul.
Edwardes, C., Hosein, I., & Whitley, I. (2007). Balance, scrutiny and identity cards in the UK. *Criminal Justice Matters, 68*, 29–30.
English, R. (2006). *Irish freedom: A history of nationalism in Ireland.* Basingstoke: Macmillan.
Ericson, R. (2007). *Crime in an insecure world.* Cambridge: Polity.
Foucault, M. (1977). *Discipline and punish.* Harmondsworth: Penguin.
Foucault, M. (1979). *The history of sexuality, an introduction.* Harmondsworth: Penguin.
Foucault, M. (1991). Governmentality. In: G. Burchell, C. Gordon & P. Miller (Eds), *The Foucault effect: Studies in governmentality.* Hemel Hempstead: Harvester Wheatsheaf.
Fukuyama, F. (2004). *State-building: Governance and world order in the twenty-first century.* London: Profile Books.
Garland, D. (1996). The limits of the sovereign state: Strategies of crime control in contemporary society. *British Journal of Criminology, 36*(4), 445–471.
Goodhart, D. (2004). Too diverse. *Prospect, 95*(February).
Haggerty, K., & Ericson, R. (Eds). (2006). *The new politics of surveillance and visibility.* Toronto: The University of Toronto Press.
Haggerty, K., & Gazso, A. (2005). Seeing beyond the ruins: Surveillance as a response to terrorist threats. *The Canadian Journal of Sociology, 30*(2), 169–187.
Hall, S., & Winlow, S. (2003). Rehabilitating leviathan: Reflections on the state, economic regulation and violence reduction. *Theoretical Criminology, 7*(2), 139–162.
Hallsworth, S. (2002). The case for a postmodern penality. *Theoretical Criminology, 6*(2), 145–163.
Hardt, M., & Negri, A. (2000). *Empire.* Cambridge: Harvard University Press.
Hirst, P., & Thompson, G. (1996). *Globalization in question.* Cambridge: Polity Press.
Hughes, G. (2007). *The politics of crime and community.* Basingstoke: Palgrave Macmillan.
Hunter, I. (1998). Uncivil society: Liberal government and the deconfessionalisation of politics. In: M. Dean & B. Hindess (Eds), *Governing Australia, studies in contemporary rationalities of government.* Cambridge: Cambridge University Press.
Kaldor, M. (1999). *New and old wars: Organized violence in a global era.* Cambridge: Polity.
Lakoff, G. (2004). *Don't think of an elephant, know your value and frame the debate.* White River Junction, VT: Chelsea Green Publishing Co.
Lakoff, G., & Johnson, M. (1980). *Metaphors we live by.* Chicago: University of Chicago Press.
Lea, J., & Stenson, K. (2007). Security, sovereignty and non-state governance 'from below'. *Canadian Journal of Law and Society, 22*(2).
Lippert, R. (2007). Open-street CCTV Canadian style. *Criminal Justice Matters, 68*, 31–32.
Loader, I. (2000). Plural policing and democratic governance. *Social and Legal Studies, 9*(3), 323–345.
Loader, I., & Mulcahy, A. (2003). *Policing and the condition of England, memory, politics and culture.* Oxford: Oxford University Press.

Lyon, D. (2003). *Surveillance after September 11.* Cambridge: Polity.
Macey, D. (1994). *The lives of Michel Foucault.* London: Vintage.
McCahill, M. (2002). *The surveillance web: The rise of visual surveillance in an English city.* Cullompton: Willan.
McCahill, M. (2007). Us and them – The social impact of surveillance technologies. *Criminal Justice Matters, 68,* 14–15.
Metcalf, C. (2007). Policing operation ore. *Criminal Justice Matters, 68,* 8–9.
Murakami Wood, D. (2007). Securing the neurocity. *Criminal Justice Matters, 68,* 37–38.
Murakami Wood, D., Coaffee, J. with K. Blareau (2007). Lockdown! Resilience, resurgence, and the stage-set city. In: R. Atkinson & G. Helms (Eds), *Securing an urban renaissance, crime, community and British urban policy.* Bristol: Policy Press.
Putnam, R. D. (2000). *Bowling alone: The collapse and revival of American community.* New York: Simon & Schuster.
Redlawsk, D. P. (Ed.) (2006). *Emotion in political informational processing.* New York: Palgrave Macmillan.
Reiner, R. (2007). *Law and order, an honest citizen's guide to crime and control.* Cambridge: Polity.
Rose, N. (1989). *Governing the soul.* London: Routledge.
Rose, N., & Miller, P. (1992). Political power beyond the state: Problematics of government. *British Journal of Sociology, 43*(2), 172–205.
Rose, N., O'Malley, P., & Valverde, M. (2006). Governmentality. *Annual Review of Law and Social Science, 2,* 83–104.
Schmitt, C. (1996). *The concept of the political.* Chicago: Chicago University Press.
Sheptycki, J. (1997). Insecurity, risk suppression and segregation: Some reflections on policing in the transnational age. *Theoretical Criminology, 1*(3), 303–315.
Spalek, B., & Lambert, B. (2007). Muslim communities under surveillance. *Criminal Justice Matters, 68,* 12–13.
Stenson, K. (1993). Community policing as a governmental technology. *Economy and Society, 22*(3), 373–389.
Stenson, K. (1998). Beyond histories of the present. *Economy and Society, 29*(4), 333–352.
Stenson, K. (2000). Someday our prince will come: Zero-tolerance policing and liberal government. In: T. Hope & R. Sparks (Eds), *Crime, risk and insecurity.* London: Routledge.
Stenson, K. (2005). Sovereignty, biopolitics and the local government of crime in Britain. *Theoretical Criminology, 9*(3), 265–287.
Stenson, K. (2007). Framing the governance of urban space. In: R. Atkinson & G. Helms (Eds), *Securing an urban renaissance, crime community and British urban policy.* Bristol: Policy Press.
Stenson, K., & Edwards, A. (2001). Crime control and liberal government: The 'third way' and the return to the local. In: K. Stenson & R. R. Sullivan (Eds), *Crime risk and justice, the politics of crime control in liberal democracies.* Cullompton: Willan.
Stenson, K., & Watt, P. (1999). Governmentality and 'the death of the social'?: A discourse analysis of local government texts in South East England. *Urban Studies, 36*(1), 189–201.
Wall, D. S. (2007). *Cybercrime.* Cambridge: Polity.
Wilkins, L. T. (1991). *Punishment, crime and market forces.* Aldershot: Dartmouth.
Wood, J., & Shearing, C. (2007). *Imagining security.* Cullompton: Willan.
Young, J. (1999). *The exclusive society.* London: Sage.

PART IV: BEYOND CRIME CONTROL

SURVEILLANCE AND EDUCATIONAL TESTING: NO CHILD LEFT BEHIND AND THE REMAKING OF AMERICAN SCHOOLS

John Gilliom

ABSTRACT

Educational testing launched under "No Child Left Behind" (NCLB) brings unprecedented levels of surveillance to public education in the U.S. The testing regime is moving American pedagogy away from types of teaching which are either politically disfavored or not easily tested. The impact of NCLB will be strongest in lower-income schools which fare poorly on such tests; these schools can expect to see sanctions, shaming, and a concomitant departure of committed families and teachers. The reshaping of American education wrought by NCLB compels us to reimagine mass surveillance as not primarily a means of watching the world, but as expressions of power capable of effecting significant changes in institutions and behaviors.

Surveillance and Governance: Crime Control and Beyond
Sociology of Crime, Law and Deviance, Volume 10, 305–325

ISSN: 1521-6136/doi:10.1016/S1521-6136(07)00214-X

INTRODUCTION

> But who will write the more general, more fluid, but also more determinant history of the "examination" – its rituals, its methods, its characters and their roles, its plays of questions and answers, its system of marking and classification? For in this slender technique are to be found a whole domain of knowledge, a whole type of power.
>
> – Foucault (1979, p. 185)

In 2002, President George W. Bush signed a reauthorization of national education law that has come to be known as No Child Left Behind (NCLB). In so doing, he unleashed what is probably the most ambitious surveillance program in the history of the nation. Under the new program, tens of millions of Americans, in every community and state, are subject to unprecedented frequencies and degrees of state monitoring. The educational testing regime set forth under NCLB is a transformative moment in the history of not just education, but government surveillance itself. It thus offers a rare opportunity to explore surveillance and the legal and political conflict over its impact and implementation.

Perhaps the most fundamental lesson to be learned is one that has been encountered in numerous prior studies of the politics of surveillance: even the most technocratic forms of surveillance tend to extend and enforce extant patterns of race and class bias, extant moralist and criminological agenda, and extant assumptions about subject populations. In *Surveillance, Privacy, and the Law: Employee Drug Testing and the Politics of Social Control* (Gilliom, 1994), I found that the employee drug testing initiative of the 1980s was embedded within the broader anti-union and anti-labor politics of the Reagan era. The testing movement could only be understood as a political–cultural bid to infuse workplace politics and ongoing labor–management struggles with the broad patina of the so-called War on Drugs. Evidence that workplace drug use was relatively minor problem outside of a few industry sectors, that drug testing was an ineffective response, and that the drug most likely to be detected, marijuana, was the least worrisome, fell by the wayside under an onslaught of pro-testing propaganda and media cooperation.

In *Overseers of the Poor* (Gilliom, 2001), a study of the computerized surveillance of welfare clients, a major surveillance initiative was launched to deal with relatively low levels of fraud and abuse in state welfare systems. Accompanying the launch of the new programs were major media campaigns highlighting anecdotal stories of welfare fraud and portraying the poor as abusers and cheats in need of close monitoring to protect the taxpayers' money. The implementation of the system itself brought salvoes of shame upon welfare clients with the constant trumpeting of the assumption that they were cheats and frauds adding to the already extreme social stigma of poverty in America. In welfare surveillance, finally, the optic and metrics of the system, created a bizarre state-centered view of poor Americans, a depiction that belied a complex reality with partial figments of the bureaucratic imagination.

Each of these studies joined a body of research and analysis finding that government surveillance and information initiatives are almost assured to miscast and misperceive their subjects in ways which not only express the political, cultural, and technological constraints on policy, but more importantly, render a body of data and knowledge that ensures poor policy and a failure of statecraft (Scott, 1998). In the present examination of the optics and implementation of surveillance of America's teachers, schools, and schoolchildren, we see another chapter in the unfolding story of the failure of surveillance in contemporary governance – budget limitations, cultural biases, technological shortcomings, and political preferences combine to render a system that will see what it is able to see and see it poorly at that.

Surveillance has, of course, always been a part of formal education. From simple things like quizzes, tests, assignments, and attendance records, classroom teachers monitor and assess the work of their students. Their goals are multifaceted. Teachers want to see if students are absorbing and retaining information, learning to use new skills, and developing the capacity for critical thinking. Teachers also use testing as a means to compel students to do the reading and coursework, to take the course seriously, and, inevitably and sometimes unconsciously, as a way of flexing bureaucratic muscle in the politics of the student–teacher relationship. Anyone reading this has at some point been a student and, probably, a teacher; we all know from those experiences that surveillance, understood as observation, monitoring, and evaluation, is a big part of the game.[1]

The new surveillance manifest in the emergent regime of standardized testing is a fundamentally different thing than the older regime of quizzes and blue books. It is a fundamentally different thing in terms of sheer size and scale; in terms of the nature and targets of sanction; in terms of the expressions of political will and the centralization of power; and in terms of the standardization of American education. The exertion of power in NCLB is evident enough that Frederick M. Hess uses the term "coercive accountability" to describe the policy and Jones et al. use the term "measurement driven reform."[2] Each of these phrases reminds us that the testing apparatus established under NCLB is not some sort of neutral or inconsequential assessment: there are policy goals, outcomes, and implications from any system of social assessment like this. And there are also, as we shall see, critically important implications tied to the nature, limitations, and use of the mechanisms of observation. In the end, this essay will argue that the testing movement achieves such a radical redrawing or reimaging of American education that it compels us to confront the question of whether the term "surveillance" and its idea of simply watching really captures the full magnitude of this initiative and its many companions. Massive new systems of social surveillance have come to be such powerful bodies of communication, depiction, and social organization that the old language of observation is obsolete.

NO CHILD LEFT BEHIND: THE BASICS

> The purpose of this subchapter is to ensure that all children have a fair, equal, and significant opportunity to obtain a high-quality education and reach, at a minimum, proficiency on challenging state academic achievement standards and state academic assessments. This purpose can be accomplished by ... ensuring that high-quality academic assessments, accountability systems, teacher preparation and training, curriculum, and instructional materials are aligned with challenging state academic standards so that students, teachers, parents, and administrators can measure progress against common expectations for student academic achievement ...
>
> – *U.S. Code: Title 20: Section 6301*

NCLB is a reauthorization and revision of the 1965 Elementary and Secondary Education Act (ESEA). The carrot and stick behind NCLB is a budgetary item called Title I which started as the 1965 Act's effort to steer federal money to disadvantaged children. It is the single largest source of federal money for education (almost $12 billion in 2003) with aid going to 90 percent of school districts in the nation (Rudalevige, 2003, p. 25). It is the threat of losing this money that serves as the leverage to bring states into compliance with federal guidelines. In short, no states have been *commanded* to obey NCLB, but nor has any state been willing and able to forego the federal dollars attached to the educational surveillance initiative in NCLB.

NCLB is a massive initiative, but for our purposes we can focus on the following key features:

- States must establish a standards-based curriculum and design tests to assess mastery of that curriculum.
- Annual tests are required in reading and math for all students in grades 3–8 and once in high school; science testing is mandated after 2007–2008.

- Test scores must be reported for schools and for significant subgroups within schools, including ESL students, disabled students, with population breakdowns by race and family income.
- Test scores for schools and subgroups are assessed to determine whether a school has made "adequate yearly progress" (AYP); failure to achieve AYP in multiple years earns escalating sanctions. While only about half of the nation's schools receive Title I funding, NCLB requires all schools to meet the standards. The actual Title I schools (meaning the poorer ones) face the strictest public accounting and sanctions of NCLB; after two to four years of inadequate progress they face such measures as state takeover, staff replacement, or conversion to charter school.
- States must have *all* children up to proficiency by 2014 and make measurable progress toward that goal during the years leading up to 2014. That progress must be evident in each identified subgroup.
- "States are free to determine their own standards, to create their own tests, and to determine for themselves the scores that individual students must receive in order to be deemed 'proficient.'" (Ryan, 2004, pp. 941–942).[3]
- NCLB also mandates that the National Assessment of Education Progress (NAEP) examinations in math and reading be given every two years to fourth and eighth graders. NAEP uses a random sample testing method to create a composite of a state's progress on a national metric of skills and content. It is frequently used as that national standard for comparisons among states and nations.

Much of what emerged in the 2001 NCLB had been circulating in the educational policy world for the last couple of decades. Many observers point to the Reagan era study *A Nation at Risk* as the most visible starting point for the modern push for standards-based educational assessment.[4] Momentum for performance standards grew during the Bush I and Clinton eras, with Bill Clinton (as both president and governor) playing a key role in the push toward setting education outcome goals. Indeed the 1994 reauthorization of the 1965 Elementary and Secondary Education Act mandated standards, testing and the AYP framework to such an extent that it "fundamentally changed the nature of Title I. Instead of providing funds to support remedial instruction for disadvantaged students, Title I funds now had to be used to create standards for all students" (Ryan, 2004, pp. 938–939). But the laws of the 1990s, though they called for state standards and testing, left enough room for flexibility and interpretation that "work progressed very slowly" (Rudalevige, 2003, p. 29). During this period a rough battle line emerged in national politics with Republicans being opposed to a national curriculum on state's right grounds and Democrats being opposed to a regime of standards-based tests (Rudalevige, 2003, p. 30).

With conservative Democrats joining Republicans on a number of fronts, a series of acts in the very late 1990s put into partial action most of the key elements of NCLB. Numerical goals, AYP, public report cards, state-set standards, and the menu of sanctions were all developed during this period. Meanwhile in Texas, tycoon Ross Perot had led the state to a system including annual testing and the disaggregation of school data to compel ethnic and economic subgroup assessments (Hess, 2003, p. 68). When the incoming Bush administration laid out the blueprint for what would become NCLB, one aide to Senator Lieberman was able to say that Bush "essentially plagiarized our plan" (Rudalevige, 2003, p. 36). But that is not quite right. What emerged in NCLB was a convergence of some longstanding Washington ideas and some of the key Texas policies developed in the 1990s. In an unprecedented convergence, this author and the Bush administration agree that the testing and accountability measures were the most important elements of NCLB. In the logrolling and compromise that resulted in NCLB, the Bush team even gave up their interests in school vouchers, and faced the wrath of cultural conservatives and the loss of many Republican votes in the Congress, in order to preserve the testing mandate.

The following analysis of the surveillance mechanisms in NCLB addresses several main points. Namely, that the methods of assessment are designed in such a way as to reduce their effectiveness for authentic educational outcome assessment; that the key metric AYP promotes lower standards and racial and economic segregation; and that there are fundamental fallacies at the heart of the idea of a single measurement of academic performance. Subsequently, the chapter explores the ways in which NCLB testing protocols are reshaping the curricula of U.S. schools, particularly those with concentrations of low income and minority students.

OPTICAL ILLUSIONS

As a tool of observation and assessment, the testing established under NCLB is profoundly flawed. Ignoring all of the questions of politics, values, and power struggles to artificially isolate the simple process of assessment, the distortions and limitations are enormously important.

The Optics of Mediocrity

The stated goal of NCLB is to improve learning. In pursuit of this, NCLB launched an ambitious program of frequent high-stakes testing, but there is considerable research indicating that state level high-stakes testing has no discernible outcome on learning. At first glance, this may seem counter-intuitive because scores have, in general, been rising on state level tests since they became popular in the 1990s and then expanded under NCLB in the early 2000s. Therefore, learning must be improving. And learning may indeed be improving, but it may just be learning about how to take the statewide tests as teachers and administrators become more familiar with the content and question design and are able to train their students to more successfully fill in the forms.[5] In this section, we begin by reviewing data indicating that more testing does not necessarily mean more learning and then examine the metrics and finances of NCLB testing to see how improving state test scores may be relatively easy and that the improvements achieved may not indicate improvements in educational outcomes.

When skeptical researchers check up on the efficacy of statewide tests by examining how students do on other examinations, they find little to suggest that rigorous state testing is associated with student improvement:

> Analyses of scores and participation rates for the NAEP, ACT, SAT and AP tests suggest that there is inadequate evidence to support the proposition that high-stakes tests and high school graduation exams increase student achievement. The data presented in this study suggest that after the implementation of high stakes tests, nothing much happens ... The data presented in this study also suggest, however, that after the implementation of high school graduation exams, academic achievement apparently decreases ... ACT, SAT and AP scores decline. Indeed, on balance, the analyses suggest that high-stakes tests and high school graduation exams may tend to inhibit the academic achievement of students, not foster academic growth. (Amrein & Berliner, 2002, pp. 57–58)

So even on one of the most basic and widely agreed upon goals of NCLB, we may actually get exactly the opposite of what was promised.

One clear area of impact, intended or not, is that the educational curricula of schools across the nation are being changed not just to meet the newly required state *content* guidelines but to teach students with *methods* that are most conducive to success on the testing instrument. We know that content is shifting more toward the tested areas of math, reading, and, increasingly, science, with corresponding cuts in arts, music, and other untested areas, particularly in high need areas. But what is also occurring is a shift of education within areas like math and English as teachers are compelled to move away from broader, more theoretically rich approaches to forms of instruction that emphasize the smaller particle and tactics that can be affordably tested. Furthermore, as a state's testing style and content become known, teachers can orient to the specific content and question styles of given tests.

Such outcomes point to one of the key vulnerabilities of standardized educational testing as a mechanism of surveillance. As testing expert W. James Popham (2005) explains, any test measures only a tiny subset of relative knowledge. An ideal test, on "everything," would take too long to complete, grade, and process, so we pick a relatively tiny subset, or sample, of materials. The fine art or science of designing that subset is the work of experts known as psychometricians. As states are unable to find or afford competent psychometricians, or overwhelm the ones they have, the test questions and question structures see more and more repetition from year to year and the tests become easier and easier to game. Teachers learn the pattern of the test, old tests are used for training, private sector preparation books enter the market; the end result is that the test itself becomes the object of study and the successful negotiation of its terms the goal of pedagogy. The training and gaming allows school districts to show improvement, but it is improvement at taking the tests, not necessarily improvement on mastering the universe of knowledge and skill that the tests seek to assess. For authentic proponents of testing as a real measure of educational progress, such outcomes are deeply problematic because they render the tests far less meaningful as tools of assessment. As Popham explains, if that tiny sample of knowledge that is to be tested becomes known in advance or over time, the entire logic of the assessment tool collapses.

The problems are compounded by the aforementioned fact that standardized educational testing in the U.S. has to be done on the cheap. The explosion of testing in the wake of NCLB was so enormous that by 2006 there was actually a test question crisis in America. As of this writing, the dramatic expansion of standardized educational testing has created a severe shortage of psychometricians and a true problem in the industry's capacity to make competently designed test questions (Toch, 2006). Part of the problem is created by the quintessentially American federalist approach to NCLB. Under the compromise brokered in the Congress, there is a national educational accountability program manifest in the demand that all children have annual tests in grades 3–8 and one more to graduate from high school. But there was also a nod to federalism in allowing each state to set its own educational standards and design its own exams. With this nod, the hope of a technically competent testing regime was destroyed: there are simply not enough qualified personnel to design a testing regime which includes not just several grade levels in each of 50 states with their

50 curricula, but the many other testing programs under state minimal competency laws *and* the classics such as the SAT, ACT, LSAT, MCAT, etc.[6] Furthermore, as has been widely noted, NCLB provided little in the way of funding to support the new testing programs. According to the U.S. General Accounting Office, multiple choice tests run about one to two dollars per student while more comprehensive performance assessments range from 35 up to 70 dollars per student (Jones, Jones, & Hargrove, 2003, pp. 16–17). This has created a budget situation in which the only feasible forms of testing are the most simple multiple choice tests – the slightest nod to an essay format or creative student responses puts the cost of evaluating the tests through the ceiling. Under economic and time pressures, many states have opted for the cheapest and most easily scored types of test questions – multiple choice questions measuring basic rote knowledge.

One of the key implications of this helps us frame and understand the recent evidence of test score improvement that we see in many school districts. With a limited capacity to create test questions that are either truly creative or well-designed, districts and teachers are readily able to "teach to the test" – essentially training students to the sorts of questions that can be expected and, therefore, over time, creating an image of improvement. Such training need not be in the form of actually teaching the direct content of the test, though this certainly occurs. Teachers can, for example, spend time teaching what they call "look-alikes." Once it is known that tests will structure the multiplication and division questions in certain formats, teachers can orient their instruction to that format. There is no necessary improvement in instruction, in fact – as we will soon see – it may be less effective. But such practices do lead to better measurements (Jones et al., 2003, p. 66). Indeed, there was a major scandal for the state of Texas when a RAND corporation analysis found disparities between student performance on state level testing and the national standards exam. The conclusion was that teachers in Texas had figured out enough about the statewide test that they were able to prep their students, who showed improvement over the years. There was also evidence that school district administrators had cheated. Once the children of Texas faced the differently formatted NAEP, the evidence of improvement vanished.

Absolute Criteria and the Reshaping of American Schools

A critically important choice was made in designing the surveillance procedures at the heart of NCLB to work with an absolute criteria scale of assessment rather than a value-added or rate-of-improvement scale. Under the current absolute criteria approach, if Nebraska sets its statewide curriculum standards to the expectation that every third grader can do long division to the fourth decimal point, then every school in the state must have a certain percentage of their students meet that standard to avoid being designated as failing. Yet under a value-added approach, if a school had, say, been able to get a given percentage of its third graders to do long division a certain measure better than they could do it in the second grade, the school would be judged successful even if the students could not make it out to the fourth decimal point. The value-added approach is more flexibly applied to the different contexts in which school and teachers function and has the potential to add accountability while avoiding a number of what have been called "perverse incentives" (Ryan, 2004) in the absolute standards approach.

> To appreciate the promise of value-added assessments, it is important to revisit why (the absolute criteria approach to) AYP is a relatively useless measure of school quality. As mentioned above, student performance is the product of a number of factors, some of which schools can control, others of which are beyond a school's ability to influence. A student's score on a standardized test is the result of both school and teacher inputs, as well as a host of exogenous factors, including innate ability, socioeconomic status, parental involvement, community stability, and peers. Because of the influence of these exogenous factors, looking to whether students in a school hit a uniform benchmark of achievement-the current approach to measuring AYP–actually tells us very little about the quality of the school itself. (Ryan, 2004, pp. 978–979)

For example, schools with relatively advantaged students typically post better test scores than those with relatively disadvantaged students. But it does not follow that the former schools are better at educating students than the latter; the scores may simply reflect the fact that the former school has students who take tests better than those at the latter. It is a well-known truism in the testing business that most assessment tests largely assess the socioeconomic background, or "social capital" of the students taking the test – known as the "Volvo effect" because, as Jones et al. (2003) summarize, "simply court the number of Volvos, BMWs, or Mercedes owned by the family and you have a good indicator of how well the child will perform on standardized test" (p. 118). Testing critic Alfie Kohn gives several examples of the Volvo effect at work:

> A study of math scores on the 1992 National Assessment of Educational Progress found that the combination of four variables unrelated to instruction (number of parents living at home, parents' educational background, type of community [e.g., "disadvantaged urban," "extreme rural"], and state poverty rate) explained a whopping 89 percent of the

> differences in state scores. In fact, one of those variables, the number of students who had one parent living at home, accounted for 71 percent of the variance all by itself… The same pattern holds within states. In Massachusetts, five factors explained 90 percent of the variance in scores on the Massachusetts Comprehensive Assessment System (MCAS) exam, leading a researcher to conclude that students' performance "has almost everything to do with parental socioeconomic backgrounds and less to do with teachers, curricula, or what the children learned in the classroom." … Another study looked just at the poverty level in each of 593 districts in Ohio and found a .80 correlation with 1997 scores on that state's proficiency test, meaning that this measure alone explained nearly two-thirds of the differences in test results … Even a quick look at the grades given to Florida schools under that state's new rating system found that "no school where less than 10% of the students qualify for free lunch scored below a C, and no school where more than 80% of the students qualify scored above a C. (Kohn, 2001)

The absolute standards approach that is currently used in the U.S. asserts that an overwhelming percentage of students must achieve certain statewide benchmarks by certain dates and at the exact same as every other child in the state. The implications of this choice are enormous. As Ryan (2004) explains, schools that do not show AYP are marked as failures and face media attention, financial sanctions, and professional shame. Staff may be let go and state agencies may take over. Parents with the financial wherewithal may move away while any parents with children in such schools are theoretically able to move their children to a better school in the district.[7] We should expect these effects to be more intense for schools with a significant number of low income and minority students. This is because, as discussed elsewhere, there is a distinct class and race bias to standardized testing outcomes, but also because such schools are most likely to be subject to the highest stakes testing (Amrein & Berliner, 2002, pp. 12–13). Schools with high numbers of lower income and minority students are more likely to be direct recipients of specific Title I funding, meaning that they face the strictest sanctions for poor performance. Higher hurdles, higher stakes.

This leads to what Ryan (2004) names the "perverse incentives" of NCLB. Working from the well-founded premise that success on absolute criteria test scores is primarily a measure of the extent to which a school is white and affluent, Ryan explains how weaker schools will slide further behind as quality teachers and educationally committed families bail out. Furthermore, racial and socioeconomic segregation will increase as affluent families flee for the better schools and the better schools begin to exercise self-protective measures to avoid taking weaker students. Schools also have an interest in urging low-performing students to drop out and to avoid accepting students who appear to be at risk of low performance. The results, argues Ryan (2004), are predictable: a further cementing of the class and race gaps in American education as a poorly designed measure of school quality becomes the vernacular in American education.

Furthermore, Ryan argues the absolute standards approach to test design fuels a race to the bottom in the setting of state standards. Because all of the children in the state must pass the same nominal threshold at the same time, educational standards and testing criteria must be pushed down to a level where it is feasible for a significant number of schools and districts to pass. State leaders and education agencies will find it politically unacceptable to have an enormous number of schools failing the test and so they will be compelled to lower the bar until enough are able to make it over.

> Indeed, some already have. Louisiana, Colorado, Connecticut, and Texas have all tinkered with their scoring systems in order to increase the number of students who will be deemed proficient for purposes of the NCLBA. In Louisiana, for example, passing scores had been divided into three categories: basic, proficient, and advanced. Last year, only 17% of eighth graders scored at the proficient or advanced level on an English test, while 31% scored at the basic category; in math, only 5% were advanced or proficient while 37% scored at the basic level. So what did Louisiana do? It deemed those who scored at the basic level "proficient" for purposes of the NCLBA. Similarly, Colorado and Connecticut have redefined categories of scores, making it easier for students to reach the newly dubbed "proficient" level. And the Texas State Board of Education, after a field trial of state tests, lowered the number of questions students must answer correctly in order to be considered proficient on the third-grade reading test. (Ryan, 2004, p. 948)

The end results of the assessment choices built into the NCLBA is that "while the Act is supposed to raise achievement across all schools, it creates incentives for states to lower academic standards. Second, while the Act is supposed to close the achievement gap, it creates incentives to increase segregation by class and race and to push low-performing students out of school entirely, which will make it even more difficult for disadvantaged students to catch up to their more affluent peers. Finally, while the Act is supposed to bring talented teachers to every classroom, it may deter some from teaching altogether and divert others away from the most challenging classrooms, where they are needed the most. In short, although the Act is supposed to promote excellence and equity, it may work against both" (Ryan, 2004, p. 934).

A New Curriculum

Several studies find evidence of a notable shift in teaching priorities in response to the NCLB testing. A national study of teachers found that

"A large majority of teachers felt that there is so much pressure for high scores on the state-mandated tests that they had little time to teach anything not covered on the test" (Pedulla et al., 2003, p. 2). Another study of teachers found that "teachers reported that after the implementation of the testing program, they spent substantially more time teaching the tested subjects of mathematics, reading, and writing and less time teaching science, social studies, the arts, and physical education and health ... This narrowing of the curriculum has been reported in virtually every state where there is high-stakes testing of only a few subjects" (Jones et al., 2003, pp. 29–30).

Jones and his colleagues report that:

> testing sharply defines the knowledge and skill that students will learn ... Prior to high stakes testing, teachers made the decision about what to teach within a broad framework of topics. Testing, however, not only defines what will be taught, but also defines the context of the knowledge. Whereas teachers may have previously embedded instruction in integrated units or taught concepts across multiple grades, testing necessitates that topics be taught in ways that can be assessed through discrete items on written tests given at very specific point of time. (Jones et al., 2003, p. 26)

As the Pedulla study team found, "Across all types of testing programs, teachers reported increased time spent on subject areas that are tested and less time on areas not tested. They also reported that testing has influenced time spent using a variety of instructional methods such as whole-group instruction, individual seat-work, cooperative learning, and using problems similar to those on the test" (Pedulla et al., 2003, p. 4).[8] They also found that teachers in states with particularly high-stakes testing programs are more apt to "engage in test preparation earlier in the school year; spend more time on such initiatives; target special groups of students for more intense preparation; use materials that closely resemble the test; use commercially or state-developed test-specific preparation materials; use released items from the state test; and try to motivate their students to do well on the state test" (2003, p. 5).

The primary educational result of the NCLB optics of surveillance may be a narrowing of the American K-12 curriculum to an outsized focus on training students to make correct choices between simple answers to questions in the three Rs: readin', 'ritin', and 'rithmetic. In many states, it will be just two of the Rs: there will be no 'ritin' because it is too expensive to score (Toch, 2006). These effects will be magnified in working class and minority schools because it is here where the tests are hardest to pass and the stakes are most dire. NCLB thus absorbs, cements, and advances our longstanding system of class and race stratification in the American education system: The surveillance optics and metrics are set in ways that largely give upper income populations an easy pass while giving lower income and ESL populations almost insurmountable hurdles. The strongest sanctions for inevitable failure are reserved for those schools that receive Title I funds which are, by definition, the lower income school populations; those schools must adapt their curriculum in order to get over the testing hurdle; the result is that lower class schools focus on testing content and practices while more affluent school systems put up with the minor nuisance of a week or so lost to filling in some bubbles.

This section began with the argument that surveillance is not really surveillance – it is not mere watching, but must, rather, be understood as a form of creative depiction and world-making. Here, we have seen how true that is and that it goes far beyond the merely symbolic terms conveyed by the idea of depiction. In an act of veritable world-making, the standardized testing movement compels teachers and students across the nation to shape the content of the educational system to match the measures of the tests. The compulsion is far stronger in lower income schools than affluent and, therefore, stronger for people of color than it is for whites. The result can only be a strengthening of the extant class biases in American education, with poorer schools being pushed to the fragmented, technical archives of information that are so readily assessed.

FUNDAMENTAL FALLACIES

Steven J. Gould's *The Mismeasure of Man* (Gould, 1996) remains the most essential and philosophically rigorous exploration of the fallacies of the testing movement. While Gould keeps his primary focus on general intelligence testing, three of the central testing fallacies are relevant for the types of standard assessment and proficiency ratings discussed here:

Reductionism. Among the most important fallacies is the error of thinking that the intelligence or proficiency of an individual, let alone a school, can be expressed as a single number or scale item. Simple rating systems necessarily belie the complexity, mutability, and incomparability of human and institutional qualities. A rating system that stamps an entire school with the label "Academic Emergency" or "Outstanding" is little more than a compilation of myths and exclusions that mocks the real and difficult process of assessing institutions.

Reification. Another fallacy is to treat a reductionist creation as a real thing; to invent concepts like "intelligence" or "school quality" and then speak of them as if it were a real thing, in a fixed location. Any attempt to measure and codify necessarily creates reified illusions of certainty and reified partial snapshots of complex realities. When we then speak of these things as knowable and manageable entities – "let's bring our quality up" – we have given false and misguiding life to a figment of our cultural imagination.

Ranking. This occurs when we hierarchically sort individuals or schools on the basis of the reductionist reifications we construct. Any state will have a variety of schools – some are poor, some are moderately affluent, some are rich; some serve rural areas, some serve university towns, some focus on a broad and tracked curriculum, and some offer a more narrow and egalitarian set of courses. To take all these unique and different programs and institutions and somehow say that one is superior to another belies the complex and multifaceted nature of not just the institutions, but their locations, our values, and our metrics.

In *Overseers of the Poor* (Gilliom, 2001), I argued that one of the great fallacies and errors of financial surveillance in welfare administration was the artificial reduction of the multitudinous and diverse people known as "the poor" to a simple set of figures and statuses. The true and important multidimensionality, variance, and depth of families and individuals were subjugated for a simple terminology that the state's rule system established and managed. Using these terminological reifications, families were then sorted and ranked according to the complex rules of eligibility. The result is a false and incomplete fiction that takes on all the immense power of state action.

My concern about this transcends aesthetic mourning for the lost richness of humanity. As James C. Scott demonstrates in *Seeing Like a State* (Scott, 1998), the rational modernist government is doomed to failure at public policy interventions because its optics, or ways of seeing, necessarily simplify, reify, and reorder from a state-centric perspective. The resulting information and informational regime will invariably err because successful policy interventions require the sorts of complicated local knowledge and wisdom that the bureaucratic state simply cannot see. As I summarized in *Overseers of the Poor, Seeing Like a State*

> suggests that the power of surveillance is often an almost bumbling power which miscasts the world and its inhabitants, overlooks essential points of information, and helps generate the seeds of its own resistance through its ongoing misreadings of local knowledge. Scott argues that modern states must produce knowledge and information to guide their various missions of social intervention and design. To do so, they must both simplify a complex social reality and rewrite the terms of that reality to fit the terms of the intervention: a mass of people becomes a list, and last names and even

street addresses are introduced as ways of organizing and knowing the population. A wilderness becomes a forest with full analysis of species and harvesting schedules. Scott argues that modern statecraft requires these systems of knowledge and that, further, it is these very systems of knowledge which doom statecraft to failure. The failure of statecraft is virtually guaranteed, he argues, because the systematic state knowledge necessarily omits or overruns the sorts of local and varied knowledge and practices that are inherent to any setting. Since these local forms of knowledge would be essential to the success of state planning, their omission essentially guarantees failure, as well as conflict and resistance from subjected peoples. (Gilliom, 2001, p. 131)

As an example, let me point to a unique high school in the Appalachian region of Southeast Ohio. It is the consolidated high school for one of the poorest, most rural, and geographically dispersed school districts in the state. The academic performance of its students is low and the teachers are poorly paid. The local taxpayers persistently refuse to pass levies while the state legislature perennially balks at reforming the property tax system of educational finance in Ohio even as the State Supreme Court has repeatedly found it unconstitutional. In the NCLB mandated rating system, the high school has only recently emerged from Academic Emergency status to hover between Continuous Improvement and Effective – low to middling scores. Most of its tests scores now run very close to or pass the state minimums, though science and social studies continue well below the norm (see: http://www.ode.state.oh.us/reportcardfiles/2006-2007/DIST/045914.pdf).

From the snapshot created by NCLB, it would be impossible to learn that the school is led by an award-winning educator or that its program of democratic education has made the school a national beacon of educational innovation. As summarized by the Center for Secondary School Redesign:

In spite of the challenges faced in this region, the school has received numerous awards under Dr. Wood's leadership including an Ohio's Best Award for the school's internship program, designation as a First Amendment School by ASCD and the Freedom Forum's First Amendment Project for the school's work in promoting active democratic citizenship, and being named one of the first five Coalition of Essential Schools 'Mentor Schools' in conjunction with the work of the Gates Foundation. (http://www.cssr.us/keynotes.htm)

This high school represents a unique and intriguing social and educational experiment which appears to be working, but all of these dynamics are invisible to the state profiles created by the NCLB surveillance technology. As the school modifies its curriculum to satisfy the commands of the state curriculum and surveillance program, there may well be catastrophic consequences for the innovations currently underway. As Jones et al. (1993) show, systems of high-stakes testing not only modify curricula, absorb time, and reduce resources for anything outside of the tested subject areas, they create pressures that reduce staff morale, push teachers out of the profession, and suppress teaching techniques that stray from the necessary pedagogy of the testing regime. That pedagogy, they show, is one that focuses on the rote acquisition of basic skills such that students can readily respond to simple, discrete, multiple choice questions in a known format. Such programs at Federal Hocking High School, which focus on democratic values, first amendment freedoms, and engagement through internships, face a hard time under the new optics of educational assessment.

CONCLUSION: EDUCATIONAL TESTING, EDUCATIONAL SURVEILLANCE

Each year, some 50 million standardized tests are administered to the children of America. The results are used to develop files on individuals, assess the work of teachers and school leaders, rate schools and districts, and, if current laws are followed, deliver severe financial and political penalties to schools that fail to "measure up." The less formal effects are equally important. The testing movement is reshaping the American school curriculum, centralizing control over educational decisions, transforming pedagogy, and shifting billions of dollars of funding into the testing industry coffers. NCLB is, arguably, the greatest single national event in the history of American education.

And it is also one of the greatest expansions of mass surveillance in American history. We may initially balk at thinking of educational testing as surveillance because when we think of surveillance the mind first turns to things like eavesdropping, spy satellites, and phone taps. But a brief reflection on the idea and processes of surveillance makes it clear that educational testing falls into this category. "Surveillance" is derived from French, with a rough translation being "to watch (veil) from above (sur)." Surveillance has been widely studied as a form of management, political domination, and social control (see Lyon, 2007). The field rightly covers a host of policies and technologies from obvious practices like the increasing use of closed circuit cameras and national identification cards to the less popularly recognized but no less important surveillance practices like insurance scoring, credit reporting, and computer monitoring of social welfare clients (Lyon, 2007; Monohan, 2006; Haggerty & Ericson, 2006; Gilliom, 2001).

Surveillance programs of varying stripes gather information through watching, measuring, and monitoring around some set of norms, rules, or expectations. That information is used, in varying ways, as a means of control. The control agenda could be nearly anything. Reducing crime, illegal drug use, or illegal immigration; reducing fraud and error in the management of social welfare and healthcare systems; reducing traffic infractions such as speeding and stop-light violations; reducing the risk of violence in air transportation, schools, and public buildings; fighting the emergence and transmission of disease; building a successful college or law school cohort; and enhancing public education. The element that unites all of these areas is that the primary method of action is observation and assessment.

In some ways, it is unfortunate that "surveillance" has come to be the accepted term for these practices, for its specific connotation of watching is an inappropriate meaning for the policies and technologies at hand. It is critical that we should not lose sight of the fact that surveillance is not *really* best understood as an act of watching – it is often far more importantly understood as an act of depiction: a creative rendering of an impression of a part of the social environment. I raise this because the idea of watching implies a simple observation of a given object. It is sort of what you might get from a junior high school level understanding of what journalists do – "we report, you decide." As if there were no decisions going into the complicated process of observing, interviewing, ordering, narrating, editing, and pitching. But, of course, there are. In newsmaking, there are choices about what to cover, which aspects to cover, where to get video, how to frame it and clip it, what to say, how to smirk, what to say next, and so on. And there are institutional structures that are not really everyday choices – the preference for strong video images, surprising or shocking information, and ongoing institutional biases that shape television news coverage (Bennett, 2006). For this reason, we come to think of something such as the news not as a window on reality, but as a subjective, limited, and constructed depiction that we can or cannot accept as telling us something useful about our social environment. Furthermore, institutions such as mass media news do more than just giving us partial pictures of the world, these enterprises *shape* the world by creating incentives and disincentives for behaviors, favoring and disfavoring people and policies, and, at least partially, defining the public understanding of reality.

And so it goes with other forms of surveillance and information gathering and dissemination. In educational testing, as we have seen, the optics are shaped by budget constraints and the limits of testing technology; they are shaped by test designer choices about what matters in education; they are shaped by political and technocratic choices about values, metrics, and criteria; and since "testing" is, of course, inextricably linked to the broader structure of American education, they are shaped by the broader histories and patterns of regionalism, racism, classism, and disparity in the American educational system. But beyond the important problems of the optics are the critical effects that such programs have in actually shaping the world they watch. As we have seen, some subjects and types of teaching become disfavored and impracticable in a pedagogical world defined my standardized testing. Testing is not just watching the classroom, it is defining it.

NOTES

1. For a historical review of the different modes of educational assessment and their transition since the medieval period (see Wilbrink, 1997).

2. But each of these authors, in my view, errs in speaking too confidently about the goals and intentions of NCLB. It is difficult, if not impossible to speak about the "intent" behind legislation that is a hodge-podge of state and federal practices hammered out in compromises between the White House, both houses of the Congress, and a very influential Conference Committee. Add to this the competing agendas created by lobbyists for teachers union and the testing industry, influential members of congress, and education policy entrepreneurs. Next, kick the whole thing out to the 50 states for implementation over a 12-year period with ongoing compliance negotiations and waivers brokered by a federal Department of Education. Then factor in thousands of school districts and schools and their administrators, principals, and teachers. In the end, I would say, it is a mistake to speak confidently about the intent of NCLB as if it were a singularly conceived policy in a hermetically sealed environment.

3. This state autonomy is a fairly bizarre turn, apparently an outgrowth of continued discomfort over a national curriculum and federal government testing regime.

4. As will become clear later, but should be hinted at now, the "standards" approach is just one way to look at educational policy and it is fraught with implications. Under the standards approach, all schools are expected to demonstrably meet the same essential norms of education. Whatever the location, challenges, or makeup of the student body, all children must be proficient at, say, long division by the fourth grade. This means that affluent schools pass the bar with nary a second glance, while schools low in social capital and high social problems may never make the bar. Alternatives, like the value-added approach, will be discussed later.

5. Indeed, many in the educational research community view the state tests as too dubious to use as benchmarks or references because they are so easily gamed by repeat players and the incentives for gaming are so high.

6. The psychometricians are spread even thinner by the use of standardized testing in many professional and military certification programs.

7. This is a joke to rural parents who have just one school per district and to urban parents who live in uniformly poor districts.

8. I can report from personal experience that the High School in my community rewards every sophomore who can pass the NCLB state achievement test the first time through by allowing them to skip all their subject area final exams during the last week of school in the late spring; English, science, math, history, and languages all take this hit in support of the statewide testing. And I can also report what every parent of school-aged children knows – there is not just one test a year, because the schools ready themselves with pre-tests, "short-cycle" assessments tests, and other tests to train and assess the students in preparation for the Big Test.

REFERENCES

Amrein, A. L., & Berliner, D. C. (2002). The impact of high-stakes tests on student academic performance: An analysis of NAEP results in states with high-stakes tests and ACT, SAT, and AP test results in states with high school graduation exams. Education Policy Studies Laboratory, Arizona State University. No. EPSL-0211-126-EPRU.

Bennett, L. (2006). *News: The politics of illusion.* New York: Longman.

Foucault, M. (1979). *Discipline and punish.* New York: Vintage.

Gilliom, J. (1994). *Surveillance, privacy, and the law: Employee drug testing and the politics of social control.* Ann Arbor: University of Michigan Press.

Gilliom, J. (2001). *Overseers of the poor: Surveillance, resistance, and the limits of privacy.* Chicago: University of Chicago Press.

Gould, S. J. (1996). *The mismeasure of man.* New York: W.W. Norton.

Haggerty, K. D., & Ericson, R. V. (Eds). (2006). *The new politics of surveillance and visibility.* Toronto: University of Toronto Press.

Hess, F. M. (2003). Refining or retreating? High stakes accountability in the states. In: M. R. West & P. E. Peterson (Eds), *No child left behind? The politics and practice of school accountability* (pp. 23–54). New York: Brookings.

Jones, M. G., Jones, B. D., & Hargrove, T. Y. (2003). *The unintended consequences of high–stakes testing.* New York: Rowman and Littlefield.

Kohn, A. (2001). Fighting the tests: A practical guide to rescuing our schools. *Phi Delta Kappan, 82*(5), 348–357.

Lyon, D. (2007). *Surveillance studies: An overview.* New York: Polity.

Monohan, T. (Ed.) (2006). *Surveillance and security: Technological politics and power in everyday life.* New York: Routledge.

Pedulla, J. J., Abrams, L. M., Madaus, G. F., Russell, M. K., Ramos, M. A., & Miao, J. (2003). *Perceived effects of state-mandated testing programs on teaching and learning: Findings from a national survey of teachers.* Boston College: National Board on Educational Testing and Public Policy.

Popham, W. J. (2005). *America's "failing" schools.* New York: Routledge.

Rudalevige, A. (2003). No child left behind: Forging a congressional compromise. In: M. R. West & P. E. Peterson (Eds), *No child left behind? The politics and practice of school accountability* (pp. 23–54). New York: Brookings.

Ryan, J. (2004). The perverse incentives of the no child left behind act. *New York University Law Review*, *79*(June), 932–989.

Scott, J. C. (1998). *Seeing like a state: How certain schemes to improve the human condition have failed.* New Haven: Yale University Press.

Toch, T. (2006). *Margins of error: The education testing industry in the no child left behind era.* Washington, DC: Education Sector.

Wilbrink, B. (1997). Assessment in historical perspective. *Studies in Educational Evaluation* *23*(1), 31–48.

GOVERNING BEYOND COMMAND AND CONTROL: A RESPONSIVE AND NODAL APPROACH TO CHILD PROTECTION

Nathan Harris and Jennifer Wood

ABSTRACT

This chapter is interested in the challenge of governing beyond crime, surveillance and control. It argues for the need to re-imagine the governance of security in ways designed to both build and enrol the capacities of different actors. The authors draw on regulatory theory and the ideas developed in the areas of 'responsive regulation' and 'nodal governance' to explore the opportunities for, and the challenges associated with designing governance institutions and processes that serve to de-centre hierarchy, command and interventionism as essential rationalities and practices. Its empirical focus is on the case of child protection, where the authors argue for the importance of nurturing the capacities of families and communities to govern both beyond and in tandem with hierarchical modalities. It is hoped that the theoretical issues raised and the agenda articulated can be engaged with across a variety of empirical domains.

Surveillance and Governance: Crime Control and Beyond
Sociology of Crime, Law and Deviance, Volume 10, 327–344

ISSN: 1521-6136/doi:10.1016/S1521-6136(07)00215-1

INTRODUCTION

This chapter is interested in the challenge of governing *beyond* crime, surveillance, and control. It is less concerned with the changing character of surveillance and crime management than it is with the need to re-imagine how to do the 'governance of security', that is, to undertake 'conscious attempts to shape and influence the conduct of individuals, groups and wide populations in furtherance of… "security"' (Wood & Dupont, 2006a, 2006b, p. 2). The first step in meeting this challenge is, in our view, an acknowledgement of 'governance deficits', whereby the knowledge and capacities of some governing actors – such as those who do statutory enforcement, intelligence gathering, and surveillance – are more privileged and better resourced than others. The second step is to re-design our formal institutions and processes in ways that more effectively enrol the capacities and resources of those who bring diverse strengths to governance. There must be a renewed commitment, in our view, to recognize, build and harness the strengths of families and communities in this process.

There are 'weak' and 'strong' actors in the field of security, as in other governmental domains (Braithwaite & Drahos, 2000). This is a situation which is being further cemented through quests to 'ratchet up' the surveillance and coercive resources of state in various realms, ranging from the control of disorder on the streets, to the social regulation of youth, to the prevention of terrorism (Garland, 2001; Crawford, 2006; McAlinden, 2007; Parton, 2006). Even inter-agency efforts to enhance 'interoperability' in intelligence gathering and actuarial risk management in areas such as child protection (Deukmedjian & Cradock, 2008) and sex offender management, are serving to bolster the capacity of (already strong) state systems to govern in ways that privilege 'hierarchy, command and interventionism' (Crawford, 2006, p. 449). Regardless of the instrumental reasons for these broad shifts, what has been lost, in our view, is a previously articulated commitment to nurturing community-based and plural forms of governance (Deukmedjian & Cradock, 2008).

It is this general 'governance deficit' problem that forms the starting point of this chapter. We draw on regulatory theory and the ideas developed in the areas of 'responsive regulation' (Ayres & Braithwaite, 1992) and 'nodal governance' (Burris, Drahos, & Shearing, 2005) to explore the opportunities for, and the challenges associated with, designing governance institutions and processes that serve to de-centre hierarchy, command and interventionism as essential rationalities and practices. We focus empirically on the area of child protection which has been gradually moving away from community-based governing modalities to renewed practices of surveillance and risk-based intervention (Parton, 2006; Deukmedjian & Cradock, 2008). While our focus is on child protection, our hope is that the theoretical issues we raise, and the agenda we articulate, can be engaged with across a variety of empirical areas.

NODAL GOVERNANCE, AND THE 'WEAK' AND THE 'STRONG'

Scholars over the world are arguing for the need to expand our governance imagination. For instance, criminologists, regulatory scholars and other social scientists are now less narrowly concerned with the problem of 'law enforcement' or 'crime control' than they are with the 'governance of security'. This broadened analytical frame makes it explicit that 'security' – and the related conceptions of 'safety' and 'well-being' –is contested in definitional terms, and more importantly, that it is a problem that can be understood and acted upon in a variety of ways and through various competencies and resources. Academics from distinct disciplinary traditions are now coming together around a broadened notion of 'governance', seeing it as a set of activities for 'structur[ing] the possible field of action of others' (Foucault, 1982, p. 220 cited in Simon, 1997, p. 174) or for 'shap[ing] the flow of events' (Parker & Braithwaite, 2003) in our social world (see also Braithwaite, 2005).

There is also greater inter-disciplinary interest in 'de-centred' (Black, 2000) or 'plural' (Loader, 2000) governance, which is seen as occurring through the activities of numerous entities or 'nodes'. In simple terms, nodes can be seen as locations for distinct forms of knowledge, capacity and resources that are assembled and channelled around particular social problems (Shearing, 2004). Of course, some nodes are more formally organized, recognized, nurtured and developed than others. Families are nodes, as are schools, as are other individuals and groups that bring to bear different ways of seeing, constructing, and acting on the problem of security (see Burris, 2004).

Curiously, in a world of 'plural' or 'nodal' governance, where different means of shaping our social world are possible, we continue to 'govern *through* crime,' to borrow a phrase from Jonathan Simon (1997, 2007), a great deal of the time. 'We govern through crime', Simon (1997) writes, 'to the extent to which crime and punishment become the occasions and the institutional contexts in which we undertake to guide the conduct of others (or even of ourselves)' (p. 174). Even in regulatory realms beyond criminal justice such as in child protection and other statutory-based sectors, hierarchical techniques for shaping our conduct remain central to our governance imagination and thus to practice. This reflects, in our view, a deep tendency to privilege particular governing capacities – the ones that we know and are comfortable with – in ways that trump the contributions that other nodes can make in shaping our social world.

In countries like the United States, Australia and the United Kingdom, politicians are devoting extensive efforts to strengthen the capacities of states to govern through new forms of risk management, monitoring and surveillance (Garland, 2001), all of which can contribute in making command and control work better. This is occurring in various empirical domains, ranging from the 'zero tolerance' of policing on the streets to a 'penal populist' approach to the management of sexual offenders post-release (Garland, 2001, Simon, 1998) to, of course, the 'war on terror' (Daniels, Macklem, & Roach, 2001). This is all happening, some suggest, at the expense of efforts to recognize and further develop the capabilities of families and communities to manage security risks (Parton, 2006; McAlinden, 2007; see also Blair (2005) on the needed link between neighbourhood policing and the management of terror).

Not everyone participates equally in the nodal governance of our world (Wood & Shearing, 2007, chapter 4). There are 'weak' and 'strong' actors, as Braithwaite and Drahos (2000) describe it, who possess varying opportunities and capacities to shore up the resources required to promote the security and protection of themselves and others. State and corporate actors have done particularly well in a nodal environment, due to their access to economic and other forms of capital which bolsters their ability to self-govern. This is now a rather well-established observation, articulated in different ways, which has generated varying normative responses (Bayley & Shearing, 1996; Loader & Walker, 2007).

Beyond this observation about the weak and the strong, there is a broader, perhaps more instrumental, problem that concerns us here, and it has to do with what regulatory scholars would describe as 'regulatory design'. While at some levels governments are captured by a vision of the world which requires greater control and surveillance, at many other levels of government and in many non-government organizations the rhetoric, at least, is one that espouses the importance of collaboration and participation. Given the seemingly contradictory requirements of these world views, the question for us pertains to how to design systems that can more effectively identify, build and enrol the governing capacities of those who are best placed to carry out functions beyond those required for governing through command (see Black, 2003). As Burris puts it, '[t]he task is to find ways to make security institutions more responsive to the least powerful, and where responsiveness, resources or ability to respond effectively are lacking, to redirect resources and power to new institutions that can do the job (Burris, 2006, p. 206)'.

In the following, we will discuss the importance of designing governance systems that can more effectively build and enrol the capacities of traditionally weak actors. We will focus empirically on the case of child protection, and explore the theoretical contribution of 'responsive regulation' and 'nodal governance' to this agenda.

RE-IMAGINING CHILD PROTECTION

Child protection agencies in a number of countries such as Australia and the United Kingdom are under increasing pressure on a number of fronts. Greater awareness of child abuse within the community, changing definitions of what constitutes abuse, and legislative moves towards mandatory reporting have led to enormous growth in the number of child protection reports (Parton, 1997). Under increased political and media pressure to prevent child abuse, governments have become increasingly focused on systems that emphasise the detection of vulnerable children and protecting them through statutory child protection agencies (Scott & Swain, 2002). But at the same time child protection has come under increasing criticism for failing to provide services that substantively improve the welfare of children and for failing to live up to the ideal of supporting families (Thorpe, 1994). The dilemma for governments is that the very tools that they have come to rely on to identify and respond to concerns are also perceived to undermine their effectiveness in improving the welfare of children.

A primary response of child protection agencies to mounting pressure has been to develop risk assessment models that deal with escalating numbers of referrals by directing scarce resources to those cases in which children are considered at high risk of harm (Scott & Swain, 2002). Statutory action has increasingly become the primary tool employed as services target risk in the worst cases: either because court orders are taken or because statutory requirements are seen as the justification for intervening. This represents a shift in emphasis away from the provision of family support services and towards surveillance and control; where surveillance is represented by investigations and increased information sharing (Parton, 2006), and control is located in court orders that authorise the assessment, supervision or removal of children. This redirection of resources has not necessarily helped those who are deemed at high risk, because it has not led to an increase in the pool of support services available, and families who are assessed as low risk are often 'screened' out of the system only to reappear later as serious neglect and abuse cases (Cashmore, 2001). Critics, therefore, have called for a rebalancing of services towards providing support (Parton, 1997; Wise, 2003).

Criticism of this trend towards an emphasis on 'protection' rather than 'support' is accepted by many (e.g. Standing Committee on Social Issues, New South Wales Parliament, 2002), but reorienting services towards providing early intervention and building the self-governing capacities of families to nurture and protect their children has proved difficult. Rapidly increasing notification rates mean that the resources of child protection services are often consumed just investigating and managing those incidents that are deemed as high risk. A second hurdle in changing the emphasis of child protection practice is that many of the cases that might benefit from an increased emphasis on support and capacity building are also ones where the child is assessed as at risk. This judgement, and the responsibility that comes with it, pushes child protection agencies quickly towards a command and control response, through statutory-based action, which is often driven by a desire to secure the immediate safety of children. While parents are often invited to cooperate, relationships between workers and parents are often antagonistic and there are clear limits to the degree of negotiation that is possible.

A growing concern expressed by scholars and practitioners is that this emphasis on statutory responses is undermining, or swamping, other important (regulatory) strategies, such as the provision of support. While the focus on child protection relies on law to coerce families to accept decisions that are seen as in the child's best interests, a focus on family support and capacity building emphasises working with families to empower them to care for their children in an acceptable way. The fact that this latter approach is dependent upon voluntary participation by parents makes it difficult to reconcile with interventions that are coercive (Trotter, 1999). Legitimate reservations are raised about whether families, subsequent to an investigation or statutory action, will be willing to engage with services in a way that improves their situation (Thorpe, 1994). Indeed, research has shown that child protection investigations often leave parents feeling alienated and angry (Cleaver & Freeman, 1995; Farmer & Owen, 1995; Parton, 1997).

As Simon (2007) points out, while politicians from different points on the political spectrum stress the importance of individual and community responsibility in the management of collective problems like security, nothing less than a 'a zero-risk environment is treated as a reasonable expectation, even a right' (p. 16). This preoccupation with risk not only trumps other objectives – especially the building of capacities in communities to self-regulate – it also necessitates a highly active and dominating state apparatus that readily wields its coercive power. Furthermore, the protection of children and communities tends to be understood in a way that bolsters efforts to reduce *immediate* risks rather than to build long-term risk management capability.

Thinking Responsively: Nurturing the Capacities of Families

Responsive regulation is of great importance, according to Burford and Adams (2004) because 'it offers us a way to reconcile empowerment practices with the context of coercion', and in doing so provides a practical way in which to expand community-based forms of governance. Reconciling the potential for coercive action and empowerment of communities, within the framework of responsive regulation (Ayres & Braithwaite, 1992) occurs through varying the kind and degree of intervention that is made in any particular case, based upon the response of individuals to concerns that are conveyed to them. While traditional models of governance, such as child protection systems, move quickly to impose predetermined solutions, responsive regulation argues that governments should initially seek to address problems through deliberative, non-coercive processes that build capacity. Only if this is unsuccessful should the level of intervention become more directive and coercive in nature.

The domain of child protection has only quite recently benefited from the theoretical insights of responsive regulation (The Allen Consulting Group, 2003; Adams & Chandler, 2004; Burford & Adams, 2004; Neff, 2004). A central tenet of restorative justice is that decisions about how to govern, and which governing actors to enrol in the process, need to be transformed into deliberative ones that, to borrow phrasing from Simon (2007), 'creat[e] new pathways to knowing and acting on ... people and relationships' (p. 4). Only by altering fundamental decision-making processes and the relationships between governing nodes will child protection systems take full advantage of responsive and nodal thinking.

Two principles can be taken from responsive regulation in relation to such decision-making processes: *firstly* that agencies should be able to vary, in terms of intrusiveness, the kinds of response they take to problems, and *secondly*, that the decision about what kind of response to take should be based upon evidence as to whether the situation has improved or deteriorated as a result of a previous intervention, with a preference for using the least intrusive intervention possible. These two principles can be imagined as existing within a 'regulatory pyramid' (Ayres & Braithwaite, 1992).

At the bottom of the pyramid are strategies that rely largely on self-governance, or supported self-governance, because they involve helping to build the capacity of parents to make changes, while also making it clear that changes are necessary. In the middle of the pyramid, interventions are directed at protecting the child through more directive means. This might initially involve requiring parents to engage in certain decision-making processes, but leaves parents and their immediate communities with considerable control within those decision-making forums. At the very top of the pyramid, statutory action results in the courts making decisions about what is in the best interests of the child.

There are two unique aspects of the way in which responsive regulation conceptualises this decision-making pyramid. The first is that decisions to move up the pyramid are made retrospectively rather than prospectively. The presumption of responsive regulation is that, unless there is a strong reason not to, concerns should initially be addressed at the bottom of the pyramid. In practice, this should mean that attempts to solve problems initially occur through deliberative processes, and that the success of these attempts should guide any further action that will be taken. While there is a preference for regulating at the bottom of the pyramid there is always the possibility of escalating to more coercive decision-making processes.

A second significant characteristic of the decision-making pyramid, as a number of researchers have pointed out (Adams & Chandler, 2004; Neff, 2004), is that escalation up the pyramid, at least in the case of child protection, does not necessarily equate with more intensive intervention, but rather a change in how decision-making is made. For instance, a family might respond to concerns about the welfare of a child by deciding to have her live with relatives for a period of time during which time they would receive intensive support. This decision might occur at the base of the pyramid, as the result of counselling with the family, where there was no immediate suggestion that the child would be removed through court action. However, the same outcome could also occur at the peak of the regulatory pyramid through a court order over which the family had little say. The variation between decision-making steps can be illustrated by three possible steps on a hypothetical decision-making pyramid.

Family Support Meetings

At the bottom of the pyramid it is possible to imagine the use of informal Family Group Decision Making (FGDM) processes that locate families as the central decision-making hub. An example can be found be in the Family Engagement Model, which has been developed and is used in one office (Cannington) of the Western Australian Department of Community Development (Harris, 2008). It involves working with families almost entirely through Family Support Meetings, which focus on identifying the issues that are of concern and developing plans to address these concerns. These meetings might be held with families up to once a week, and on each occasion previous plans will be reviewed or new plans will be formulated. Private family time, in which the family is left alone to discuss how they would like to address an issue, is sometimes used, and extended family and professionals can be invited to attend where this is seen as useful. The meetings are fairly informal, do not involve independent coordinators, involve limited preparation, and the outcomes of these meetings do not have any formal (legal) status.

Family Group Conferencing

In the field of child protection the primary decision-making forum that has been associated with responsive regulation is family group conferencing (FGC) (Marsh & Crow, 1998). A number of authors (Adams & Chandler, 2004; Pennell, 2004; Neff, 2004; Merkel-Holguin, 2004) have argued that FGC's provide a logical mechanism for translating the principles of restorative justice because they provide child protection services with an alternative to taking court orders, which provides the family with much greater power in decision-making.

While there is considerable variation between jurisdictions in how FGCs are used, they are primarily convened as a response to an assessment that a child is in need of care and protection. A facilitator in consultation will invite the family, their immediate community of friends and other possible supporters, professionals who might be able to offer assistance and the child protection worker to the conference. The meeting itself will discuss the child protection concerns, hear what kinds of service might be relevant to the family and provide the family and their community with private time in which they can come up with a plan to address the child protection concerns. Usually a conference concludes with an agreement by the family and the child protection worker or facilitator on a plan. In some jurisdictions, such as New Zealand, the plan is automatically implemented unless it proves to be impractical or breaches legislation in some way.

Conferences do not occur entirely in the absence of coercion because they are usually instigated by statutory services, and while empowering the family do not allow unrestricted decision-making. Nevertheless, they might be considered a regulatory option that governments have towards the middle of the pyramid; that is, they emphasize the family's ability to self-regulate with assistance from a broadened community network. Conferences now represent an increasingly important component of child welfare services in many parts of the world such as New Zealand, where they were first used, as well as the United States, the United Kingdom, and in many states of Australia.

Court

Responsive regulation does not argue that we should entirely abandon the use of coercion. Indeed, Braithwaite (2002) argues that strong interventions at the peak of the decision-making pyramid are necessary when other approaches have failed. In the case of child protection the use of court orders to impose assessment, supervision, or alternative guardianship on families is a common outcome in a large proportion of those cases that are the subject of ongoing attention. The aim of a responsive pyramid is to significantly reduce this proportion and to increase the legitimacy of such action when it is used.

Many child protection workers already employ aspects of responsive regulation in their work. They use their professional judgement to vary their approach to individual families, and place a priority on working with families cooperatively wherever that is possible. The model that is outlined here extends this approach to the formal decision-making processes of child protection. An important aim is to increase the confidence of both families and workers to engage in collaborative problem solving by establishing a transparent decision-making structure. This should increase confidence of workers because they know that if a collaborative approach fails, they have the capacity to call on a more directive approach to safeguarding children.

Responsive Appraisals: Building Hope and Limiting the Harm of Interventionism

These fairly simple changes in the way that decisions are made could have profound implications for child protection systems, but that outcome would also require significant changes across child protection practice. Current preoccupations with risk and statutory action would need to shift towards a focus on creating contexts in which families and communities have genuine opportunities to develop their own capacity. This is most apparent in investigatory practices, which are currently shaped by overriding requirements to determine whether a child has been abused or is at future risk and to document this in possible preparation for a court case. Our intention is not to imply that all investigations are conducted in a confronting or unsympathetic way, but rather that it is likely that these overriding requirements are accurately understood by many parents, and that this has a profound impact on the outcome of investigations (Trotter, 2004). For example, research conducted in the United Kingdom by Farmer and Owen (1995) found that 70 percent of parents felt marginalised and badly treated following an investigation. In 40 percent of these cases feelings of anger and violation were long-lived, and meant that families kept workers at arm's length (p. 315).

Engaging families and communities in capacity building requires a model of investigation that is able to reflect the way in which child protection services want to work with families. Increased use of assessments by child protection services for some lower risk cases in the United Kingdom, United States, and some other countries reflects recognition of the importance of working with families in a more collaborative manner (Platt, 2006). However, a more fundamental transformation of practice requires a more flexible model of investigation that can be applied to all families. A practical way of doing this is possible by drawing on the same basic principle used in the decision-making pyramid above to develop a model of responsive assessments. In the vast majority of child protection cases, investigation could have a much stronger focus on voluntary provision of support, while leaving the door open to escalating to more coercive or forensic approaches if this proves unsuccessful, and is necessary because of risks to a child. Responding to social problems such as socioeconomic disadvantage, mental illness, drug dependencies, and relationship dysfunction, which are often the underlying causes behind reports of abuse or neglect, should not be based on the assumption that surveillance, control and statutory action will provide the most effective solutions.

Nodal Risk Management: Building Social Networks of Concern

It can be seen that a responsive model of child protection would also have implications for how risk is conceived and managed in child protection. Current practices have moved towards the development of actuarial based systems that are used to guide the actions taken by departments in response to reported concerns. The responsive decision-making pyramid and responsive assessments, outlined above, would not entirely displace such an approach, but would enrich the framework for decision-making by informing subsequent decisions about how risks should be managed.

Current conceptions of risk management would be broadened through recognising the contributions made by social networks to the management of risk. An assumption of the model proposed here is that it involves significant engagement with various actors, or nodes, in the community, such as families, schools, doctors, child care centres, and many others. These nodes would take on a significant role in risk management, through contributing to responsive assessments and the various decision-making processes (e.g. FGCs). As a result, risk management would shift from being understood as an actuarial process, based on risk factors and performed by a professional, to a dynamic process that draws on the insights and concerns of multiple nodes that are directly interested in the well-being of a family.

Enroling Wider Nodes of Community Capacity

The responsive model that has been outlined so far has implications for how decision-making, risk assessment and risk management is understood within a responsive framework that endeavours to enrol families in child protection. This model poses a new set of challenges for child protection workers, the police and other statutory agencies with whom they work. One of these challenges, which may be a key to realizing the model in practice, is that child protection workers and agencies need to maintain and foster reciprocal relationships with community nodes. The model proposed here relies on such relationships because children's social networks, such as extended family, friends, professionals, and community organizations are clearly essential to the nodal governance of child security. For example, members of a child's community may be important in convincing their parents to engage with workers, so that they can conduct an assessment collaboratively. They may be even more essential in guiding a worker's subsequent strategy if the parents refuse to engage with workers. Being able to locate and draw on such nodes of capacity is both central to implementing the model described and also one of the greatest challenges.

One positive effect of pluralism is that governments have the opportunity to expand their capacity to provide services through enrolling actors that have additional capacities (and indeed rationalities) to offer. In child protection there are a multitude of actors, both organizations and individuals, who possess different forms of knowledge and skills that can contribute to the welfare of children. These nodes represent major resources that can be harnessed in ways that compensate for the weaknesses of command and control governance.

A responsive approach to child protection provides a framework in which these actors can be utilized more effectively because the greatest contribution that most of them can make is at the deliberative end of the pyramid, which is where the capacity of statutory agencies is probably weakest.

A primary difficulty for child protection services in locating and drawing on these nodal capacities is that workers tend to be located in organizations that are remote from the social networks that are most important to children and their families. Contact with professionals and families in organizations, such as schools, is likely to be infrequent and tends to occur as a result of interventionist actions. In some cases, particular workers will develop strong relationships with particular organizations due to their involvement in particular cases, but these are likely to occur in a fairly unsystematic way. This means that individual workers, or even teams, are much less likely to form systematic nodal relationships. This may become a significant impediment to the responsive model that is proposed, because experiences of inter-agency collaboration suggest that interpersonal relationships of trust and reciprocity form a crucial foundation for robust nodal linkages (Rhodes, 2006). It might be argued that this will be even more so where the focus is linkages within and across professional and community-based nodes. Child protection workers would be, in effect, seeking to enrol others to assist them. While each of these organizations will have their own motivation to safeguard children, the degree of cooperation with child protection services will depend greatly on the ability of workers to develop trust in these organizations and vice-versa.

Processes of enrolment, however, need not be one-way and state-driven. It has been argued elsewhere that more energy needs to be placed in creating the conditions in which traditionally weak actors can do the enroling as well (Shearing, 2001; Burris, 2006; Wood & Shearing, 2007). This could be one avenue towards greater self-direction on the part of various nodes (Shearing & Wood, 2003) whilst also creating opportunities for nodes to negotiate competing interests and build trust. Efforts to build responsive and nodal models that place families and communities in decision-making hubs and in roles requiring them to enroll the knowledge and capacities of themselves and others has been expressed, for example, in the contexts of security governance in shantytown communities (Shearing & Wood, 2003; Johnston & Shearing, 2003; Burris, 2006) and youth safety and well-being promotion in schools (Marks & Wood, 2008).

If child protection agencies are to forge sustained relationships of mutual enrolment with other nodes, they may need to restructure their organizations around maximizing their ability to nurture networks within various communities associated with children and families. In practice, this probably means that child protection workers need to be located within, or spend considerable amounts of time working from organizations that are responsible for the safety and well-being of children. Particular workers, or possibly small teams of workers, could be responsible for a single organization, such as a large school, so that they could develop and maintain relationships based on the mutual exchange of knowledge, capacities and resources. As part of this, child protection workers could undertake activities in relation to early intervention, outreach and advocacy for families in need. A child protection worker at a high school might conduct workshops on how to parent teenagers. This has the potential to give the worker a meaningful role within that community that helps them to nurture and build networks with teachers and parents. When child protection issues arise, such as a child protection report, the same worker would be able to investigate, and provide ongoing assistance to particular families where that was necessary.

A significant advantage of locating workers within organizational nodes like schools is that it would provide a useful mechanism for shifting the resources of child protection systems from the current emphasis on reactive and crisis-driven intervention to early intervention. Locating workers within such organizations would mean that they would be aware of families who were struggling before reports were made, and would have the opportunity to offer assistance. In some respects, such a model would move towards the public health model advocated by Dorothy Scott (2006), which proposes a child protection system that places greater emphasis on strategies that improve the well-being of children across the population by tackling the underlying causes of problems associated with neglect or abuse. In this way, child protection workers would play an increased role in addressing universal needs of children and their families. Rather than being entirely focused on addressing abuse in a specific population, they would also be increasingly concerned with building capacity within community-based nodes to address such issues as mental illness, inadequate housing, drug problems and social isolation.

In summary, child protection workers would both enrol, and be enrolled by a wide range of nodes, including health and education services, as vehicles for safeguarding the welfare of children and their success would need to be seen not just in reducing child abuse and neglect, but in advancing the social opportunities for all children (Parton, 2006). As an ultimate result, we could potentially move beyond the governance of security *through* security (Valverde, 2001) towards the governance of security through health

(Burris, 2006) and other rationalities that can serve to de-centre hierarchy and control as central modes of shaping our social world.

CONCLUSION

Despite what many observe to be the markers of plural or nodal governance, there are equally clear indications that established ways of constructing and managing our social world remain central to our imagination. It would appear that in many areas that pertain to the 'security' of our lives, nodal governance has simply offered new opportunities for modalities of hierarchy, command and interventionism to intensify and diffuse (Crawford, 2006). At the same time, we suggest there is potential in a nodal world to reinvent governance, especially in ways designed to both build and enrol the capacities of different nodal actors. Our empirical focus has been on the case of child protection and we have argued for the importance of nurturing the capacities of families and communities to govern both beyond and in tandem with hierarchical modalities.

This enterprise we are committed to is not pre-occupied with abandoning rationalities and practices of command and control, but rather to de-centre them whilst building on the latent strength of other governance forms and nodes. In so doing, we are engaging with the theory and model of 'responsive regulation', which conceptualizes coercive interventionism as a last resort and locates it figuratively at the top of a 'pyramid' of other (preferred) options for shaping the conduct of individuals and collectives. In our analysis we also draw from existing work on 'nodal' and 'network' governance which places both instrumental and normative emphasis on drawing on the knowledge and capacities of numerous entities, especially those who are currently at the margins of governance efforts; what Braithwaite and Drahos refer to as 'weak actors' in the system.

By weak actors we mean those whose capacities are not recognized, nurtured, developed, or are otherwise used sub-optimally or ineffectively. It could be argued that all governing nodes are 'weak' in some way, including state ones. While we accept this to be true, our discussion focussed on the need to better recognize, build, and harness the capabilities of nodes contained in families and communities. We may never wish to abandon crime and control as central to how we shape our social world, but we can certainly take better advantage of opportunities for governing nodally and governing beyond what our imagination might currently allow. There are obviously numerous challenges associated with stretching our imagination and putting new ideas into practice. We are committed therefore to facing the hard questions as we 'sow the seeds of innovation' (Scott, 2006).

ACKNOWLEDGMENT

This chapter was developed with the support of two grants from the Australian Research Council (grants LP0348682 and LP0669230). Jennifer Wood was a Fellow in the Regulatory Institutions Network, Australian National University, when this chapter was initially drafted.

REFERENCES

Adams, P., & Chandler, S. (2004). Responsive regulation in child welfare: Systemic challenges to mainstreaming the family group conference. *Journal of Sociology and Social Welfare*, *31*(1), 93–116.

Ayres, I., & Braithwaite, J. (1992). *Responsive regulation: Transcending the deregulation debate*. New York: Oxford University Press.

Bayley, D., & Shearing, C. (1996). The future of policing. *Law and Society Review*, *30*, 585–606.

Black, J. (2000). Decentring regulation: Understanding the role of regulation and self-regulation in a 'post-regulatory' world. *Current Legal Problems*, *54*, 103–146.

Black, J. (2003). Enrolling actors in regulatory systems: Examples from UK financial services regulation. *Public Law* (Spring), 63–91.

Blair, I. (2005).'The Richard Dimbleby lecture', *BBC News*. November 16. Available at http://www.news.bbc.co.uk/1/hi/uk/4443386.stm

Braithwaite, J. (2002). *Restorative justice and responsive regulation*. New York: Oxford University Press.

Braithwaite, J., & Drahos, P. (2000). *Global business regulation*. Cambridge: Cambridge University Press.

Braithwaite, V. (2005). Ten things you need to know about regulation and never wanted to ask. *Australian Law Librarian*, *14*(3), 19–28.

Burford, G., & Adams, P. (2004). Restorative justice, responsive regulation and social work. *Journal of Sociology and Social Welfare*, *31*(1), 7–26.

Burris, S. (2004). Governance, microgovernance and health. *Temple Law Review*, *77*, 335–359.

Burris, S. (2006). From security to health. In: J. Wood & B. Dupont (Eds), *Democracy, society and the governance of security* (pp. 196–216). Cambridge: Cambridge University Press.

Burris, S., Drahos, P., & Shearing, C. (2005). Nodal governance. *Australian Journal of Legal Philosophy*, *30*, 30–58.

Cashmore, J. (2001). Child protection in the new millennium. *Social Policy Research Centre Newsletter*, *79*, 1–5.

Cleaver, H., & Freeman, P. (1995). *Parental perspectives in cases of suspected child abuse*. London: HMSO.

Crawford, A. (2006). Networked governance and the post-regulatory state? *Theoretical Criminology*, *10*(4), 449–479.

Daniels, R. J., Macklem, P., & Roach, K. (2001). *The security of freedom: Essays on Canada's anti-terrorism bill*. Toronto: University of Toronto Press.

Deukmedjian, J.E. and Cradock, G. (2008). From community to public safety interoperability in policing and child protection. *Canadian Review of Sociology*, (in press).

Farmer, E., & Owen, M. (1995). *Child protection practice: Private risks and public remedies*. London: HMSO.

Foucault, M. (1982). The subject and power. In: H. L. Dreyfus & P. Rabinow (Eds), *Michel Foucault: Beyond structuralism and hermeneutics* (pp. 208–226). Chicago: University of Chicago Press.

Garland, D. (2001). *The culture of control: Crime and social order in contemporary society*. Chicago: University of Chicago Press.

Harris, N. (2008). Family group conferencing in Australia 15 years on. *Child Abuse Prevention Issues*, *27*, 1–19.

Johnston, L., & Shearing, C. (2003). *Governing security: Explorations in policing and justice*. London: Routledge.

Loader, I. (2000). Plural policing and democratic governance. *Social and Legal Studies*, *9*(3), 323–345.

Loader, I., & Walker, N. (2007). *Civilizing security*. Cambridge: Cambridge University Press.

Marks, M. and Wood, J. (2008). Generating youth safety from below: Situating young people at the centre of knowledge-based policing. In: T. Williamson (Ed.), *Handbook of knowledge based policing: Current conceptions and future direction*. West Sussex: Wiley. Available in http://www.wiley.com/WileyCDA/WileyTitle/productCd 0470028998,descCd-tableOfContents.html

Marsh, P., & Crow, G. (1998). *Family group conferencing in child welfare*. Oxford: Blackwell Science Ltd.

McAlinden, A. (2007). *The shaming of sexual offenders: Risk, retribution and reintegration*. Oxford: Hart.

Merkel-Holguin, L. (2004). Sharing power with the people: Family group conferencing as a democratic experiment. *Journal of Sociology and Social Welfare*, *31*(1), 155–174.

Neff, R. (2004). Achieving justice in child protection. *Journal of Sociology and Social Welfare*, *31*(1), 137–154.

Parker, C., & Braithwaite, J. (2003). Regulation. In: P. Cane & M. Tushnet (Eds), *Oxford handbook of legal studies*. Oxford: Oxford University Press.

Parton, N. (1997). Child protection and family support: Current debates and future prospects. In: N. Parton (Ed.), *Child protection and family support: Tensions, contradictions and possibilities*. London: Routledge.

Parton, P. (2006). *Safeguarding childhood: Early intervention and surveillance in a late modern society*. Houndmills: Palgrave Macmillan.

Pennell, J. (2004). Family group conferencing in child welfare: Responsive and regulatory interfaces. *Journal of Sociology and Social Welfare*, *31*(1).

Platt, D. (2006). Investigation or initial assessment of child concerns? The impact of the refocusing initiative on social work practice. *British Journal of Social Work*, *36*(2), 267–281.

Rhodes, R. (2006). The unholy trinity and network governance. In: J. Fleming & J. Wood (Eds), *Fighting crime together: The challenges of policing and security networks* (pp. 15–34). Sydney: University of New South Wales Press.

Scott, D. (2006). Sowing the Seeds of Innovation in Child Protection. Paper presented at the 10th Australasian Child Abuse and Neglect Conference, Wellington, New Zealand.

Scott, D., & Swain, S. (2002). *Confronting cruelty: Historical perspectives on child abuse.* Melbourne: Melbourne University Press.

Shearing, C. (2001). Transforming security: A South African experiment. In: H. Strang & J. Braithwaite (Eds), *Restorative justice and civil society* (pp. 14–34). Cambridge: Cambridge University Press.

Shearing, C. (2004). Thoughts on sovereignty. *Policing and Society*, *14*(1), 5–12.

Shearing, C., & Wood, J. (2003). Governing security for common goods. *International Journal of the Sociology of Law*, *31*(1), 205–225.

Simon, J. (1997). Governing through crime. In: L. M. Friedman & G. Fisher (Eds), *The crime conundrum: Essays in criminal justice* (pp. 171–189). Boulder, Colorado: Westview Press.

Simon, J. (1998). Managing the monstrous: Sex offenders and the new penology. *In Psychology, Public Policy, and Law*, *4*, 452–467.

Simon, J. (2007). *Governing through crime: How the war on crime transformed American democracy and created a culture of fear.* Oxford: Oxford University Press.

Standing Committee on Social Issues (New South Wales Parliament). (2002). *Final Report on Child Protection Services* (Parliamentary paper, no. 408), Sydney.

The Allen Consulting Group. (2003). *Protecting children: The child protection outcomes project.* Sydney: The Allen Consulting Group.

Thorpe, D. (1994). *Evaluating child protection.* Buckingham: Open University Press.

Trotter, C. (1999). *Working with involuntary clients.* Crows Nest, Australia: Allen & Unwin.

Trotter, C. (2004). *Helping abused children and their families.* Crows Nest, Australia: Allen & Unwin.

Valverde, M. (2001). Governing security, governing through security. In: R. J. Daniels, P. Macklem & K. Roach (Eds), *The security of freedom: Essays on Canada's anti-terrorism bill.* Toronto: University of Toronto Press.

Wise, S. (2003). The child in family services: Expanding child abuse prevention. *Australian Social Work*, *56*(3), 183–196.

Wood, J., & Dupont, B. (Eds). (2006a). *Democracy, society and the governance of security.* Cambridge: Cambridge University Press.

Wood, J., Dupont, B. (2006b). Introduction: Understanding the governance of security. In: J. Wood and B. Dupont (Eds), *Democracy, society and the governance of security* (pp. 1–10). Cambridge: Cambridge University Press.

Wood, J., & Shearing, C. (2007). *Imagining security.* Cullompton: Willan.

FROM THOUGHT CONTROL TO TRAFFIC CONTROL: CCTV POLITICS OF EXPANSION AND RESISTANCE IN POST-OLYMPICS GREECE

Minas Samatas

ABSTRACT

This chapter demonstrates that while in most late modern societies there is a neoliberal hegemony to expand police Closed-Circuit Television (CCTV) surveillance for crime control and antiterrorism, in Greece there is serious controversy and resistance against the post-Olympic use of more than 1,200 Olympic CCTV cameras. Drawing on the interesting politics of CCTV expansion and resistance, the chapter traces the reasons why, in the Greek context, this very expensive Olympic surveillance "dowry" has been opposed, even for traffic control. It critically attributes Greek citizens' fear and mistrust primarily to their past police-state experience of authoritarian, thought-control surveillance.

Surveillance and Governance: Crime Control and Beyond
Sociology of Crime, Law and Deviance, Volume 10, 345–369

ISSN: 1521-6136/doi:10.1016/S1521-6136(07)00216-3

INTRODUCTION

In our post 9/11 era a hegemonic, neoliberal security logic promotes high-tech panoptic surveillance systems throughout the world, especially video surveillance or Closed-Circuit Television (CCTV) as a normal way of life[1] to enhance traffic and crime control including antiterrorism (Lyon, 2001, 2003a; Marx, 2002; Schulhofer, 2002). In most late modern societies, especially in those which produce such technologies, governments follow neoliberal policies promoting the securitization of all social issues, adopting techno-centric solutions instead of social policies, at the expense of privacy and civil liberties (Coleman, 2004; Monahan, 2006). Especially the US and British governments promote high-tech security and surveillance products of their corporations, which, looking for new markets after the end of the Cold War, propagate panoptic surveillance systems as a panacea to the social problems (Norris & Armstrong, 1999, pp. 27–32; Ball & Webster, 2003; Lyon, 2003b). However, in all democratic surveillance societies, including Australia, Canada, the European Union (EU) and the United States there is concern for the new surveillance technologies such as the street cameras or CCTV systems linked with databases, which can cause "the death of privacy" (Coleman, 2004; Garfinkel, 2000; NYCLU, 1998; Sidel, 2004; Walby, 2006). "Fortified urban space" with continuous street monitoring and potential facial recognition (Coaffee, 2003, p. 15; Gray, 2003; Warren, 2004) is causing worries about what happens to the images once they would be used by the police and who else would have access to them. Especially in the EU, privacy protection laws and Data Protection Authorities (DPA) attempt the effective regulation of public and private use of CCTV (Gras, 2004; Dubbeld, 2005).

Unlike the hegemonic acceptance for CCTV use in most parts of the world, in Greece there is serious controversy and resistance against police CCTV cameras. This controversy was escalated after the Athens 2004 Olympics when Greece acquired a surveillance system of more than 1200 CCTV cameras, part of a 1-billion-euro security program, which also aimed for post-Olympic use against crime and terrorism. This chapter inquires the reasons why in post-Olympics Greece there is serious controversy and resistance over the post-Games use of the Olympic cameras, which might be welcome in most other countries. Drawing on the interesting politics of CCTV expansion and resistance in the Greek context, the chapter traces the reasons why this very expensive Olympic surveillance "dowry" has been opposed, even for traffic control. Rejecting Greek "exceptionalism" or uniqueness, we attribute Greek citizens' fear and mistrust towards police and security cameras primarily to their past police-state experience (1950–1974) of authoritarian, thought-control surveillance. In an effort to substantiate this argument, the chapter tries also to give answers to the following related questions:

- Why do Greek citizens fear police CCTV cameras and ignore or do not mind private cameras?
- Why have CCTV systems failed in Greece for traffic control and as a deterrent for car accidents and urban violence?
- Is the Hellenic DPA ruling an adequate privacy safeguard against CCTV monitoring?
- Are the police cameras a threat only to those who break the law or potentially for all Greek and European citizens?
- Can we generalize from the Greek resistance to police CCTV cameras to similar attitudes in other post-authoritarian societies?

To answer these questions, we trace:

(a) the politics of the last two Greek governments, which have used the Olympic Games for CCTV expansion and legitimation, as a panacea for all problems from traffic control to crime and terrorism, urban and sports violence;
(b) the Hellenic (Greek) DPA CCTV ruling based the whole Greek and European privacy legislation, including the Greek constitution and the European Convention on Human Rights;
(c) a case study of 10 Olympic cameras, deployed across the Maliakos highway "horse-shoe" to reduce car accidents;
(d) the public debate over the CCTV antiterrorist efficiency, after a rocket-attack launched by a terrorist group on January 12, 2007 against the US embassy in Athens; and
(e) the multifaceted resistance to police CCTV cameras, ranging from legal actions to vandalism and civil disobedience by mayors, civil society groups, academics, leftist, and antiauthority groups, even by the police union.

We substantiate this CCTV resistance in Greece by drawing data from the press and web pages, but also from informal discussions with academics, activists, police officers, and concerned citizens. We also take under consideration the initial results of our ongoing survey on CCTV use and privacy protection, and the rich international bibliography on CCTV.

THE ATHENS 2004 OLYMPIC SURVEILLANCE "DOWRY:" OVER 1,200 CCTV CAMERAS

The Athens 2004 Olympic Games, the first Summer Olympic Games after the terrorist attacks of 9/11/2001, were a catalyst for surveillance explosion in Greece. The country had to acquire very expensive security and surveillance systems, including hundreds of CCTV cameras to cope with real and perceived threats. Two years ahead of the 2004 Games, i.e., beginning January 14, 2002, the first day of CCTV monitoring of Athens' roads, more and more cameras were put up on Athens streets to record traffic violations in bus lanes as well as for general security purposes (*Athens News,* February. 15, 2002, p. A07). The "Panopticon" (Foucault, 1977; Simon, 2005) in Athens 2004 Olympics was based on a "super-surveillance project" which prescribed a network of interconnected 1,250–1,600 CCTV installed throughout the Athens metropolitan area, running 24 hours/day. The 1,283 microphone-equipped cameras, one camera every 50 meters installed at Olympic venues and on the streets of central Athens, were part of an elaborate security system that cost $255 million Euros. CCTV systems were linked with a whole surveillance network of mobile sets (TETRA, Terrestial Trunk Radio), which broadcast images and sound in real-time to 22, 160 security staff members organized in five operation centers, coordinated by a central information security station (*Ta NEA*, March 20, 2004, p. 31). Besides the Olympic Village and the Olympic Stadium, which were the primary surveillance focus, broad coverage also included the new Athens airport "Eleftherios Venizelos," the Athens Metro system, the major seaport of Piraeus, and the national highways; hundreds of CCTV were in constant operation, interconnected as a network with the other Greek cities hosting Games events (*Ta NEA*, March 20, 2004, p. 31, and *Eleftherotypia*, June 30, 2004, p. 20).

The Olympics central information security system, known as C4I, which stands for "command, control, communications, and integration," as the latest antiterrorist super-panoptic technology consisting of 30 subsystems, became a technical nightmare and actually failed because of its enormous complexity and could not be implemented on time for the Games. Thus, the security of the Athens Summer Olympics was dependent on conventional means (Samatas, 2007a).

Nevertheless, much of what the Greek governments sought in the costly C4I system was for post-Games applications of the security system. The then Public Order Minister Yiorgos Voulgarakis justified the extreme cost with promises for post-Games long-term security benefits has stated: "What matters is that after the Olympic Games we will find ourselves with very sophisticated systems that will allow us to police Olympic cities very well, for example, in matters of traffic control. Through cameras, we will be able to intervene without men in the field" (*Athens News*, July 23, 2004, p. A05). The Greek Police (EL.AS) was pleased with their new digital, wireless encrypted, multi-channel communication system, TETRA , the only C4I subsystem that has actually worked ever since the Olympics (*Eleftherotypia*, September 11, 2004, p. 56).

In addition to TETRA, the Olympic surveillance dowry includes Subsystem 17 of C4I, consisting of the aforementioned 1,200 CCTV surveillance cameras on the roads of Attica, equipped with Intelligent Traffic System (ITS) or "Traficon", monitored and controlled by the Traffic Police Headquarters of Attica (TPHA).[2] "Traficon" has the potential to watch and evaluate the following categories of incidents of traffic control: traffic congestion, circulation load, circulation speed, locating an accident, unlawful stop and parking of a vehicle, unlawful circulation of a vehicle, etc.[3]

The excessive cost of the Olympic CCTV system pushed the Government to insist on its use, not only for traffic control but also for security and antiterrorism. The EU DPA has, however, suggested the Olympic panoptic systems should be either sold or rented to other states, which organize mega-events (*To Vima*, May 23, 2006).

THE GREEK DPA RESTRICTIONS ON THE CCTV USE AND THE CONTROVERSY OVER THEIR MIXED PURPOSE FOR TRAFFIC CONTROL AND SECURITY

Since September 2000, long before the Athens 2004 Olympics, the Greek (Hellenic) DPA had set out guidelines restricting the use of CCTV in public and private places. DPA has prohibited the recording, use, monitoring, and retention of personal information through CCTV on a regular, continuous, or permanent basis. CCTV recording is only lawful when it is done for the protection of individuals or goods or for traffic regulation and only under "the principles of necessity and proportionality." According to the principle of necessity, the processing is allowed "if its purpose cannot be achieved by any other equally effective method but less troubling to the individual (such as detectors at the entrance and exit of indoor premises);" the principle of proportionality, implies that "the legitimate interest of the controller must prevail over the rights and interests of the individuals to whom the data relate, provided that their fundamental rights are not violated." In these exceptional cases, and others for crime prevention, the DPA must grant special permission to judicial and legal authorities to use cameras, with strict guidelines for use and retention (Hellenic DPA, Directive on CCTV systems, 1122-September 26, 2000).

In May 2004, the DPA approved a police request to operate CCTV cameras on the streets during the "operational phase" of the Olympics, as long as the cameras work under the basic terms of the aforementioned DPA directive 1122 of 2000, and they are not used after the Games. Hence, the cameras were legally operated during the Olympics from July 1 to October 4, 2004. Other conditions were that the cameras not be set up in such a way that they film the entrances or interiors of homes or that they record the conversations of passers-by, and that there be adequate signage informing citizens they are entering surveillance areas (Hellenic DPA Opinion 4/ May 3, 2004).

After the end of the Olympics and Paralympics, the Government urged legalizing the extension of the function of the CCTV, used for the security of the Olympic Games; the Ministry of Public Order with its Ref. No PN 101/ October 4, 2004 document asked DPA for the post-Olympic extension because, as it claimed, "its function is necessary for reasons of public interest and specifically for the purpose of traffic management."

In November 2004, the DPA allowed the continuation of the Olympic CCTV on the streets for a period of six months, as long as it was used only for car circulation and not for any other reason including the ascertainment of illegal acts other than those related to the traffic control. The use of cameras was allowed only in the roads of high circulation and not in the roads of low traffic or at places, squares, parks, pedestrian-precincts, and public assembly areas (e.g. entrance of theaters). The cameras were to be set in such a way that they did not film the entrances or interiors of homes, and microphones must be taken off the poles. Transmitting data to third parties is prohibited, and the data will be kept for seven days at the most. Also, warning signage should advise of the existence of these cameras (Decision 63/ November 24, 2004).

DPA's annual report for 2004 (p. 45) stated that DPA inspection of police CCTV cameras found that there were no warning signage of the existence of these cameras and also several CCTV systems were deployed in inappropriate locations. After DPA inspection, the police put up signage but did not dislocate nor incapacitate these cameras. Hence, critics argued that the DPA has no way to monitor effectively whether the restrictions it issues are actually initiated by police.

The Ministry of Public Order with its Ref. no. 62/24-5-2005 document, requested again more from the Authority to extend the time period of the operation of the CCTV system that is used for the purpose of traffic control, that expired on May 18, 2005. This time the Ministry's request included extending the processing purpose for the personal data that are received through the CCTV system for traffic control, that is declared as the primary purpose of the processing, and for the protection of individuals and goods, that is declared as the secondary purpose of the processing.

According to DPA, the aim of the Ministry, as embodied in its request, was to install a general electronic surveillance system that would cover, besides traffic management, all the possible activities that take place in public places, where CCTV have been installed (circulation of vehicles and pedestrians, transportation of important people, marches and assemblies, police patrols, etc.). Further, there would be coverage of as many targets as possible (traffic roads, squares, parks, public buildings, embassies, etc.). Through this extension the Ministry aims to make the electronic surveillance system as efficient as possible with regard to the ulterior purpose – to prevent and repress possible criminal or terrorist acts.

To decide on the Ministry's request the DPA took into account the following legislation:

- "Article 8 of the European Treaty on Human Rights that refers to the right to respect for private and family life (Gallagher, 2004).
- Convention 108/1981 of the Council of Europe for the protection of the individual with regard to computer processing of personal data.
- Articles 7 (respect for private and family life) and article 8 (protection of personal data) of the Charter of Fundamental Rights of the EU (see Gras, 2004).
- DPA Directive 95/46/EC of October 24, 1995 for the protection of human beings with regard to the processing of personal data and also for the free circulation of these data.
- Articles 9 and 9A of the Greek Constitution that stipulate, respectively, that "the private and family life of the individual is inviolable" and that "all persons have the right to be protected from the collection, processing and use, especially by electronic means, of their personal data, as specified by law ..."
- The Greek Law 2472/1997 on the protection of individuals with regard to the processing of personal data."

As a result, and after Public Order Ministry pressures, DPA accepted the use of hundreds of Olympic security cameras for traffic control only, deeming them unconstitutional for security and counter-terrorism, even for law enforcement. Thus, on August 2, 2005, when DPA (Decision 58/2005, August 12, 2005) issued rules that severely restrict the use of surveillance cameras in public, DPA dealt police counter-terrorism efforts a potential blow. In fact, DPA permitted the operation of the system, including 293 street cameras that were part of the C4I system and 49 police cameras that pre-dated the Games, only *for the primary purpose concerning traffic management* for the requested period of time, that is until May 24, 2006, according to the terms and conditions of decision 63/2004. As to the secondary demand of the Ministry, DPA rejected to extend the purpose of the CCTV processing for the protection of individuals, which is related to the public safety. DPA cited the principle of proportionality between means and ends, and a 2005 *Home Office Research Study* (No. 292), suggesting that CCTV is ineffective in guaranteeing security in large public spaces, and extremely discouraging as to the efficiency of CCTV systems to assure the safety of the citizens.

Moreover, DPA cited the First Instance Court of Patras, which, after five lawyers' petition, decided through the process of a restraining order as to the operation of CCTV systems in public places (decision 2765/2005) that the operation of the cameras is against the Greek Constitution articles 5.1 and 9A, and "unlawfully violates the right to personality [...] of the citizens, because it puts them under control and unjustifiably restricts their freedom, ... and hinders the free development of their social and political activity." Consequently, DPA rejected the secondary demand of the Ministry prohibits CCTV use for any other purposes besides traffic management, and while allowing nearly 350 cameras for traffic control, it banned use of their microphones and ordered the removal of 32 cameras located in low-traffic roads, public spaces such as parks and squares, pedestrian roads, and theatres' entrances. In this DPA decision, however, there is a "window" loophole that gives permit for monitoring for other purposes than traffic control in case of emergency or other special occasions, called "security events" (*Rizospastis*, August 23, 2005).

That DPA's decision on August 12, 2005, disappointed everyone – the police, the government, the leftist opposition, and antisurveillance NGOs. Police spokesman insisted that the massive deployment of cameras as quite helpful in handling a wide range of security issues, from robberies to terrorism. In Parliament the Left Coalition (*Synaspismos*) and the Greek Communist party KKE accused the Greek government of blatantly violating its pledge to take down the cameras after the Games, and that the DPA had abandoned any pretense of independence, bowing to government pressure to allow sweeping surveillance. The Left parties claimed that the cameras were trained on protesters during demonstrations so as to keep records, and asserted also that US and British law enforcement authorities have access to Greek police surveillance data.

On the other hand, the government contested the DPA ruling, saying national security trumps the privacy of citizens, and took the issue to the Council of State, the country's highest administrative court to annul that DPA decision. The Council of State, according to a ruling made public in August 7, 2006, approved the use of CCTV cameras by police until May 2007, but only for monitoring traffic in Athens. Yet, its ruling on use of other cameras installed as part of the one-billion-euro security system for the Athens Olympics is still pending (*Kathimerini*, August 8, 2006).

Meanwhile, a snap inspection by the DPA in May 12 and 24, 2006 found that 13 out of the 32 illegal Olympic CCTV cameras, which had been placed in low-traffic roads, public squares, parks, pedestrian ways, and open spaces for public gatherings and therefore should be taken away as illegal, were working. Hence DPA imposed a 3,000 Euro fine on the Ministry of Public Order (*Athens News*, January 26, 2007, p. A08). This fine, however symbolic, ratified by the Council of State, was annoying to the Government, which has not hidden its annoyance against DPA on that and other occasions as well.[4]

CCTV FAILURE TO PROMOTE ROAD SAFETY AND DETER CAR ACCIDENTS

A widespread application of CCTV systems, which is very legitimate and welcome in many countries all over the world, is their deployment in roads for traffic monitoring and prevention of car accidents. Moreover CCTV could be very useful in Greece, which is the champion in deaths from accidents in the EU. In fact, Greece is first among the EU 15 member-states in vehicular/car deaths, having a rate (9.1%) triple the 3.1% EU average, according to "Eurostat" 2006 data published in *Eleftherotypia* (February 23, 2006, p. 61). Throughout the country there are "red-blood miles" where for the last 15 years (1990–2005) there are data for 31,246 deaths; still, an average of 1,500 people are killed annually and 4,000 are seriously injured (*Kathimerini*, March 9, 2006, p. 3).

The "champion province" in serious car accidents is the Maliakos Gulf road in central Greece, the so-called Maliakos horse-shoe. It is a road of 78 deadly kilometres length, with average daily traffic of 50,000 vehicles.

During 2001–2005, in 214 car accidents 130 people were killed and more than 445 were injured, of which 91 were seriously (*To Vima*, March 6, 2005, p. A34). That is why, almost immediately after the end of the Athens 2004 Olympics, 10 Olympic cameras were deployed across the Maliakos "horseshoe" to reduce car accidents. All ten CCTV were used as speed cameras to monitor car violators, issuing tickets of 53 Euros as penalty for those exceeding the maximum speed limit of 50 km. Soon after their placement most of these CCTV cameras were used as phoney speed cameras, because of the expensive operation and maintenance cost, and the inability of the road police to staff them with trained personnel. Thus the CCTV cameras failed to reduce car accidents there since most drivers, who have to wait in car lines for long time due to endless construction works, had soon realized they were phoney. Instead of CCTV monitoring, the construction of a tunnel 4.5 km long, which would have saved many victims and shortened the Athens–Thessalonica road distance by 45 km, has been repeatedly cancelled by local tourist interest groups and competing construction firms (*Eleftherotypia*, March 10, 2006, p. 45).

Despite CCTV failure at Maliakos, additional 390 more CCTV, totalling 400 road cameras nation-wide in the whole 3,500 km, were planned to be installed for traffic control and the prevention of car accidents. That means 1 CCTV every 10 km (*Eleftherotypia*, March 3, 2006, p. 12). The project is "all Greek roads with Speed Cameras," although camera systems alone cannot reduce car accidents on dangerous roads. Instead of expensive and inefficient CCTV systems, road safety requires better road systems, street lighting, police presence, and driver education (Samatas, 2006).

THE POLICE CCTV FAILURE AGAINST URBAN VIOLENCE IN ATHENS AND THE DEBATE OVER THE BALANCE BETWEEN SECURITY AND CIVIL LIBERTIES

Frequent violent incidents in central Athens during the last years have proved the inefficiency of hundreds of police CCTV cameras installed after the Olympics. In fact, central Athens, the most police-protected area, is routinely burnt chaotically by a group of about 100 hooded, so-called "known–unknown" anarchists, while riot police MAT units keep an eye from a distance (see pertinent report and photo at *Kathimerini*, March 9, 2007, p. 3). Both the Public Order Ministry and the police have proved unable to enforce law and order by using the post-Olympics surveillance system. They blame DPA restrictions on the CCTV use, although it is clear that police cameras alone are useless to enforce law and order unless there is political will and determination to do it.

Even the Athens US embassy, the super guarded place in central Athens, has proved a vulnerable fortress, after a rocket-propelled grenade was launched by a new Greek terrorist group on January 12, 2007 against the embassy. This rocket demythologized the efficiency of the security systems and the CCTV cameras monitoring the area around the embassy. As the US and London terrorist attacks have proved, security cameras cannot stop terrorists even if they can assist to identify them ex-post-facto (Monahan 2006, pp. 6–7). All evidence collected in connection with the embassy attack, both by Greek police and the FBI officials who had examined CCTV footage from the embassy's cameras did not help to identify the rocket launchers (*Kathimerini*, January 17, 2007).

After the US embassy rocket attack, the New Democracy (ND) government felt it was an opportune moment to push for the police's right to have greater powers of surveillance, and took the opportunity to intensify its call for hundreds of CCTV Olympic microphone-equipped cameras, installed for the Olympics in the Games venues and on the streets of central Athens, to be activated. This government's plan to introduce broad security surveillance sparked fierce political debate over the controversial issue of the balance between security and civil liberties. Foreign Minister Dora Bakoyannis, and the Public Order Minister Vyron Polydoras defended government's desire to start using hundreds of CCTV cameras in the capital for surveillance purposes; but Public Works Minister Giorgos Souflias and the ND Mayor of Athens Nikitas Kaklamanis said that they were not in favor of creating Big Brother with the cameras (*Kathimerini*, January 23, 2007).

Opposition PASOK leader George Papandreou said that his party would not support the use of the cameras to monitor people because this would infringe upon their human rights, and declared that when in power his party had addressed security issues without limiting civil liberties. PASOK's former public order minister, Mihalis Chrysohoidis, and MP Theodore Pangalos broke ranks and expressed support for use of the cameras for security, but with DPA permission (*Kathimerini*, January 17, 2007). Nevertheless, Greece's two largest parties, the ruling conservative ND and the so-called socialist PASOK, are in agreement in the use of CCTV cameras by police to combat crime and terrorism. Only the Greek Communist Party (KKE) and Synaspismos Left Coalition have always categorically opposed the use of police CCTV cameras for security purposes.[5]

In an editorial in the daily *Kathimerini* (January 22, 2007), entitled "Who's afraid of CCTV cameras?" we can read:

> We really should not kid ourselves. The privacy of citizens in this country is being violated on a daily basis on many levels and using various methods ... the problem regarding the lack of protection of our personal data has not been created by CCTV cameras. It is already long-established. The cameras are a threat to those who break the law ... These cameras can actually protect the citizen and the democratic state, as long as we ensure that what they record is handled responsibly. Here, we all agree, there must be some guarantees.

Yet, in a poll conducted for Sunday's *Kathimerini* (January 21, 2007) a week after the embassy attack and violent clashes between anarchists and riot police in central Athens, nearly two-thirds of those surveyed favored police using CCTV on roads in central Athens for security purposes. Only one in three respondents said that they feel CCTV systems intrude more than they protect, and only a quarter of people questioned said that the cameras installed in 2004 should be ripped down whereas 61% said they should be used. The results of this poll were interpreted by the newspaper as "a substantial shift in public opinion which until now has viewed the surveillance system as an intrusion on people's privacy" and generalized to suggest that 57% of *all* Greeks now think that cameras protect people rather than violate their personal or political rights. It is important to note though that some 70% of the respondents said that in Greece rights and freedoms are often not protected.[6]

CCTV EXPANSION: CAMERAS AS PANACEA FOR ALL PUBLIC AND PRIVATE TROUBLES

Despite DPA and police controversy over the issues of CCTV cameras, in Greece as in all late modern societies (Monahan, 2006), CCTV cameras are also deployed in public and private places as a panacea to all troubles and social problems (Samatas, 2007b). Hence, in early 2007 CCTV was advertised by the Greek Government and its supporting media as a remedy against sports violence, so to legitimize their deployment in most stadiums against hooligans (*Sunday Eleftherotypia*, April 7, 2007). Also, the Transportation Ministry and the Greek Trains Organization have announced that CCTV will be installed in all new trains to monitor passengers. CCTV have been also installed in some private and public schools against vandalism, even in churches and monasteries against thefts.

Recently, DPA has ordered for the ripping down a CCTV system of four cameras, deployed against vandalism at a municipality square of the broader Athens area, on the basis that citizens' rights of privacy and freedom at public open spaces are proportionally more precious and should be more protected than the few benches and a pergola at this square (Decision 51/2007 June 20, 2007). DPA also rejected the petition of someone who requested a permit to legalize the already-installed CCTV cameras at his own house for video recording the babysitter during his absence, for the protection of his children (Decision 3/2007, January 12, 2007). Yet, DPA ordered the removal of a CCTV system in a hotel's corridors, targeting mainly its personnel. (Decision 31/2007 April 4, 2007).

POLITICS OF RESISTANCE TO POLICE CCTV CAMERAS: FROM LEGAL ACTIONS TO VANDALISM AND CIVIL DISOBEDIENCE

The Greek people had understood there was a national interest for extra security and surveillance systems during the Olympics. However, after the games they see no serious reason to continue sacrifying their privacy, human rights, and civil liberties. Senior Greek legal experts, civil liberties organizations, left-wing opposition parties, and various other civil society groups have strongly condemned the post-Olympic use of the surveillance system and the Olympic cameras. Protests against the Olympic surveillance systems started before the Summer 2004 Olympics, continued during the Games, and escalated immediately after the end of the Paralympics in September 2004, when the Government made public its intention to use them for security purposes after the Games, to justify their tremendous cost. Citizens' concern against police surveillance cameras were reinforced as participants in demonstrations in Athens and Piraeus have accused police CCTV cameras of monitoring them, despite DPA banning. Government spokesman B. Antonaros justified the cameras' operation "as monitoring bulk events, which may affect car traffic" (*Rizospastis*, August 6, 2004). During the Games, Amnesty International and the Human Rights Union had protested the Olympic CCTV systems, the surveillance blimp, and the C4I system. The Athens Prefecture Council voted unanimously against the perpetual use of Olympic cameras in September 6, 2004, like other organizations as the Imprisoned and Exiled Fighters during the last dictatorship (SFEA), and the Greek Women Federation (OGE)

(*Rizospastis*, September 30, 2004). Since September 2004, immediately after the Games, several labor unions and the Panlabour Movement (PAME), in Athens, Piraeus, Thessalonica, Larissa, and Heraklion had organized protests against the post-Games use of the Olympic cameras, with very little media coverage, except in the leftist press. (*Rizospastis*, September 9 & 11, 2004; see activists photo at *Eleftherotypia*, May 5, 2005, p. 49). Interestingly, even the Attica Police Union has accused the Public Order Ministry and the State Security Agency of monitoring the police union activities since 2003 and provided evidence of using the Olympic CCTV system, including the blimp during the Games, to watch if the policemen at various posts were doing their duty (*Rizospastis*, November 3, 2004).

Eminent academic professors, like N. Alivizatos, G.A. Mangakis and A. Marangopoulou, have vigorously protested against the perpetuating operation of the Olympic cameras after the Games.[7] Also the Heads of all Greek Lawyers' associations declared their opposition to police CCTV use (*Rizospastis*, January 23, 2007). Several mayors, like of Athens Nikea, Galatsi, and Halandri, backed by citizens who had suffered from the authoritarian police state, disabled surveillance cameras, protesting for violations of privacy and civil liberties. Mayor Stelios Benetatos of Nikea city, who ordered crews to vandalize security cameras is facing Court sanctions; he has claimed the actions did not violate any laws because the police and the ministry had not sought the town's approval to install the cameras (*Rizospastis*, January 25, 2007).

NGO's human Rights groups such as the "Marangopoulos Foundation for Human Rights" and the "Democratic Rally for Freedom and Solidarity," (D.R.) in 2005 petitioned the Council of State to stop the use of cameras in public places even for traffic control.[8] In a press conference in May 5, 2006 they presented their petition's arguments; they claimed that CCTV use constitutes a violation of the European Convention for Human Rights, and distributed a DVD of police CCTV surveillance of various demonstrations of workers, employees, students, and teachers at central Athens under the pretext of traffic regulation (*Rizospastis*, May 26, 2006). The Council of State in its ruling of August 7, 2006, ignored the protests of Democratic Rally and the objections of 41 citizens who filed complaints against the use of cameras by police to monitor buildings and people. The Public Order Ministry in a document given to the Court, accepted that the operation of CCTV for traffic control includes "security events" like demonstrations, and other various social events, VIP and soccer fan club transit, etc. (*Rizospastis*, November 15, 2006).

After all these accusations against the police cameras DPA inflicted a 5,000 euro fine on Greek Police (EL.AS) for illegally monitoring the rally by members of the whole educational system in March 22, 2006 (*Rizospastis*, April 20, 2007). This fine, however small, is the second DPA fine on police for illegal CCTV use.

The opposition to the Olympic cameras has contributed to opposition to CCTV systems in other public places, especially in schools. High school students and parents associations have protested at the Ministry of Education against CCTV cameras' deployment in a public school of Piraeus, Thessalonica and Karditsa and forced the school chancellors to rip them down (*Rizospastis*, November 1, 2005).

This antisurveillance struggle targeting especially the Olympic CCTV cameras has taken various forms, like: blinding the cameras with black hoods; ripping off their cables; spray painting of CCTV lenses; throwing down the CCTV poles as the "antiauthority" movement.[9] Yet, the most radical blow against CCTV systems has been delivered by anarchist, "antiauthority" groups in Athens, with an extensive arson campaign using Molotov cocktails against police cameras "of electronic spying (hafiedismos)." In the last week of November 2006, arsonists had put on fire 20 CCTV cameras and their operations box in various places of Athens. (*Rizospastis*, November 29, 2006 & *Kathimerini*, November 30, 2006). According to official police data, through the end of November 2006, 180 CCTV cameras or their electronic operations boxes had been burnt by such radical groups (*Sunday Eleftherotypia*, January 7, 2007). The same radical groups have also put up posters and made graffiti with antisurveillance messages all over central Athens justifying CCTV destruction as a civil disobedience act defending civil rights against illegal police monitoring; they equalize their anti CCTV fight with the struggle against the last dictatorship (*IOS* report, in *Sunday Eleftherotypia*, January 7, 2007 and Papaioannou, 2007).

CONCLUDING REMARKS

Based on the documented analysis above, we can now provide some answers to the questions we have asked in the introduction.

(a) Why in post-Olympics Greece is there serious controversy over the post-Games use of the Olympic cameras, which could be welcome in most other countries? The basic reason for CCTV skepticism, opposition, and resistance

in Greece relates to the legacies of past authoritarian surveillance. Greek citizens, especially the older who retain vivid memories of the post-civil war police state (1950–1974), distrust police monitoring even for traffic control. As police surveillance files were kept in the past for all citizens and they had been used to enforce social and political controls, punishing and excluding non-loyal citizens from public benefits, police cameras now can also be used to identify citizens – especially protesters, active unionists, and visible minorities – and violate their rights and liberties. Therefore, the memory of authoritarian and repressive surveillance used in the past for thought control, and the fact that police CCTV cameras represent the central state and the security forces, now cause fear and mistrust that police electronic cameras are not only used for traffic control but can be also used for socio-political control. That is why, unlike the police video surveillance in the UK that is widely accepted and viewed as a fact of life, in Greece memories of the past police surveillance cause distrust of police use of CCTV cameras. Whereas mayors in the UK and Australia compete for deployment of more open street CCTV systems for traffic and crime control (Sutton & Wilson, 2004), in Greece most mayors, not only those backed by leftists, but even the Athens conservative party's mayor, are vehemently opposed to police cameras. The left parties and rights groups claimed that the cameras are trained on protesters during demonstrations so as to keep records, and asserted that US and British security authorities have access to the Greek police surveillance data. Civil liberties advocates are worried about misuse of technology that tracks the movement of thousands of cars and people.

(b) Why are Greek citizens afraid of police CCTV cameras and ignore or do not mind private cameras? Greeks are not afraid of any visual surveillance, for instance that by their neighbors' gaze; they do not care for about private CCTV in banks, kiosks, hotels, malls, etc.; they are seriously concerned, however, about police surveillance and files ("fake-loma"), which, as in the past, can now keep personal socio-political information in electronic databanks, to be used for discrimination and exclusion from public sector benefits. There is still vivid post-civil- war memories in older generations of the security police and the military surveillance network of police informers throughout the whole country, which systematically watched, collected, stored, and updated information in special police surveillance files. The post-civil war police state kept these notorious surveillance records ("fakeloi") on sociopolitical ideology and activity of every Greek citizen, family, group, and organization, classifying

them according to their "national loyalty" (*ethnikofrosyni*). This bureaucratic surveillance procedure, expressed by the popular Greek term *fakeloma* (literally, filing), was the basic mechanism of the oppressive sociopolitical control system, which I have called "authoritarian bureaucratism" (Samatas, 1986, 2004).

The problem is that the legacy of this mass political surveillance has produced a negative surveillance culture, which does not permit any positive perception and trust of any legitimate monitoring or personal data collection, even for traffic control (Samatas, 2004, 2005). Every legitimate "institutional surveillance" (Lianos, 2003) presupposes trust in the state-public and private institutions; this institutional trust has never really existed in Greece, since even in the present liberal era several public agencies have often violated the privacy rights of Greek citizens, especially of the weak minorities. Corruption or negligence characterizes several public services, including the police. In brief, if we can accept that in a democratic state police cameras can actually protect the citizens, as long as there are institutional guarantees that their record is handled responsibly, in Greece those guarantees are still missing.

(c) Why have CCTV systems failed in Greece for traffic control and as a deterrent for car accidents and urban violence? Technical and financial problems like installation and maintenance cost, but mainly police dysfunction, are the basic reasons. Because, the way "new surveillance" technologies are applied in practice depends upon how they fit in with existing organizational, cultural, and political practices in different locales and institutional settings (Marx, 2002; McCahill, 2002).

As we have seen, the decentralisation of Olympic cameras on the dangerous Maliakos road, plus the governmental project to deploy speed cameras on all Greek roads reflect a techno-fix solution which cannot succeed in a country with a terrible road system and inefficient police. Aside from the high installation and maintenance cost, CCTV systems alone cannot succeed in traffic control and reduction of car accidents on very dangerous roads. Instead of spending for expensive and inefficient CCTV systems, better, safer roads should be constructed with street lighting and continuous police presence. In the Greek case both ruling parties should be blamed for the inability in the entire 30-year post-dictatorial period to build a safe road system in a small country, despite the generous infrastructure finance from the EU.

Both the Public Order Ministry and the police have also proved unable to enforce law and order in central Athens by using the post-Olympics surveillance system. They blame DPA restrictions on the CCTV use, although it is clear that police cameras alone are useless to enforce law unless there is political will and determination to do it. As the US and UK experience teach, CCTV as deterrent for crime and terrorism is no longer accepted (Monahan, 2006, pp. 4–8); video surveillance is better at unraveling crimes than deterring criminals, particularly terrorists determined to sacrifice their lives.

Thanks to the Olympics, Greek police (EL.AS) has been electronically modernized to fight the "antiterrorist war," using American and British technology and know-how. After the Games however, the police in several cases proved unusual dysfunctional, inefficient, brutal, and corrupted behavior, failing to gain the trust of Greek citizens. Greek police dysfunction is firstly due to its political domination by each ruling party and the Public Order Minister that each year appoints a new police leadership, promoting or moving out a serious number of high-ranking police officers, using mostly political criteria. Career insecurity fosters political dependency on the ruling parties and political factionalism: police rank and file are divided between those supporting the ruling party and those supporting the main opposition party. The police unions' leaders are affiliated with the opposition political parties; hence, they usually clash with the police leadership, which is appointed by the Public Order Minister. Frequently, the police unions' leaders have publicly condemned or resisting the governmental policies. Greek Police leadership, like every police, is usually enthusiastic about CCTV cameras – unless they are designed to watch for police wrongdoing, e.g. when they expose police brutality, or humiliation of police victims (*Athens News*, November. 24, 2006, p. A08). Several high-ranking police officers, however, are skeptical of using new technology, not only due to low funding and training, but also because of government's expectations to combat crime and terrorism basically by panoptic electronic systems.[10]

(d) Is the Hellenic DPA ruling an adequate privacy safeguard against CCTV monitoring? Of course not, due to several reasons. DPA's work is very difficult given the fact that in Greek society there is no privacy protection culture and the Greek state traditionally violates the citizens privacy, using personal data for sociopolitical control. (Samatas 1986, 2004, pp. 135–142). Aside from its battle against Olympic CCTV cameras, DPA also won the battle of identity cards against the powerful Greek Orthodox Church when it ruled for removing the religious affiliations from the new state identity cards, causing massive protests and challenges to that ruling

(Decision no. 510/May 15, 2000). Its defamation by the Orthodox Church deprived the Greek DPA of wide popular support as a citizens' privacy protector. Also, due to operational problems and limited personnel the Greek DPA has no way to monitor effectively whether the restrictions it issues are actually initiated by police or private CCTV operators.

Nevertheless, the Greek DPA fights hard to justify its independence from Greek authorities and enforce Greek and European privacy protection laws in Greece. Its whole anti CCTV work in favor of citizens' privacy and civil liberties is esteemed as positive, but it is simply unable to control the CCTV cameras mushrooming almost everywhere in public and private places. This positive assessment for the Greek DPA is reinforced by the embarrassment it causes to the Government and police with its rulings and fines. In November 19, 2007, the President and five members of the Greek DPA have resigned in protest for the Government to allow the police CCTV cameras to monitor the political rallies and demonstrations during the commemoration of November 17, 1973 student revolt against the military dictatorship.

(e) Are the police cameras a threat only to those who break the law or potentially for all Greek and European citizens? Not only leftist parties and movements but also rights groups and independent academics fear that police CCTV cameras can be a threat for all citizens in the context of the Greek, European, and the Atlantic "surveillance assemblage," (Haggerty & Ericson, 2000) based on the EU–US security cooperation against terrorism. Police CCTV cameras and electronic databases can be internationally linked with a variety of networked databases for continuous online linking, processing, evaluation, classification, and identification of personal data, and the production, even simulation, of various personal information profiles for a variety of purposes. These technological possibilities will be soon immensely enhanced, according to STOA report for the European Parliament (2007). This massive sharing of data between the EU, the US, and other wealthy nations is based on a "counter-terrorism" net, comprised of a number of treaties and surveillance systems with interlinked databases to introduce the surveillance of the movements of *everyone* in the EU – citizens, legal resident third-country nationals, visa entrants, and irregular migrants – and the storage of their personal data on an unprecedented scale (Bunyan, 2005).[11] Thus, EU–US security cooperation heralds a global identification system, the global surveillance of movement and a global police information system, where there is no place for "free movement" and privacy (*Statewatch*, 2004; Maas, 2005).

The Greek government is on board in the Schengen zone with the new EU-wide surveillance measures, based on the unanimous decision to keep files on Europeans (Samatas, 2003, 2008). Hence criticism against police CCTV systems emphasizes that London-style security in Athens with Olympic cameras is the Greek government's way of introducing in a more moderate fashion the planned EU surveillance and communication data retention after EU justice and interior ministers approved plans for extensive surveillance of European citizens. However, such surveillance is in breach of the Greek constitution (*Eleftherotypia*, July 14, 2005).

Nevertheless, Greece with adequate privacy legislation, DPA decisions, and citizens' resistance to police cameras has, compared to other societies, a relative good privacy record. According to the Electronic Privacy Information Center (EPIC) and Privacy International (PI) (2006), global study on "Leading surveillance societies in the EU and the World" (February 11, 2006), an assessment of the state of technology, surveillance and privacy protection in 36 countries, Greece belongs to the third best category of "adequate safeguards against privacy abuse," scoring and ranking a pretty good privacy protection record.[12] On the category of "visual surveillance" which measures electronic surveillance, in particular by CCTV cameras, included availability of the technology, safeguards and limitations, and use of the technology in private and semi-private locations, Greece scores 3 after Germany, France, and Poland which score 4, indicating that it has some safeguards, and relatively limited practice of surveillance.

Furthermore, Greece is a member of the Council of Europe, which recently concluded against the compatibility of video surveillance of public places with the protection of fundamental freedoms. On July 11, 2007 the European Commission for Democracy (2007) through Law or the "Venice Commission," which is the Council of Europe's advisory body on constitutional matters, made public an opinion: the practice of video surveillance of public places by public and private operators is a threat to the fundamental rights to respect for private life and freedom of movement and touches on specific issues of protection of the personal data gathered in this way, and invited the member states to take privacy protection steps (Press release 496, 2007).

Greece formally has already taken most of these steps, although there is serious problem with their enforcement.[13]

(f) Can we generalize from the Greek resistance to police CCTV cameras for similar attitudes in other post-authoritarian societies? Without

comparative surveillance studies, we can only assume that acceptance, opposition, or resistance of new electronic surveillance in post-authoritarian societies varies greatly; presumably it is much different in southern European states with embedded democratic institutions, as EU member states since the eighties, and much different in the former Soviet block central European states without democratic culture and stable institutions.

Hopefully, the above answers to questions about police CCTV cameras in Greece will enrich the ongoing debate about the expansion of CCTV systems around the world, and especially in post-authoritarian societies.

NOTES

1. For "naturalization effect," the promotion of a natural attitude, taking surveillance as granted in the new post 9/11 life world see Kim (2004, p. 211).

2. Subsystem 17 can give operators access to video and sound from each one of 293 CCTV positions, and also video from 49 police cameras with Panoramic Optical Zoom "PTZ," sound box and microphone, and an electric board with codifiers, network transmitters, UPS, and the detection cards of the "Traficon."

3. "Traficon" like the N-System in Japan (Abe, 2004) is based on algorithms video process detection cards, so as to define pre-arranged alarm conditions (traffic violations, traffic flow problems, and accidents). All technical information about C4I Subsystem 17 is taken from the Hellenic DPA Decision No 63 / 2004.

4. Justice Minister Papalegouras protested against DPA' s Decision 33 of May 4, 2007, and a ban of 5,000 Euro for not protecting the privacy databank of an Orphanage, that was closed, violating the privacy Law 2472/1997.

5. KKE, commenting on the rocket launch against the US embassy on January 12, noted that if the Americans want to avoid being targeted in the future they had better change their foreign policy in Iraq and elsewhere and not rely on street cameras that are of no help anyway (*Athens News*, January 26, 2007, p. A99).

6. The results of this survey conducted by VPRC are, however, very questionable because: (a) the poll was conducted within a week following the embassy attack and violent clashes between anarchists and riot police in central Athens, and following the subsequent propaganda blitz by TV programmes broadcasting "experts" in favor of CCTV; (b) it did not specify the procedures and rules under which data would be collected and processed; and (c) the poll results based on only 606 questioned people were generalized as representing *all Greeks*.

7. Nikos Alivizatos, constitutional law Professor and former member of DPA's Board, has accused the then-Public Order Minister of "gimmick CCTV politics." He argued that the tremendous surveillance possibilities of these cameras, to focus on faces and car plates even from 2 km distance away, and recording of dialogues such as drivers' quarrels after a car accident, exceed the role of CCTV for traffic control (Alivizatos, 2004). Such a video of two drivers' quarrel after a car accident recorded

by police cameras is put on the website of special squads forces as a fun video! (available at http://www.sefeaa.gr/fun/videos/boom.zip).

8. Aliki Marangopoulou, head of the eponymous foundation, sees the push for broader surveillance as part of the US-led global war on terror. "We must fear terrorism, be obedient and accept new surveillance measures," she said (*Athens News*, January 26, 2007, p. A08).

9. All this information is taken from *IOS* reporters at *Sunday Eleftherotypia* of January 7, 2007 entitled "Why cameras are burnt?" and *Rizospastis*, September. 9, 2004; May 12, 2005, July 12, 2005; November 1&15, 2005; June 2, 2006; January 16, 2007.

10. Greece still has a low crime rate despite an increase in the number of reported offences in 2006. Lately investigations for police corruption have been increased but penalties and sanctions are still very limited (*Sunday Vima*, March 11, 2007, p. A37). Recently, video footage has exposed police brutality against students; yet, videos made for fun by policemen recorded their immigrants' serious abuse and humiliation.

11. This "counter-terrorism" net is comprised by a number of treaties (like Prüm Treaty) and various systems (like SIS II, VIS, and PNR) to be used for speculative surveillance, general intelligence gathering, and "fishing expeditions;" more importantly, individual records will increasingly result in coercive sanctions, such as the refusal for travel, visa, or asylum applications, detention pending extradition, even deportation. (Privacy International, 2004, Samatas, 2008.)

12. Greece's total score is 3.1, i.e., a 4th ranking in EU and 5th in toto, after Germany, Canada, Belgium, and Austria which have the best records on privacy protection and safeguards; in the bottom black category are USA and UK, the worst performing western democracy, while Malaysia and China are the worst.

13. The most common violation by both Greek ministries of public order, which operates 293 police CCTV cameras, and the public works ministry, that operates 208 CCTV cameras, is lack of any signage in most of the monitoring zones (*Ta Nea*, April 2 & 12, 2007).

ACKNOWLEDGMENT

A first version of this chapter was presented at the "Crime, Justice and Surveillance Conference" Sheffield, April 5–6, 2006, thanks to the Graduate Programme "Sociology" of University of Crete.

REFERENCES

Abe, K. (2004). Everyday policing in Japan: Surveillance, media, Government and public opinion. *International Sociology*, *19*(2), 215–231.

Alivizatos, N. (2004). Gimmicks for the cameras. *Ta Nea*, September 27, p. 6.

Ball, K., & Webster, Fr. (Eds). (2003). *The intensification of surveillance*. London: Pluto.

Bunyan, T. (2005). While Europe sleeps. Spectrezine, December 5, 2005. Available at http://www.spectrezine.org

Coaffee, J. (2003). *Terrorism, risk and the city: The making of a contemporary urban landscape*. Aldershot: Ashgate.

Coleman, R. (2004). Reclaiming the streets: CCTV, neoliberalism and the mystification of social division in Liverpool, UK. *Surveillance & Society*, *2*(2/3), 293–309. Available at http://www.surveillance-and-society.org

Dubbeld, L. (2005). Protecting personal data in camera surveillance practices. *Surveillance & Society*, *2*(4), 546–563. Available at http://www.surveillance-and-society.org

Electronic Privacy Information Center (EPIC) and Privacy International (PI). (2006). Leading surveillance societies in the EU and the World, February 11. Available at http://www.privacyinternational.org/article.shtml, and http://www.privacyinternational.org/survey/phr2005/aboutphrtable.pdf

European Commission for Democracy through Law or the "Venice Commission" Press release–496. (2007). Available at http://www.coe.int/t/dc/press/

European Parliament. (2007). STOA report for Radio Frequency Identification Systems (RFID). Available at http://www.statewatch.org/news/2007/aug/ep-stoa-report-on-rfid.pdf

Foucault, M. (1977). *Discipline and punish: The birth of prison*. New York: Vintage.

Gallagher, C. (2004). CCTV and human rights. *Surveillance & Society*, *2*(2/3), 270–292. Available at http://www.surveillance-and-society.org

Garfinkel, S. (2000). *Database nation: The death of privacy in the 21st century*. New York: O'Reilly.

Gras, M. (2004). The legal regulation of CCTV in Europe. *Surveillance & Society*, *2*(2/3), 216–229. Available at http://www.surveillance-and-society.org

Gray, M. (2003). Urban surveillance and panopticism: Will we recognize the facial recognition society? *Surveillance & Society*, *1*(3), 314–330. Available at http://www.surveillance-and-society.org

Haggerty, K., & Ericson, R. (2000). The surveillant assemblage. *The British Journal of Sociology*, *51*(4), 605–622.

Hellenic Data Protection Authority (DPA). (2000). Directive on CCTV systems, 1122-September 26. Available at http://www.dpa.gr/Documents/Eng/DIR.CCTV1122-26.09.2000.doc

Hellenic Data Protection Authority (DPA) (2004). Opinion 4/ May 3, and Decision 63/2004 on CCTV cameras on the Attica road network. Available at http://www.dpa.gr/Documents/Eng/CCTV%20cameras%20on%20the%20Atttica%20road%20network.doc

Hellenic Data Protection Authority (DPA). (2005). CCTV Decisions: 58/2005, Aug. 12, 2005; No 3/2007, Jan. 12. 2007; No 31/2007 April 4, 2007; No 51/2007 June 20. Available at http://www.dpa.gr/decision_eng.htm

Home Office Research Study. (2005). Assessing the impact of CCTV. Written by Gill M. & Spriggs A. No 292, February.

IOS reporters. (2007). Why cameras are burnt? The new constitutional article 1-1-4. *Sunday Eleftherotypia,* January 7.

Kim, K. J. (2004). Surveillance technology, privacy and social control. With reference to the case of electronic National Identification Card in South Korea. *International Sociology*, *19*(2), 193–213.

Lianos, M. (2003). Social control after Foucault. *Surveillance & Society*, *1*(3), 412–430. Available at http://www.surveillance-and-society.org
Lyon, D. (2001). *Surveillance society: Monitoring everyday life*. London: Open University Press.
Lyon, D. (Ed.) (2003a). *Surveillance as social sorting*. London: Routledge.
Lyon, D. (2003b). *Surveillance after September 11*. London: Polity.
Maas, W. (2005). Freedom of movement inside 'fortress Europe'. In: E. Zureik & M. Salter (Eds), *Global surveillance and policing*. Devon: Willan Publish.
Marx, G. (2002). What's new about the 'new surveillance'? Classifying for Change and Continuity. *Surveillance & Society*, *1*(1), 9–29. Available at http://www.surveillance-and-society.org
McCahill, M. (2002). *The Surveillance web*. Devon: Willan Publishing.
Monahan, T. (Ed.). (2006). *Surveillance and security*. London: Routledge.
New York Civil Liberties Union (NYCLU). (1998). New York City: A surveillance camera town. NYCLU Special report, New York. December 13.
Norris, C., & Armstrong, G. (1999). *The maximum surveillance society: The rise of CCTV*. Oxford: Berg.
Papaioannou, K. (2007). Cameras and the arsonists of legality. *Ta Nea*, June 4, p. 6.
Privacy International. (2004). Condemnation of E.U. data sharing with US (February 1). Available at http://www.privacyinternational.org/article.shtml?cmd[347]=x-347-72181
Samatas, M. (1986). Greek McCarthyism: A comparative assessment of Greek post-civil war repressive anticommunism and the US Truman-McCarthy era. *Journal of the Hellenic Diaspora*, *13*(3 & 4), 5–75.
Samatas, M. (2003). Greece in 'Schengenland': Blessing or anathema for citizens and foreigners' rights? *Journal of Ethnic and Migration Studies*, *29*(1), 141–156.
Samatas, M. (2004). *Surveillance in Greece: From anticommunist to the consumer surveillance*. New York: Pella.
Samatas, M. (2005). Studying surveillance in Greece: Methodological and other problems related to an authoritarian surveillance culture. *Surveillance & Society*, *3*(2/3), 181–197. Available at http://www.surveillance-and-society.org
Samatas, M. (2006). From thought-control to the traffic-control: A comparison of traditional and new surveillance in Greece. Paper presented at the Crime, Justice and Surveillance Conference. Sheffield, April 5–6.
Samatas, M. (2007a). Security and surveillance in the Athens 2004 Olympics: Some lessons from a troubled story. *International Criminal Justice Review*, *17*(3), 220–238.
Samatas, M. (2007b). Cameras for every illness… *Eleftherotypia*, April 20, 2007, p. 10.
Samatas, M. (2008). A'Securitization' Policies of European Union and the Southern 'Fortress-Europe. In: T. Selwyn & M. Kousis (Eds), *Mediterranean contention*, London: Pinter.
Schulhofer, St. J. (2002). *The enemy within: Intelligence gathering, law enforcement, and civil liberties in the wake of September 1*. New York: Twentieth Century Fund.
Sidel, M. (2004). *More secure, less free? Antiterrorist policy & liberties after September 11*. Ann Arbor: The University of Michigan Press.
Simon, B. (2005). The return of panopticism: Supervision, subjection and the new surveillance. *Surveillance & Society*, *3*(1), 1–20. Available at http://www.surveillance-and-society.org
Statewatch News Online. (2004). EU-U.S.: The Passenger Name Record (PNR) Deal. April 2004. Available at http://www.statewatch.org/news/archive2004.htm

Sutton, A., & Wilson, D. (2004). Open-Street CCTV in Australia: The politics of Resistance and Expansion. *Surveillance & Society*, *2*(2/3), 310–322. Available at http://www.surveillance-and-society.org

Walby, K. (2006). Little England? The rise of open-street CCTV surveillance in Canada. *Surveillance & Society*, *4*(1/2), 29–51. Available at http://www.surveillance-and-society.org

Warren, R. (2004). City Streets: The war zones of globalization: democracy and military operation on urban terrain in the early 21st century. In: S. Graham (Ed.), *Cities, War and Terrorism*. London: Blackwell.

SUBJECT INDEX

www.ingramcontent.com/pod-product-compliance
Ingram Content Group UK Ltd.
Pitfield, Milton Keynes, MK11 3LW, UK
UKHW020232290926
15470UKWH00002B/497

9 780762 314164